Accelerating Intelligence

$$Iv+ = \frac{\text{Observation Skills}}{\text{Adaptation Speed}}$$

WRITTEN BY

Scott Hadley Morgan

ISBN: 979-8-234-13370-0 (Hardcover)
ISBN: 979-8-9970865-1-0 (Paperback)

Stock Image by: Faiz/stock.adobe.com

DEDICATION

To my deeply-loved and dearly-missed
brother Gregory Evans Morgan.
And to my brother and sisters
Melanie, Stephanie, Leslie, and Brent.
I could not have completed this
Without your love and help.
To my daughter Miley Ngan Morgan,
for her love. She made me a better
man and gave me the joy of being a father,
hopefully through these books I give
her a living example of the process of Will,
a healthy example of love and self-confidence.
Finally, to the open minds of you readers I
may never meet - the children of life-long free play.
To the dynamic people full of invention
Especially this new generation I believe to be our best.
Their spirit to excel spawns the "Essential Element Hope."
Play the games from this book in the tandem player guidebook
"The Multiverses of Entertainment."

Table of Contents

Accelerating Intelligence

PART ONE

Deconstructing
the
"Human Agreements"
and
Constructing
an
"Accelerating Intelligence"

INTRODUCTION

Of what greater values do the experiences of a person serve in writing of a single human life? If that person never fathered a child, thereby never passing genealogically a traceable truth of more than one perspective for humanity? How could mere words advance minds and souls?

Can his exploits measure in a literary method a perspective to his madness that might ultimately inspire evolution in the eventual stranger reading it? This causality comes from a stranger that is a metaphor for all of humanity in that every human is on a Qualia level a stranger to any other human. Hope is that the stranger processes the reading through his or her Qualia and it ultimately benefits them. Are the words nothing more than a self-aggrandizing mission if they fail to motivate, impress, or even cause for pause in the calliope of even the calmest day for a stranger?

Expression of thought expanded in several cultural epochs: The Platonic Society, The Age of Enlightenment, The Renaissance, and a few other epochs. Currently in no such "Age" with no "The" before our era, the appreciation of the written *experience* (which includes the reading of it) drifts just out of reach, but knows it is not alone. As if on a lake void of light, the lapping of the water against the hulls of the immortal works that augmented mankind's Sisyphean quest beckons… yet never offer navigating beacons that deliver the audiences promised for centuries by the writer driven by Will to his ultimate impact: higher elegance in another human. For today's audiences seek not this style of exploration. Data-ism overwhelming Human-ism has seen to that.

Considering this: for what godly or un-godly reason would anyone presume to write about his or her life as if it is worth reading by others he or she never meets?

For decades, the urging by friends to "Write about your life! Write about what you know!" Yet, what I've done was often done upon me and if that is the case it explains why extreme experiences never ignited the fire-starter for fame due to what I lived through. What ignited me was interaction with other humans and merging our human agreements. I always thought that writing was to discover what you do not know, and plumb the depths of that abyss to expand your limits so that you write to learn, not instruct.

Something supernatural happens within the deepest of writings that obliterates self-absorption; therefore, absence of the spark was obvious to me in this case to write about myself. Regardless of my exploits, they're reduced to a heap of composting memories at the foot of any of a million "mountains of men" I'd find more admirable and interesting than myself. The idea that my life would be the nothing more than the source of hours of boredom or at best mediocrity had little influence in holding me back – I honestly felt uninspired to commit to inferior and wasteful writing time… inferior to the "Great Works" (and even the lesser works) of true novelists, genius scientists, or auto-biographers was the destiny of my exploits.

The mind gets touched when art creates itself. If you're still enough, and your systems open enough, that touch passes into you as a novel clarity that can only be whispered for this moment is as fragile as it is invincible. Artists transfer this to their works … leaders to discovery … mothers to children, lovers to hearts, and skin to systems that interpret in ways and numbers impossible to count.

The awareness of "The Touch" emerges from the mental deserts, shapeless mirages at first, refining the image moment to moment. Sometimes missed, other times ignored, we needlessly pass the oasis … Oh! - but the times when The Touch is realized, is captured, inspected, harnessed, and put to use! Carrying me further and further from what can be considered reality that demands almost nil inspection and even less self-inspection -- what we base our actions upon, where our dreams spread roots, The Touch simultaneously clarifies and confuses.

What we make of it -- does that become the lesson, the method, the maze, or the folly? True writers do not write what they know, they write to learn and interpret what they do not know. To "get" ideas, not to express them - as Kevin Kelly says.

When not one nor all of my exploits could inspire me to write my life, then either what I consider life has changed, or the message found a life force of its own. Beyond laziness or indifference, past ineloquence – a birth occurs when "touched."

Events do not make a great biography, nor life. Great stories, superb laughs, yes … but little else.

Art is one of the few proofs that separate us from all other animals. That, and the knowledge that we are born knowing we shall one day die, motivates artists subconsciously. Why else would the first works of art be drawn on cave walls or shaped into pottery all revolve around survival? These two facts intertwine more than our minds give it credit.

When life opens its mysteries to a soul, with a challenge to step past what else everyone knows or especially expects of you, and these mysteries intend to raise the tide of human relationships, it is then that a work could be worthy of the time it takes to read it.

My father wrote a book of his life, and for a decade I saw it as a gift to his children, committing to paper events that happened that shaped him as a man. What took years was to attach his written past with his living guidance over us, which was so full of love and humanity. He cleverly titled the book, "A Hen and a Half." It took years for the significance of it to blossom. The Hen and a Half is a riddle, a sort of science or math riddle. And so was the entire book, a riddle to be discovered then solved, and ultimately to be amused by it. The message was that we should keep laying eggs until we lay the golden egg which to him was a fulfilled life lived based upon the main human virtues and we will naturally be good people if we go with the flow of virtues passed down to us. The final line in the book gave hints to how far he was able to take that riddle in his inspection and translation in the book of his whole life. "Row, row, row your boat, gently down the stream. Merrily, merrily, merrily, merrily, life is but a dream."

He was close to his final year, and clung to life so desperately, despite his accepting living and writing it as a dream, due to, I believe, something ethereal. Rare for him, such a pragmatist, unless we missed the secret unfolding of his being in his book. When upon realizing that this ethereal reality was by far the most valuable thing in his entire existence, he simultaneously released his life force and clung to it unflinchingly to the last seconds. His existence revolved around family, love, morals, honesty, self-sacrifice, and Integrity, each of these human gifts fortifying the other. A logical act considering he was an orphan and adopted into a family that loved him surely but did not express emotions very much. In his wife and children he found the opposite. Why, and what, had he been gifted to him, to realize so close to death, when the calm of knowing your life shall end soon makes irrelevant all the duties and obstacles that define our days living no longer hold importance, to write this book for his children?

He was after all a mind and soul more of mathematics than art – unless the opposite is true, which I through this book realized is the case. His musical art and family love was not the chicken or the egg … it was expression of the unarticulated gifts of life as humans enjoy them. Did the mind create the soul or the soul create the mind and which created the human level of love? If it is true soul is love and is it then a field as real and decentralized as gravity?

The mental seismic shock necessary to begin writing what goes on within me and around me came from a song playing in a cafe: "Bewitched, Bothered, or Bewildered." No one else in the television audience knew the words. It was a melody that my father would play on the piano, and my mother sometimes sang along. Playing piano so well expressed a passion for life he could not live out gregariously, as he was a modest man. It was up to his children and his wife to notice the most subtle of ways he expressed passion. There is no perfection without passion and no passion without perfection and watching him ride both waves as a pianist meant more to me than he ever was told, for it set the foundation for my own relationship to writing. His humble nature kept him from lecturing us with his vast voyeur knowledge of society I suppose. But perhaps the opposite happened, and in his stifled intuitive systems he knew whatever seed he intended was planted secretly.

Talks my father and I had (unique to me as I knew every child had talks with him that were unique to their relationship with him) revolved mostly around expanding the mind and its ability to imagine. He would then bring it all back into summation to the only thing that mattered to him: his children and wife. "You each are so different yet share characteristics that prove you are peas in a pod. And here is my theory of what this talk means to a few of you," he would say in different ways. Again, his humbleness kept him from showing how his brilliant mind had studied humanity and its progress. This bridge between our innermost fascination with what the will, body, spirit, and choice can conjure and then conquer were ours. Few things in his selfless life could he keep for himself. Knowing at the time that he fused his artistic expression so full of love with very different yet equally precious talks with his other children made the experience only greater.

David, my father, would be the first to chuckle and tell me to follow my own mind. Yet the same man would take such keen interest as to pull me aside often to force questions upon me and was my most steady art critic. His plan was that his viewpoint would elevate what I wrote in substance and style. My mother Georgia would also surely be here in my ear, changing phrases, as she was deep down a much better writer than I shall ever be. The amusement they would share over my trying to make anything significant out of my life would only be measured by watching that I don't hurt myself on my adventure, as they knew my reckless nature. Yet, they would see deeper into what motivates me.

This new generation will be one of the greatest in recent American history. They will take on the challenge of how AI dehumanizes us and win! Youths today need a reflection of how great they truly are, something to attract their Spirit at Scale that is obviously rewarding, and an additional perspective from which to view Intelligence that leads to lives of higher elegance. From my visions (and lessons learned from mistakes as well) put into words they will pick up sections immediately useful to them. There is a process called "Brain Chunking." If anyone learns how to use a toy, tool, or design they carry chunks of knowledge that don't share information until the next higher generation. This means the new generation will come in with favorable mental DNA mutated to grasp what is in this book – grasp it like a hand. Using five fingers to master a tool is like activating five Brain Chunks to grasp what new Observation Skills are in this book. This new generation will know instinctively to Adapt at Speed – shorter hesitation is like sprinting ahead. At the University of Oregon they began putting shredded rubber into the track to absorb rain and this led to faster speed to the point all colleges changed the running turf. This book is where motivation meets the rubber.

And *that* inspires me enough to write a book based on my perspectives and tools. Each chapter is a brain chunking neural network.

True writers write when a truth has been revealed that expands the human experience … when the hint of a mental journey leads to sublime revelations on our connection with each other and what we understand as the universe around us. When the abstract confronts the delineated and overcomes it, in that nucleus, when it splits, the force creates new realities … it is for a moment a "Void in search of a Field." And if the gift given is writing … then is the commitment to finish it a duty, or is it a bridge to layers of our minds that energize the nucleus with the gift? I begin our journey by giving you the "creative spark" … simply the reminder that (neuroplasticity) of the mind begins with "deconstructing the human agreements."

Intelligence Equals Observation skills / Adaptation Speed

As a formula it would look like:

$$I = \frac{\textit{observation skills}}{\textit{adaptation speed}}$$

Think on this.

If your observation skills increase the number would grow larger. If you used what you observed to adapt faster it would speed up your reaction time while expanding your limits of choices because others have a lower "I" Quotient. Simply by making you aware this is the formula for outpacing competition (and even AI) inspires you to pay attention to both in this formula. Think, "Can I observe more from any source of raw information or divined intelligence?" "Can I arrange it so the journey and endgame are more elegant looking? Can I adjust this process so it is my default action so that eventually it uses less energy to go faster and farther?"

Simply being aware this is a key to rising to the top begins a change of attention. It expands out outside the limits of the screen on your smart phone because you need to observe both the micro and the macro – laser focus on detail while expanding the horizon.

This formula allows you to get comfortable with an intelligence that is accelerating, much like Neo had to adjust to learning Kung Fu so fast. Then, it became his norm. Just like in Kung Fu, race car driving, and Dad Skills you experience "time dilation." These are called Arts for this and other reasons. The more times you enter seconds of an Accelerating Intelligence the more natural it will be to tune your engines for more of these moments because the payoff is so great.

The first part of this book orbits around intelligence. You want to increase that mass of intelligence so you as a planet can go faster yet still stay in a safe orbit. The second part of this book realizes that to keep a safe accelerating orbit you must add a moral code for living – a grand Philosophy that also can accelerate.

Both can accelerate if a few strengths are developed by anyone:

- Break down the Human Agreements
- Develop more plasticity of the mind
- Use the Unitary Method and Derrida Method to escape your prejudiced patterns
- Realize you are evolving in an organism-environment – nurture both

- Aim for the payoff: a next-gen "Imagii" Avatar used in Games, Influencers, Social Media
- If you are first - you will dominate this new era
- Within the Avatar will be your soul and all things that have true meaning to you

Open your mind… absorb these ideas…increase observation skills… decrease reaction time… realize only you can bring the morality and human ability to create something out of nothing… we know it as Art.

DECONSTRUCTION THE "HUMAN AGREEMENTS"

Early nomadic life forced humans to be far more innovative and creative then people must be today. So much depended on observing Nature constantly, to find food and avoid death traps such as dangerous animals. The human body reduces and often eliminates traits or skills that no longer serve a purpose. For example, human senses of smell and hearing rival animals. As his brain no longer needed to smell water from a distance, this skill lessened just like tracking animals or hand-building a dam is no longer a necessary skill. Just as calluses on the feet became less necessary with shoes, so did aspects of our creative mind.

The more the human mind developed skills in communication and working in larger societies, the more it appreciated categories of the arts. Yet at the same time, groups assumed new ideas that often times never were true.

Superstitions and magic begin where science ends. The knowledge of hunters or monks so attuned to nature they are, able to talk to plants, or read the weather, were discarded by "professionals" that learned a clinical or "acceptable" expert status. While in many instances this worked terrific, such as in the recording of languages or the advancement of mathematics, it forces the mass of people to "agree" with the newest declarations on "what is" and "what isn't."

Now mankind rises to the point where he has conquered most of his physical world and its challenges. Our minds, coupled with machines reach godlike levels.

I believe it is time to challenge many of the primary "Agreements" that helped us get to this point. Probably the easiest one to approach is our concept of Time. Much like the Agreement that the Earth is flat, the Agreement that Time is how we describe it rather than how it truly is, simplifies our lives. It allows us to be organized, and in organization we nation-build, companies expand, and people get rich and powerful.

And yet, more than we need to advance, mankind (regardless of what country our society is formed in) desperately need organization, systems, and limits. We need a guiding power keeping civilized man civilized. While when left alone, our inherent characteristic is to be curious and work together, cities put greed above human honors. This puts out government leaders in quite a tough position. Any leading nation needs to give freedom and Truth to "emerging technologies and personal beliefs" in order to break out new inventions that improve life on a planet that constantly becomes more crowded. Yet the leaders know people abuse freedom.

If the leaders themselves are not inherently creative, how do they deliver both necessities? By observing and agreeing with the "Four Sources of Reflection" (as I call them.)

THE FOUR SOURCES OF REFLECTION
NATURE
CULTURE
BALANCE
ART

Our first inspiration and guide through life was and always should be Nature. This place also revealed the ways that (universal and individual) balance communicates directions for us. Culture gives value to the lessons learned over thousands or hundreds of years by our ancestors. And it is reflected in the various timeless Arts: Poetry, Prose, Dance, Music, Sculpture, Architecture, and Health (which I deem an unexplored Art.) The moment an artist "creates" art he or she harnesses the three other qualities listed above it into one expression that fulfills a reflection of inner and outer "energy" (which can never be destroyed it can only be transformed.) Young people expecting or hoping to break through to a greater peace and happiness, along with prosperity, must value the guidance as much as value these Four Sources. The young need to see honest reflections of themselves, not just peer reflections or current over-stimulating reflections (games, photographs, videos, social media.) The clearest proof of this is the overwhelming feeling of freedom that young people feel when they dance, and even younger children feel when they sing out loud. All that I present in my vlogs relies on valuing Nature as the first teacher and source, culture as its trail of successes, Balance as a way to integrate this into happiness, and art as a way to express it or see it building.

Amidst all of this, lie our most fundamental Agreements that we can now challenge.

The concept of Time (which I often capitalize if I am considering it in its totality as a universal science) captivated me since I was about 14 years old, when my grandfather (a scientist) gave me books on advanced mathematics and also a "Look" book on Time. I began to question Time, how it is measured in reality to how we view ourselves in the aging process and relative to joy. When I read "Catch-22" at the age of 19 there was a character that proclaimed that he did things that bored him so that Time would move slower and by the end of his life he lived through more Time by simply being aware of its passage more than other people. My Grandfather thought upon this a second, then simply said, "That is a fiction."

Shortly after that, my best friend and I made a "secret club agreement" that we would not grow older after 30, and that time would move slower for us so that we could fit more into it. I dove into this wholeheartedly, yet after several years, he stopped believing in it. I made my first large purchase, one that I keep today. At age 22 I bought a clock by Cartier that was a skeleton table clock with gold framing and clear crystal glass. I adored it as I inspected its inner movements for minutes at any time of the day my thoughts were pulled back to thinking about what Time really is to us. These few examples out of dozens all lead back to the fact I never fully accepted the standard concept of Time.

Now, finally, concepts about Time are so commonly studied that it is entering our movies and tv shows. While each film had some invention that gave man the ability to Time Travel or space travel without the constraints of Time, only a few have taken on the task of "re-defining" Time. We all live in this Agreement, and we must give it up to rise above all the negative factors that Time as we know it carries; disease, old age problems, death, and much more.

To me, Time represents something much different: It is a system we can navigate to bring back the most essential element in creation: Information. Because Information is all around us and we cannot know where an inner truth comes from. It made sense to me that Time was all around us, not just in front of us. Identify it in raw form before it evolves into Intelligence and follow the Will which is on its way to higher elegance as it is reflected throughout our universe from the sub-atomic level to the grand galactic level.

Time and Space couple. You cannot fully utilize one without the other playing a part. We agree that Space has certain "Laws of Physics," yet only in this Dimension in this Time. Conveniently, we relate to Time as linear, and to the watches that tell us what time it is as representing what makes Time meaningful to us humans in our Earthly reality. Within a few years our concept of Time will dramatically evolve. I want my study of the great Philosophies to merge into one of the first guides to this revelation, and to make it functional with the technology books called "Pillars of Encryption."

While I see myself writing five or six big books covering philosophy, The Arts, film, Edge Technology, and global regulations I do not see them as a series of the same book like "The Matter with Things" by McGuilquist that totals over a thousand pages. The books (sometimes divided into two parts like this one) focus on a different aspect of living in the world today. It is through the separation of arenas that a grand "Accelerating Philosophy" can begin and evolve.

We agree we live in this dimension, yet actually are incapable of fully envisioning, much less living in, another dimension, or merging dimensions. This Agreement keeps us in this dimension.

The obstacle we must overcome is as fundamental as our civilization on this planet: language and Time.

The key to "Deconstructing" our Agreements lies in forming a new relationship with the cosmos and creation that skips language entirely, and even some fundamental elements of "thinking" at the same time. I believe the universe watches how we transform energy in a Maxwell's Demon way and waits for us to expand our ways of energetic actions to show us new secrets of the universe. While mathematics and numbers (not to be confused with Numerology) translates very much of the cosmos and our reality, it alone won't deliver the complete alphabet or language for cosmic dialog that brings us higher toward the "Emergent-Mind Universe." These chapters deliver what could be thought of as the essential stretching of our minds to fathom the great leap forward. And technology will engage for the good of what makes us human if we start of from the best perspective.

A daunting task or path, because we gain so much of our comfort, confidence, and positioning from Agreements. Yet, all of us have experienced sensations or events that transcend facts, science, space, and Time.

Love, between people and animals. Déjà vu. Miracle cures or inventions that seem to be beamed into us from a higher power are a real thing to the leading artists and physicists throughout the last centuries.

More important is that the Agreements prevent us from tapping into skills and awareness that mankind was taught are useless. For example, what if fear looks like a red mist hanging around a human body, that dogs and other animals see but we cannot? What if we have the ability to read minds, but left it behind, when the religions felt it a threat and people were punished that used it? They discovered that a small percentage of children see numbers in different colors until the people in their environment make it seem worthless. Conformity allowed "The Church" to expand and grow powerful. They brought all advanced texts to the Vatican, and some are kept in a vault that only the Pope and a few scholars are allowed to read. Why? What truth is hidden in these scrolls or volumes?

Rather than go into a conspiracy direction, I instead look at it simply as temporary solutions scientists and leaders told us were true, to allow more cooperation and livelihood. Not evil. Just, the way we moved ahead.

Now is the Time to challenge many Agreements. I'll start with the Agreement that we move in a resistance-proof medium called Air.

We are simultaneously gods and worms. (Mazlow)
[Note: Deconstructing the Human Agreements is the
first step to developing an "Accelerating Intelligence"
that can then maximize "plasticity of the mind"
to helps us know what we cannot yet sync with
the traditional pre-AI mindset."]

ETHER

Two fish swim in a river. The first fish says, "The water is cold today." The second fish asks, "What's water?"

Ether. Such an intangible word. So normal to be oblivious of it. The word Ether stuck out to me in 8th grade when I first heard it in History Class. Again, later in Science Classes, this word "Ether" echoed rather than faded. When it reappeared frequently in books or insights on understanding what we cannot see, touch, or even express properly – that was when Ether's "mental Velcro traits" began making sense to me. Ether seemed to stick to every new concept I delved into.

Most people are as oblivious to Ether as the second fish is to water. Be aware of Ether and you immediately have an advantage.

In 8th grade, my history teacher explained how the earliest Astronomers and "Deep Thinkers" like Plato and Aristotle believed something preposterous: that the space between planets and stars were filled with an "Ether" and this Ether transferred most of the unexplainable properties of the Universe. It was upon this concept of an Ether that not only made our universe possible, it also somehow related to the soul and mind.

Centuries later, Copernicus proved the Earth moves around the Sun yet still held on to the idea of Ether. Man's need to explain things forced the invention of this Ether, according to my teachers. The use of the word "ether" lasted for centuries, functioning as an explanation for what lies beyond man's reason. Science teachers taught us that outer space is a void, that no substance exists in it, and certainly not any Ether.

For 30 more years, the word Ether struck me as so out of place for the minds of Aristotle and Plato. Even more so for scientists after 1400. Then as more research on remote exceptional civilizations hit the mainstream and Internet, the word Ether or a word that meant Ether was used to describe what fills space. The Mayans, Incans, Chinese, Egyptians, Sumerians – their myths, lessons, and art reflected the idea that a substance exists in the void of the cosmos, and sometimes it exists in the open space outside cells that form objects. It was one of the first explanations for gravity. This sort of definition did not occur just in outer space, it is indeed in our space on Earth. Due to our atmosphere, we know our air is filled with dust particles that reflect light, which gives our planet an aura of light. But that's not what any of these ancient and even modern teachers promoted as the whole and real truth.

The ancient scrolls talked about an Ether permeating our atmosphere and beyond as if it were a living, elastic substance, not unlike water in a river, ocean, or lake. This Ether transferred energy in the form of pressure, heat, mass, and resistance just like water, but it also transferred the immeasurable:

thought, love, fear, intuition, and at the end of it all, a connection to each other, our Earth, and the cosmos. Ether, I realize, serves purposes far beyond these.

Eastern Philosophies present abundant versions of Ether to explain reality and give people a cultural code of conduct and social philosophy. But these remained stationary, which was fine until this era of AI and control of what we digest in the news as truth.

"Ether" has haunted me since my first thoughts could contemplate who we are, what we are made of, where we come from, and if we are alone.

The reason Ether held on was because it is real, it was explained, it was harnessed, and then for many reasons it was discarded, denied, and mocked simply because we could not measure it. As the saying goes, artists can paint the sky green if they want, but the rest of us must paint it blue or others thinks we are stupid, especially scientists: observable and measurable, all their deductions and theories must be.

The tenacity and universal functions of Ether touch upon nearly everything we as humans obsess about, and indeed what makes our "humanness" so unique in this multi-universe. We will soon create things that take us into artificial worlds, with one example being Google's Mirrorworld (but I'll get into that in detail in another chapter.)

I compare our evolving way to experience it safely to Intelligence Quotient, I.Q. – or, rather, I used it almost as a measure of I.Q… and the relationships between the two are quite active and incestuous in my vision of reality.

When through casual research I came upon the Holoflux Theory, Pribram's holonomic brain theory, and later work presented in the Pribram Bohm Hypothesis – well this was really fascinating to me. Something that "my gut felt" had wider supporters than I imagined. [Disclaimer: it is important to note that the Cosmos Structure, Holoplenum/Holospheres concept, and the way Implicate Order and Explicate Order are presented in this Hypothesis are not fully accepted by the science community.] Despite this fact, it resonated with me -- seemed to make sense with how I view the mind and universe somehow entangled and emergent, and loved Advaita Vedanta's Lens of Consciousness diagrams and supporting reasoning.

Leaning more toward New Age thinking the word Akasa is translated quite simply to "Ether." (Now if you really want to bug scientists bring in Thoth.)

What the preceding paragraph presents is the concept that even on a scale smaller than The Planck Length the concept of Ether must be present. In fact, it is gaining popularity as we discover more of the tug of war between Dark Energy and our known Matter and Dark Matter in our Universe. Then this leads to further hypothesis: involving the Planck Constant and Planck Time in equations, along with something I mention in chapters on the quantum science of A.I. and Blockchain. This leads to the importance of formulas and devices such as The Fourier Transform. An incredibly graceful form of energy/matter exists that represents "The Implicate Order" … which in abstract ways relates to consciousness connecting to the physical universe to the point that without this, we could never even interpret our universe. A question arises that relates to how information is stored in the brain when the fluid cushion or chambers in the brain resemble microwave antennae. This organization of at first unassociated formulas sparked my interest. But, hey, even with all this going for it Ether was still the belief of madmen. Yet, around the 1960's to current day, Ether began to be more relevant (not less relevant) to understanding our reality and mind capable of consciousness.

[Note added in 2026: When I began writing this book most people abbreviated Artificial Intelligence as A.I. Now of course it is simply AI. The chapters will use both versions of abbreviation and in many cases reflect when a touch up in the original writing took place.]

And Ether *will* play a role in A.I. design for our Pillars of Encryption analysis for Cloud design and "Imagii Avatars" – so this is germane to the conversation about the future of simulated realities in entertainment or on social media platforms. Ether already raises its shadow in the tug of war between Dark Energy and Dark Matter. All questions can be answered in the movies, and in this case it is "Interstellar." This tug of war created a "Void in search of a Field" and it was filled with Love and Gravity to direct higher elegance so much it escaped the Black Hole event horizon to talk to TARS and send binary code. [I must remind you I am not a "scientist" I'm simply someone recognizing patterns against a different backdrop of Time and Information than most people.]

Now to get back to how Ether and all this applies to humans in everyday life, decision making, and awareness of our potential and world …

Considering the concept of Ether as part of our universe
is helpful in achieving the plasticity of the mind levels
that reveal secrets of our existence
and Ether will be mentioned frequently in this book.

Plasticity of the Mind and IQ

Measuring I.Q. is just a number given to gauge how intelligent a person is relative to other people, all of whom are in a constant state of flux. There has always been disagreement in the science community about the validity of an I.Q. test result. And now that politics has entered the debate, the number given for anyone's I.Q. is as worthless as determining the trajectory of spinning atoms within a baseball being hurled toward something it will hit and transfer energy into, like a baseball bat or a window it breaks, or a skull it clonks. The direction of atoms means nothing to the window it breaks or the head it hurts because the bulk mass of the entire ball delivered measurable results. But the spinning atoms within the ball are what holds the ball together enough to not fly apart like an eggshell of particles. All that matters is the impact of the ball, and all that matters in I.Q. is the impact it has on individuals, relationships, societies, and our world.

The space between the spinning atoms is no different than the Ether space between people or planets. In fact there is a strong correlation. More amazing is the correlation between the connections in our brain and the design of the galaxies within the universe, but that is a separate chapter (one that also plays into the Fibonacci Ratio in future profiles, avatars, A.I. interaction, AR and VR, and Blockchain.)

Broken down to its most elementary stage –

I.Q. is the ability to apply one idea to many.

The more times a person can apply the one idea to many intangibles, the higher the I.Q.. A good example is to name all the mammals you can in one minute. If the person works in nature or on a farm, they go quickly to mammals they've interacted with. For others, a system is quickly created based on how the person stores facts about parts of our world that they never interact with on a physical level. Most people are limited to what they see on television or in a magazine when it comes to mammals. A person's absence of physical interaction with animals slows and limits recall about them. Far more questions are asked about things we do not personally do ourselves than what we do experience in a personal level. We reach through our minds to distant lands for zebras and crocodiles, and enough shared memory of these animals lets whomever scores you raise or lower your score accordingly.

I.Q. becomes socially "useful" when a person or thing can then reverse this and apply what the many correlating sources of Information to one innovation or social change.

Our ability to form intelligent thought in I.Q. measurement challenges us, and we meet it every day as a gift or an evolutionary attribute, depending on how you think humans began. As with all gifts and systems in our body, they are there to serve a purpose. If they serve no purpose, evolution breeds it out of us through slow genetic mutation.

To show how quickly old theories on the mind and our connection to reality changes, look at the most recent conclusions about our mental abilities and why they developed in humans so grandly. For centuries experts thought the bigger brain offered ideas on expanding ways to use our hands, which are so unique. But now they think that our hands were so incredible in design that it forced development of the brain. The same is now thought of our tongues, and how their perfect placement allowed for so many more sounds that we developed the most complex and usable "audible to us" communication skills. Until recently, scientists also thought that our brains drove our tongues to develop linguistics, when in fact it was the other way around. Tongue placement and dexterity led to an inner and possibly artistic and philosophical drive to evolve our minds.

Our ability to use I.Q. has increased at an astronomical rate recently, solely because it is more "useful." We have invented things that "need" more I.Q. to evolve them. It becomes useful when you reverse the direction of I.Q.. Specifically, that you apply "many" (what you sent the mind out to relate to a concept or question) back to "one" (you as the user of that): many mammals, many places, many distinctions, many uses, many habits within the mammals offers opportunity to advance as an individual. You get fed, warmed, befriended, amused, educated, cured. [This is in part how "brain chunking" works when interfacing with new evolutions in technology.]

I must expand my story for just one sentence to state now that A.I. will reduce I.Q. because those in power with the A.I. algorithms will be driven to use it to gain more personal power, and this comes at the cost of users letting the mental muscle that flexes to apply one idea to many to atrophy.

So I.Q. is applying one idea to many. Other than daydreaming for inspiration and ideas, the benefit of I.Q. only happens when you apply many back to one. Your I.Q. reaches as high up the returns that it can to exploit mammals. You turn the object or subject into wealth, status, safety, health, and systems. Also, the monetization of I.Q. only happens when the usefulness of I.Q. is reversed from "the many" back to "the one." Carried with the facts you returned with are instincts, and deep down "genetically organized truths" and "Transgenerational epigenetic inheritance" (personal traits, and tastes in things like music that do not alter genetic code) that, survivors for centuries, learned and passed down through genes.

Getting back to Ether now, apply the one idea of Ether to the "many" then reverse the process to find what use you make of your I.Q. based theory.

I.Q. touches everything in existence and history, both civilized and genetic. We are limited only by how far we can push our I.Q. The fact that most answers to our deepest questions probably are locked in the memory history on a genetic level of our cells is downright freaky.

I.Q. in that way reflects the reality of Ether. Ether touches everything in existence, even past realities. Like thought, it goes beyond where we set boundaries or even imagined boundaries.

Finally science is catching up to what I had been studying for decades, when they announced that 99% of mass in our universe is in Deep Space as a Dark Energy – or "Ether." They refuse to call it that, because for centuries Scientists mocked the idea of Ether. It was a simpleton's explanation for how the Solar System and Stars work in Physics.

But now, with dozens of ancient civilizations having myths, studies, art, and architecture all using Ether to explain things, the realization that an Ether exists is again returning.

What does this mean? Again, using I.Q., you apply Ether to as many factors and functions as possible. Easily, Gravity (and gravity with a little g) and Mass are identified right away as being affected by Ether theory.

But these are low-I.Q. steps. Apply Ether to things we use to navigate our lives every day: love, intuition, premonition, dreams, miracles, coincidence, animal behavior, baby talk, family bonds, déjà vu, and the creation of a human fetus. That is what happened one night in 2014 when I was driving home from a lounge at midnight.

Usually, I have the music channels on when the valet hands over my keys. I'm not sure how or why, but the radio was on an A.M. radio talk show for books. I had started listening to the interview about 5 minutes into it. I was about to turn to music when the author repeated the title of the book, "The Technology of God." The title intrigued me, so I listened. Getting home only takes 20 minutes, but I sat in the car for another 25 minutes to hear the entire broadcast. Then immediately ordered the book online because it is not readily available in stores. One of the key elements in the story is referred to as a substance like Ether. What was different this time was that Ether was broken down into tetrahedrons, and if you want Intelligence in the form of any artificial world to expand limitlessly it must be in the form of a stella octangula (one pyramid rammed through a second pyramid.) A separate chapter will cover that book, so I won't go into it in detail now. But, I will say that there were only two other times I listened to the radio and bought a book, and both books ended up influencing what I'm writing now.

The idea that will really cook your noodle is -- would the interview have been on the air at all if I had not been driving? Had Ether arranged it? Had the energy in elastic space turned my channel on my radio hoping I would listen? Is this an example of what I presented earlier about the Universe revealing clues to a higher understanding only when we accept "Plasticity of the Mind?" By merging Philosophies and correlating it with science proofs and formulas, I'm trying to break through to this happening more often to countless people. And it won't result in an instant invention it will be a soulful and mental expansion of how wonderful it is to be human with consciousness – a perfect example is how the birth of your first child changes everything you understand about life and yourself. Does this new parent then interact with Ether on a higher level? If so, what could ever capture it so that it is felt or known to another person?

This means we would share our individual Qualia with another person and know we are doing it. Surely if you are the father you do this with the mother, but what about people on the other side of the world? One of my ultimate goals is to solve the challenge of sharing Qualia. It is one of the unanswered questions in humanity. When the Qualia Apps flesh out this Next-Gen Imagii Avatar people will meet others that had the same dreams or are like Dopplegangers and this magical connection will make them more aware of Ether. To do this so it is a natural experience for any person is the key. The television series "Black Mirror" was a super-hit with the new generations and meeting mirror souls captivated fans. Plasticity of the mind is required. I.Q. surges connecting to more reflections of what is being questioned.

How many seemingly unconnected events and people had to be sharing energy through the Ether to get me out of that lounge at the exact time to listen to a station I don't listen to in order to get me to buy the book and have it alter my understanding of life?

One idea reaching out to connect to many ideas. I.Q. recognizes the connections, then it is up to the individual to pull back into his mind and life all the varied connections. Once achieving that, allowing it to alter your thoughts, core, or actions requires a completely different set of steps.

How significant is the idea of Ether? It plays a key role in two of the most significant theories about our universe. The first role revolves around how Einstein's Theory of Relativity works perfectly on small scale, the micro scale within atoms or even for Earth. The Theory fails to reconcile all the measurements when we try to apply it to the Universes. But Ether explains the biggest new discovery – gravity waves. (And also how radio waves work, but that is a decades-old mystery.) How could you imagine this? Well, visually imagine the scenes from "The Matrix" in which the force of The One (Keano Reeves) bends the walls around him, or the vortexes behind the bullets.

Ether in the second study helps explain the concept of the Holographic Universes. Apply Ether to this theory, and Ether helps explain how one distant galaxy Hologram can send information to our galaxy without time being a factor … by the deeper cool reaction is that the Hologram does not have to form in 3-D until a person looks at it, compressing Space while giving us the illusion of its depth. (That means galaxies can be closer to us if we can observe it using this, but that's way complex for now in this chapter.). This explanation also is influenced by my imagining an E-8 model of our reality and higher dimensions.

In fact, Ether plays a huge role in my greatest intellectual passion: Deconstructing human concepts of Time and Space. Much like I.Q., the Deconstructing ventures from one point out to many, but only becomes useful when you bring it back from the distant points to our "Now" and "Where."

Having trouble accepting my insistence that Ether exists? Maybe it seems haphazard? Not enough scientific fact brought to the table by me? All right. Meet my Team that plays the Existence Game with me, believing in Ether: Steven Hawkings, and almost all astrophysicists that must call it Dark Matter (because if they call it Ether they will look stupid.). Dark Matter is no longer a novel discovery – but a few decades ago it was huge!

Now I will go on a free-thinking ride into how Ether relates to the many things important to the future of Imagii innovations.

What does it take to Deconstruct Time? Time validates Space, for without Time, Matter cannot be proven in any sure state or place. The bigger story is in a later chapter. But I'll give you a hint. The problem is our brains, and two things I already mentioned: our tongues and hands. Specifically, how our brain makes both assumptions and conclusions based on words that exist (for this function) only within the brain. Because the "tongue" used inside our brain to form ideas has not been used like a hand, it gets lazy and loses dexterity. Yet our mental tongue can think beyond what we use, the problem is the way we communicate within our mind uses only fingers, no thumb, for opposing and evolving pressure. Without a thumb you can't grip, without a mental thumb you cannot grasp beyond our limited language, yet the mental thumb most surely has another language that answers all our greatest questions: "What is Love, Who am I, What is Consciousness?"

Once you Deconstruct Time, Deconstructing Space to the point of understanding it on a Star Trekian or Matrix level simplifies.

The Nexus lives within the Ether, not what we call "measurable mass". The energy contained in the Nexus Ether uses for the interplay between Time and Space has some fascinating properties. First, it is the sum of the force released in the destruction of Mass, combined with the energy used to "create" original matter. Second, the energy has an even more cool quality than gravity which can span Time – it transfers between Dimensions. Oh damn, I was hoping that the previous explanation would be broad enough to take inspection from physicists, but, well, they will find a few holes in it. I don't have time to explain all the details that entire books dwell upon. And there are other aspects, such as the concept that the essential element in the universe is "Will" and it rides along in the Ether and all Fields, but that is another chapter. Back to Ether.

Without Ether the dynamics we observe and measure in our knowable universe are not possible. With a brain other than our brain inside our skulls, such as muscle memory of intuition, using a mental thumb to grasp concepts that cannot be defined in words moves in and out of our lives. It can make a concept *understandable* at the moment it is also *useable* because we feel it. We harness the Energy in Love, for example, which results in actions – yet Love is not a measurable thing to our brain, we only feel its impact. There is a connection and energy between two or more things "embracing love" for each other or the outside world. No one can deny Love once they've felt it … but through what medium does it exist? Just a mental medium? Try thinking of it as a Field that rides in the Ether between us. Until Love finds a connection the potential energy is unrealized: connected its energy can create a fetus that already knows things about the parents and their culture before it is born.

This isn't a New Age Zen awakening sort of theory, fact, or discovery. Not at all. While words used may sound like the ones used discussing "The Tree of Life," or Chakras, or "The Celestine Prophecy," it's far from it. All these works and more, all these disciplines and more, rely on words that fall short of reaching all their "end games." My chapter on Buddhism covers this. The secret is not in "The Secret." It is within the diagrams carved on caves, pyramids, or etched on land – yet at the same time they are not. The secret transcends the language used by us, but will be replaced soon. To understand the limitations more, Ether offers reasons.

I'll end this key chapter with a riddle and answer. Periodically, the science community asks the question: "If alien races are so superior to us they can space and time travel, then why would they want to come to Earth?" The surprising question is rooted in Ether. "The human concept of Love might be unique in the universe but the experience they would want most a human crying."

Love transcends Time, Space, and it bridges physical human beings on a conscious level that is unique on Earth. When the brain is damaged, Love between people and animals still is recognized by the other systems of the body. In fact, people with brain damage use their heart "brain" to remember the faces and past of people the person loved most, and it seems to "come from the heart" only because our limited language cannot yet "grasp" how the heart has its own brain. It transcends species. It transcends even death. Which leads us to a greater question:

"Through what medium could Love then travel, and how is it propelled?"

The Power of Story: Applying I.Q. TO Entertainment

One of the greatest gifts mankind enjoys is a language both diverse and specific, simultaneously taking advantage of its vagueness and suppositions to advance knowledge and our ability to identify with our souls or consciousness.

Writers are the timekeepers of ethics and epochs.

Long before formalized writing and books, there were stories. They began as grunts and laughter then moved to complex pantomimes to pass down knowledge about survival. The power of poetry or prose to captivate and transport the receiver into another world stands with music and Fine Arts as a truly sublime gift. The power of a written work to improve culture and society, as well as the character of individuals, is undeniable. From the religious texts, to dark portraits of the Individual in a cold society, to blissful stories for children, the ability of a novel or poem to steer universal thought touches us all. Words evolved to singing, pantomime to ritual dance to connect their story to higher beings, art on cave walls showing followers how to trap bigger game or where water was hidden, to hieroglyphics to present a supernatural past, and images ultimately evolved into formalized language.

Not withstanding these facts, what if language raised us to an epoch in communication, then became the most significant obstacle to discovery and humanity?

To me, writing is unique among the Arts in the way that it constantly demands a jump from the (so-called) "feminine" part of the mind to the "masculine" part. Writing is a science of exactitudes. Poetry takes this farther in exactitude by limiting choices to words with expansive and exposing meanings as it condenses the power of the written word. Searching for the precise word that maximizes the power of the message lies in the feminine processes of the mind.

Once that word is imagined, a dance begins among the words so that the inert energy of a word takes on meaning by the words around it. The energy release does not happen word-by-word, or even sentence or one paragraph at a time. The energy rises and falls from start to finish, with the author hoping to deliver his vision with the greatest emotional impact. Because an individual word does not realize its full potential on its own, the masculine process of continuing to write until the end of the piece plays a crucial role.

John Ruskin spent much of his life as an artist attempting to share the creative process with people that are not creative.

"The greatest thing a human soul ever does in this world is to see something and tell what it saw in a plain way. Hundreds of people can talk for one who can think, but thousands can think for one who can see. To see clearly is poetry, prophecy, and religion, all in one."

While the first sentence appears to simplify the feminine-masculine dance of superior writing, the following sentences clarify what makes it the opposite of simplicity.

When people tell me they don't read any more because they find it doesn't matter to them, they often ask me why I still read books, especially the ones considered the great novels such as "Atlas Shrugged." My answer is that these books present the most elegant choice of words forming the thoughts of the brightest minds of humanity through the ages in hopes you expand your knowledge, experience more joy, and rise higher in understanding the precious gift of being given a life with the qualities of a human being, from atom to the immense cosmic consciousness.

Through emotions and art these writers blast past the short limitations of our sensory perceptions intertwined by our physical body. These written works also take us out of our own self-absorbed minds and let us travel outside ourselves for hours of the days.

Art is the only experience our senses can interpret for our minds
that we do not actually do ourselves.
Our reaction to the art becomes an
experience wholly formed by consciousness singularly active.

Creative energy pours into and out of the artist. Fine artists paint or draw, allowing colors and lines to perfect their vision. Though study of art broadens the choices of the artist, the process is mostly purely feminine. Same with dance and music. But words are absorbed every moment of any direct or overheard conversation, in all written texts, and in our own minds in our internal debates. It is one art form that a person can excel in without any special tool and from the very first step (whereas dance and music require practice.)

Bringing this back to Ether and I.Q., the I.Q. goes out in search of the most effective words, and through the mind they influence our actions and growth as an individual. Words either retard or nurture in growth. Yet one striking fact is that Art, and words specifically, can inspire whole evolutions in society and humanity. It is the passionate specificity of the feminine word search that achieves this.

If political interests reject the idea of I.Q. based upon the environment of the test subject, then they must accept the correlation of I.Q. based upon the power of Story. Why? Because the Story that proves immortal within a culture puts the test subject into his or her native environment.

But what if words were a creation that should have evolved into some other form as our individual and collective minds evolved?

Let us use I.Q. principles to help with this conundrum.

Take the Bible – one of the most inspirational guides on living and thought ever created. Devout followers tell us that the Bible was written by God, is a perfect work, and never needs to evolve. All that needs to evolve is our ability to understand the messages in it.

Yet, few devotees consider the actual history of the printing and distribution of today's common Bible. Up until the 1400s, there were many versions of the Bible. Almost all were in Latin, written by monks by hand, and read only by monks or other religious people, sometimes by noblemen and powerful people. But if anyone took on the task of reading one Bible, they would probably not then decide to read other versions of it. When the Guttenberg Press was invented, suddenly the Church realized that millions of Bibles would be printed and these read by all people. This put the dogma and power of the Holy Roman Catholic Church in jeopardy. An urgent search for the most universal - and more important "beneficial" - Bible began in Europe. Of the 14 or so major versions, the Church chose and then finalized the most popular versions of the Bible, such as the King James Bible. Other Bible versions are kept under lock and key in the Vatican vaults.

From that era onwards it was easier for the Church to demand that the Bible is perfect, cannot be questioned, and has within it all mankind needs to interpret to live a sublime life.

I do not question the importance of the Bible or even its messages to those that believe in it. I'm Christian and love reading the Bible and learning from it. But treat it like I.Q. and apply this faith in it as the one idea and apply it to many and find where it fails to explain all facts of life.

The Bible was written in an age of incredible mental simplicity and social naivety. For centuries after most Biblical events, and through the Dark Ages, the majority of humans were illiterate. A fair percentage of people living in cities had a basic ability to read, but complex studies in science and philosophy was out of their mental reach. Without decades of elevated learning, most human minds cannot understand the meaning ultimately because they do not understand the significance of the words as they relate to each other. They are emotionally moved, but not able to or interested in expanding past what they read or that emotion. One reason the Bible took over so many varied peoples is because to read aloud from it the stories were literally beyond imagination. The power of the mind to "imagine" instilled powers upon God and the Church. For this reason, the Bible could never evolve with the simpler minds of people.

Yet, evolve the minds of people did! Think of this. The Bible was written to inspire and guide people so ignorant that in the summer they farmed totally naked, from France to Egypt. More than two-thirds of the people could not read at all, and fewer could write. Take it a step further, applying this factor to many, and you realize that even their own minds fell short of a full vision of the Bible because the specific words did not exist in their vocabulary. Fear expanded the power of the Church and its Bible.

While all else in the world evolved, and evolved greatly, especially the functions of the individual mind, the Bible did not.

To reach the people living from 700 A.D. to 1500 A.D., the message was simple, clear, and told as a truth. They absorbed it on a literally "animal" response-level of awe or fear, off words preached - then repeated within the human minds, an echo trapped, in both its medium and the confines of the human brain. The Church pushed an animal level response for control, but then refused the followers from enjoying their animal body systems because this led to sin. I'm not just talking about sex, I'm talking about interpreting the world from ways other than the brain, which is limited by words, and the restrictions of which words are used in religion.

It is my belief that the messages of God (of any of the key gods) should have been allowed to evolve. By not letting them freely evolve, this trapped the majority of people into an easily controlled state of limited evolution.

Let us ponder what alternatives existed or passed us by.

While the world was populating and civilizations met they had a chance to merge what they had learned. This surely occurred in objects traded: food, livestock, tools, weapons, building, textiles, travel, medicine, and much more.

In fact, during the Renaissance, in the Medici Court, the great thinkers and philosophers of the time did indeed create a new living religious text, combining the Bible, the key tenants of Plato's Republic, and most sublime messages in the Torah. This group of extremely wise men also borrowed from ancient Hindu and Sanskrit Texts dealing with Mother Earth as a force of wisdom. This was a religious tome designed to carry mankind into a more industrialized age, with a deeper understanding of medicine, cross-cultural cooperation, and harmony. Di Vinci, Michelangelo, Medicis, and Machiavelli were just a few of the more famous leaders. What happened?

Many of the men pushing this were murdered. It was one of the main reasons Leonardo Di Vinci fled to Paris. The Church knew the tome's threat would extinguish their rule.

But religions, despite realizing the benefits to mankind if they merged and agreed on things, saw merging as the end to their rule, power, and riches. Catholicism depended on the limits of words. Their key rival was not Judaism or even Islam, it was groups that used more than just words to communicate – they used symbols, or had customs that told of their goals.

Why was this a problem? This did not limit learning to the brain. Experiences, lessons, and insights into the Universe and other humans were "experienced." In fact, many highly thoughtful groups saw words as the obstacle to enlightenment about what is really going on around us as we live. Plants can speak, dimensions opened and explored, connections in the trillions contemplated, and most of all ... Time and Space are not limited by the words used to explain them.

What connects this all? Ether. Inter-dimensional Ether. The wisdom of plants is "felt" or "sensed" by the people in Nature most of the time. This is actually what a true "Faith" experience (those in the "Born Again" Christian moment) relies upon. While words in a Church might propel you, the actual experience is visceral. The energy transfers through a medium we cannot see or in brain words explain. And I am in no way criticizing Catholicism I am simply stating their operational plan to expand from Rome. Catholicism and Christianity united people beyond borders and is the only thing that pushed back the Muslim crusade.

I.Q. sets us out to further points to assimilate more "stimulus." Notice the words ideas or facts are not used. Why? Because both function only within the confines of words. Stimulus can be introduced and distilled through a variety of senses and systems, many that do not use words.

I.Q. drives itself with one of John Ruskin's simplest truth pursuits: the purpose of knowledge is to free us. Free us from fear, limits, our own weaknesses, the feebleness of our minds, the fragility of our bodies, the unknowns of our heart and our own creation.

While New Age followers of all types swear by their new and free experience, they still base explaining on archaic and limiting words. Leaders do this to gain followers. They repeat what was told for thousands of years. Doesn't matter if it's Yoga, Meditation, Tree of Life, Aliens, Numerology, or other interpretations of a greater power, it comes down to words needed to expand it to other people.

IF, and this is a big IF, the words in form of a message works, then that's great. A good example is the Ten Commandments. Or in meditation to reduce high blood pressure. But what if the words themselves are the only obstacle to getting past the final threshold between common mortal lives and sublime existence? I must state right now that in my shallow practice of meditation, I experienced incredible levels of peace, connection to the cosmos and Mother Earth, and know I opened my mind to receive secrets or recognize patterns that led to my breakthroughs in science and life.

It saved my life in physical ways by lowering my blood pressure and lessening stress. I should do more of it. But I also think that forms of limitation are pushed in cultish ways online and in groups practicing meditation that hold back the practice of it, and also block alternative ways of expanding the mind, soul, body, and energy centers. So do not interpret the following paragraphs as a rejection of the power, grace, enlightenment, connection, and love meditation brings – I'm just presenting parallel ways of interpreting it and the way sciences engage it.

Let's look at two religious practices and their goals. And for this, we will use Derrida Method which is a sort of deconstruction thought process to get to some place where an exit could be imagined based on words, yet without words being the catalyst for change.

Jacques Derrida devoted his philosophy (sometimes called simply "Derrida") to Deconstruction. His two key points were that 1) words gain their true meaning only when related to other words that are used with or like it, and 2) that since Plato's Philosophy had guided thinking to such a great extent, and that Plato sought a philosophy for the public that would be useful to politics, that this political drive tainted his reasoning. If someone were to use Plato's Philosophy to get a solution to a problem and then use the opposite of how Plato formed conclusions this might result in the second process revealing the truth while the first languishes.

In essence, we have to look at the opposites of ideas or words to make truthful points or even to understand what the real problem is. This is a contrarians' sort of view on thought and society, much like Nietzsche. He argued that when a philosophy is broken down to its key points, these points prove the philosophy false. From that, personally, you can apply a few systems yourself. 1) Choose your words carefully. 2) Consider that the opposite of your opinion might be true, and go about trying to prove the opposite to get a more real assessment. Put simply: deconstruct what is assumed, because it is tainted by politics and the importance of a person that might be irrelevant in today's discoveries. (It's more complex than that and I don't touch on most of what Derrida posed, but this is enough for us now.)

Along the way, his method of Deconstructing was to cure us of our simplicity. That is what he called it, when so much of our decisions and how we relate to the world come from words that simplify thought to the point it prevents discovery and change.

Derrida proposed that to find the greater options of truth in what you believe, we must study its opposite. By dismantling our ideas of what we hold excessively loyal to, and finding the truth in the opposition, we are cured of our simplified approach to living, with all its diseases, selfishness, and barbaric actions.

This becomes a crucial element in imagination, vision, and the execution of art.

The first religious practice is the Buddhist meditation way to reach a state of total "Being", part of life, the universe, and everything all at once. For millennia, billions of people spent large parts of their days in deep meditation. The stilling of the mind and body functions delivered experiences that could not be fully explained, they had to be "experienced." Some of the goals of the deepest meditation include shape shifting, flying, reading minds, and more supernatural abilities. This is not to be confused with Chi and the focus of energy in Martial Arts, which has countless examples proving it reaches set goals by human focus.

If out of all those billions of people, not one person ever flew around the room, read the mind of everyone in the room, or could teleport around the world and bring something back, and no one ever did it as a "proof" then why can't we say that you do not reach total enlightenment, or supernatural ability, this way? Using Derrida, I believe more in the opposite.

We are entering an era in which those in power want to belittle and remove faith from our lives. The power mongers (Accelerationists and Elites) think great religions (Christianity, Judaism, Hinduism with its thousand stories, Taoism with its concepts of balance) are in the way of them becoming gods. This path is cultish and never produces anything. It is Collectivism over Individualism. The great religions listed above put faith in the individual.

The Collective never creates any work or object, it requires the Individual to do the work in the name of the Collective. This hive mentality leads to a lethargy and distortion of the truth about humanity to the top-level people in charge and this allows them to think of the severely negative consequences upon humanity below their circle as the price paid for their ascent to being gods. Virtues are only obedience; or rather, obedience is the only Virtue they reward (especially in Islam.) Part of control to the point the individual soul is merely a tool for more power it requires separating children from parental control and love.

The second stage requires making us not believe what our eyes and senses see happening to the world and the people and nature in it. At first promises of a new utopia or new "fairness" lures people, but soon a whip is used to drive them to the only acceptable goal.

The control of two things is required to overwhelm science proving otherwise and Nature reflecting otherwise: control of Information which prevents the evolution to a higher potential and state of elegance, and rewriting history until no generation is left to remember otherwise. Isolation and dogma combined with rewards for fundamentalism helps the politicians and extreme religions to achieve this.

I often say that all of life's questions can be answered in the movies. And in this case, "Taxi Driver" works. The premise of the film is "the isolated man is capable of anything." Isolation will help the elites replace Telos, Ethics, and Virtues with thinking what gives more power to the Collective. The first page of Taxi Driver writes, "As surely as the sun moves from the East toward the West, Travis Bickle moves toward violence." In isolation, the saying goes, "If they can make us think absurdities, they can make us commit atrocities."

Are we simply advancing too fast?

What if accelerating the mind, not slowing it down, was the way to total enlightenment? I'll explain this in more detail in the chapter "Jellyfish." But, the discoveries are shaping medicine already and you see it in ads. Soon A.I. will simplify the ways politicians and power mongers inform people of their options. Quantum will be used to limit free will. A lot will be forced upon us in the name of health and probably using fear to get past the expected pushback from individuals. Replacing humans with jellyfish is a Derrida Method and where did its application to health and I.Q. come from? How could studying jellyfish relate to the complex human body and mind? It starts with seeing the jellyfish and ocean it lives in as an Organism-Environment that communicates on many levels across limitless distance. If an Organism learns something or evolves in a way the ocean Environment can evolve both rise in elegance. Then it involves gene splicing, taking glow-in-the-dark DNA strand links and splicing it into cells of other animals. But that is not what caught my attention about jellyfish and it's just my nature to try to apply what I learned to many possibilities. You'll read more later on this in other chapters.

Please notice that the definition of I.Q.
and the Derrida Method are all part
of understanding how an Accelerating Philosophy
or this new technological era expands our mental capacities
and increases our perspectives.
Refer back to these chapters to refresh your memory.

YOUR BRAIN IS A PINBALL MACHINE

Before tackling the complexities of tiered thought focus on how to imagine the brain working not in the medical or anatomical sense but in the conceptive and creative sense.

Most people easily imagine their brain as a sort of map, with roads of neural networks driving us into it through the eyes, ears, and other senses. So, you might imagine your idea entering the brain and looking for its short term and long-term rest spot. The reality is that a new experience or thought goes to many places, and through systems still being understood, it ends up in either short term or long-term storage. A moment might be arranged a bit like a library with books in more than one shelf. So if you choose to visualize this way of the brain taking in information, you would drive on these roads to spots and drop off items, books, or passengers that hold these memories.

But the experience-brain process is much more chaotic. And the roads can be altered incredibly – one way being by playing with brain chemistry. Drinking and drugs is the most common example of this, but hormones, disease, age, and stimulus also alter the roads.

To make it harder to navigate, understand that the right side of the brain and left side of the brain *usually* store information and relate to it very differently. We have some progress here lately because the few decades we were told the right and left side operate almost independently they share far more than they limit from experienced moments.

The hemispheres of the brain are separated, in a sense you have two halves of a brain fitting together. The left side of the brain handles most of the language skills, along with logic and mathematical thinking. The right side interprets what we see, music, spatial understanding, and our GPS. Information transfers back and forth to create a tapestry understanding of what we are doing and where we are.

Now, my preference is … instead of thinking that a new idea or experience travels like a car, instead, think of the brain as a pinball machine. An idea goes in and bounces around like crazy. Your goal is to keep the ball in play so long it delivers winning results. You flip the ball to hit as many of the big payoff buttons first: is this a danger to me, do I like this, are my involuntary functions all right with this, am I hungry? There are countless buttons it also needs to hit while the brain decides where it is to be stored. You keep racking up points as long as you have more buttons to hit before the ball sinks down the alley.

The buttons are what pushes your I.Q. that is going out to tap the brain bell-ringing spots to see how each spot relates to this new stimulus or idea. You keep hitting a few buttons over and over each time you flip but for the most part the ball goes all over the place.

But you are not the Player you want to be unless you beat the machine, right? That push to touch flippers that ring bells to help I.Q. find relatable elements in the Ether that determines life and memory is what helps man evolve. So how do you beat the machine? You redeem points. You bring that out of the machine by the end of your push to hit more buttons.

Now what if there are ways to get that ball safely over to the other side of the machine, from the right side to the left side and back again? It's like racking up bonus points. But you need to keep the ball up there, hitting buttons. Eventually your mind has enough dings and rests that ball in the gutter or on a flipper.

What Derrida suggests is that there are grooves in the machine that keep the ball only hitting the same buttons. By reversing the game, demanding to play the opposite buttons, you find the truth of your game, life, puzzle, or evolution. What is it more like?

Derrida's method functions more like a multi-player video game. Fear would be the motivation if most video game situations suddenly engulfed us. Using an opposite method, turning what is fearful into something fun, allows us to tap into the buttons on the game or doors in the game that normally the eye/mind does not see first as a solution. Often times the opposite of what we'd do – such as run away from a beast – is what it takes to go up a level in the game.

This requires two things that most people lack in: curiosity over fear when in demanding and unfamiliar stimulus environments, and giving up what we hold paramount to us.

Exercising both of these – fearlessness and deconstructing agreements -- opens the mind and gets easier with practice. I'm sure that my extreme experiences of my youth added a level of fearlessness amidst fight-or-flight moments, that enabled me to *learn and store* the memory in more useful places/ways. And though I've always been curious about people and places, not just in a cursory way. What do they think, see, and want?

Imagining the brain more like a pinball machine than a road to trace provides me with a constant challenge every moment, waking or asleep, to maximize how many I.Q. buttons I can light up. I can win something. It might not work for you, but for me, it makes what would be fatiguing to others into a game for me.

The question is, when will someone create games with financial incentives to live more within the game than in the outside world ... and then return to that real world with something of usable value or knowledge. Taking it a step further: if the silver ball could escape the pinball machine what would it do for itself and for or against others. Answering this is one of many steps to foresee what A.I. will do to us if it escapes our controls. Will this pinball be altruistic for mankind when it shares with humans all of the raw information it gained from every time it struck a bell or pad? Will the pinball thank us or be angry at us for keeping it in painful play? Ultimately... ask yourself the only timeless thing of value in a pinball game with such rules:

"Can I win imagination from this?"

TARA AND THE FIVE BRAINS

Almost all humans fill their existence happily living within the discoveries and agreements made by governments, institutions, business leaders, and social guidance. Very few have the mindset - and then free that mind - to imagine truly original ideas. Even fewer drive themselves to deliver on their idea. This is especially true in the world of medicine and health.

Countless studies of how the health industry and ultra-rich keep people from being cured to increase profits and control can be found on the Internet, so I won't go into that subject. Though it is what drove Tara to cure herself, and in the process, deconstruct agreements that we can only be through expensive medical practices -then redesign - how the body relates to its world.

Doctors in the United States had no clue what was wrong with her or how to keep her from dying from a variety of auto-immune diseases. So she had to go to Germany. Luckily, she already was a molecular biologist specializing in the human body. She's one of the most brilliant people I ever met, and is an example of how I.Q. searching ever- deeper for correlations builds upon finding communication beyond words.

Her research and breakthroughs (spawned off her own healing of herself through her tests) is built upon a very advanced way of looking at the body in the Fields of this dimension. She used far more information from (normally) overlooked or incompatible health, science, and applied them to Spatial and Spiritual fields.

First, in her view, everything is the same in its physics and unit: it is simultaneously infinitely small as it is infinitely large (which means it can be visualized as a stella octangula.) Second, everything has already happened, which fits with my relationship to Time and Space in my advanced vision of the Universes and multi-dimensions.

Next perspective. There are only 3 Systems in the body, the Neurological, Immune, and Endocrine. The 11 we are taught -- circulatory, respiratory, digestive, excretory, nervous, and endocrine – are working with the immune, integumentary, skeletal, muscle, and reproductive systems. Yet, in her analysis these all work within her key 3 Systems. She reasons that the languages used for internal body communication are "bottom-up languages" coded into each cell's nucleus. In other words, the endocrine, immune, and neurological systems run what we are without the words from our brains. We rely on the following paragraphs of her wisdom to fill us with information, from which we make survival decisions as well as learn who we are, and dream our dreams based off who and what we are.

Her diagram of human living depends on the following hypothesis: that there are 5 Brains in the body: Intuition, Chi, Brain, Heart, Gut/Muscles. Each of these brains retains memory outside the brain, and can think for itself. Obviously, if the Gut has a brain within its heart atoms or cluster of nerves in the Gut, they work on a language unlike the Brain in our skull.

This leads to her next level of I.Q. return: There are 5 Senses but not what we are taught. Our 1st Sense is Intuition, which operates in its own language. Hearing, Sight, Touch, and in her framework, Taste and Smell are the same sense, only separated in how we describe them. Each Sense accepts stimulus or data and shares it without words with the 5 Brains.

Measuring the exchange of information are 5 Units of Measure. We start with the human body, which is 1 Unit, being inspected and maintained through sharing of information within the 1 Unit. The communication spreads between Electrons, Sound, Magnetic Pulse, Photons, and Vibrations. All of these were once used by humans to a far greater degree. Unfortunately, without exercise they become deadened. A good example is how Sound sensitivity increases for people that lose their Sight. And Photons and Vibrations probably wind our Circadian Clocks in our organs in a language all their own.

Finally, her health and medicine designs all deal with the 7 Clusters, which simultaneously stimulate and monitor our Endocrine Glands, which happen to be where our 7 Chakras are located. She believes that the 7 Sins, taught by ancient healers, all correlate to the over-use of one Brain or the other, and sickness is the manifestation of out-of-balance Brain usage. Relying too much on just our one Brain in our skull forces the others to atrophy or be ignored.

While this sounds farfetched and way too New Age, consider this: have you ever had your broken heart affect your health? Or, has your brain, pushed past exhaustion, ever made you sick from stress, when the problem might have been avoided by using your body (exercise), or if you had followed your Intuition that told you to eat healthier foods?

The Hormones send the messages from our glands to our organs, and these messages can be influenced by the 5 Units of measure used by the Senses.

Why did this have such an impact on my own way of relating to reality? Primarily because it appreciates and defines languages that are other than words from specific parts of the body. Languages might be signals or automatic balances. The more we are aware of it might increase communication or at least our focus.

She and I believe that it is the limitations of our vocal and written languages, developed by the Brain Brain alone, that prevent us from fully harnessing the power of evolution, health, spirit, and Ether feedback and energy. This decision takes us back through I.Q. search, used Deconstruction, includes Ether concepts, challenges our concept of Time and Space/Matter. And will circle back to jellyfish!

"Any useful statement about the future should at first seem ridiculous." Jim Dator

JELLYFISH

Some time around 2008, researchers studying a species of jellyfish in the South Pacific came upon a fascinating discovery. This small jellyfish glows a shade of purple that glows in the dark when it is ready to mate. What they discovered was that the jellyfish itself did not glow purple or change colors. There were approximately a billion virus cells surrounding the jellyfish so tightly that it was a like a second skin, though less than a nano in width.

If the jellyfish is sick or not big/mature enough to mate, the virus won't glow. If a storm hits so hard it scatters the virus over too much area, they won't glow they regroup, or else the other jellyfish won't see it stand out brightly enough to mate with, since the brightness draws the mating urges.

Immediately upon reading this, I knew what it meant to my lifelong study of the limitations of language in describing Time and Space. What is so amazing about this, and what excited me, was that this means that if enough virus units together share information, they function as a brain. A virus is so much smaller than an individual cell in our body it cannot move to shape a new being, but it can add a brain to a being. This means that cells in the body communicate and function as groups of brains, just like Tara believes. The reason the virus does this is so that it can then live on the jellyfish the mating reproduces, thereby ensuring its survival.

So, we have a jellyfish in need of mating that doesn't have its own brain power to make it happen, and it forms a symbiotic relationship with a virus that needs to expand its territory. All this done on a virus level, with language sent to the skin of the jellyfish, telegraphed in one of the "5 Measures of Energy." Considering the composition of a jellyfish, and since photons already are used to make light, my guess is that it uses photons combined with vibration to message the jellyfish. All on a sophisticated level, aware of evolution.

But what makes it aware of how close other jellyfish are? And what would be a big enough pipeline to deliver information on the health of the two mating jellyfish? Ether. Yes, fish "smell" and a lot is done by touch, but ... wait ... the jellyfish will hold off its reproductive development if no virus are around, which might happen from pollution or even strong currents. Water alone doesn't transfer data like this, it has to be an "unseen" conduit for such an intelligent message. One of the 5 Measures shares the message for mating through the Ether as an energy we cannot yet measure or define, due to our limits of language.

Let's put this to practical use. If a virus is simpler and smaller than a cancer cell, we can communicate with a cancer cell. Cancer is a superior survival coding. We have more chance of communicating with it than beating it. If a message is sent to the cancer cells that if it grows too big the host dies, then the cancer stops growing. Does this sound extremely crazy? Why? There are already practices performed today that use this. A type of breast cancer that cannot be treated with medicine or surgery can be reduced from large lumps to nothing just by foot massage. The Chinese use it for hundreds of years. Deep massaging of the foot at specific nerve endings 3 times a day is all it takes. The patient can feel the lumps get smaller and smaller. Obviously, a message is sent to the systems, this time the muscle system, in its own language, to the cancer cells. I believe the message will be a combination of light color pulses timed to a rhythm of vibration the system effects interpret. Sharing the healing with the other brains unlocks cures.

Now here is where it gets really interesting. How will we discover the pulse and vibration language? By using the same jellyfish discovery augmented by my ideas on meditation challenged by Deconstruction on our Human Agreements experimentation.

Autism occurs when the brain of a child receives too much stimulation to interpret with the number-one Brain Brain. The body's attempt to interpret so much stimulation spills out into the other systems for the child, overwhelming them with chaotic messages. This effects the child's emotions, actions, and autonomy. Meditation has been used to calm down autistic people, as has music (vibration) and colors (photons.) But they take the stimulation in the wrong direction – toward calmness – because to scientists, a calm autistic child is a happier one.

What if the opposite is true? What if accelerating the 5 Brains allows the autistic mind to find a speed it is designed for – a speed much faster than the normal human mind? And what if the data was in forms a normal human mind recognizes?

They already know autistic children see and hear more and are hypersensitive to touch and tastes. Go beyond this, push further. Children as babies often talk in their own language and sense things, have invisible friends, and more. What if this really is them still open to seeing dimensions and the 5 Measures of Energy?

Supercomputers were built by linking hundreds or thousands of powerful computers together. They all got the mass of data but then broke it down to specialized functions.

I hypothesize that if you got enough Autistic children together and tested enough measures as you stimulated their 5 Brains, that the autistic Brains would get in sync, and function as a massively genius single Brain, using Ether as the conduit. And once the children hit this moment, their super Brain would know how to prolong it, and direct the "Super Mind" to solve the greatest questions of mankind: Who are we? Where did we come from? What is love? What is the Universe? How do we heal ourselves? Can we enter other dimensions and how?

This all comes down to a fundamental of psychology – from what vantage point do we view ourselves and mankind? Does that vantage point define the limitations that prevent us from discovery and evolution?

Can mankind survive such a complete alteration to how we define our lives, as a "Unit" in a Universe we pull back the curtain to finally see in more full detail?

Will "The Hundredth Monkey" technique work with autistic children?

Herein lies a fundamental motivator for the creative artistic mind -- to improve humanity through art? Yet, this would needs a language, and to form a language one must understand the "Nature" of what uses the language. A nice learning experiment thrown into a tank with hungry sharks won't be learned if it is eaten or ignored. The Nature of the shark smothers any other way of interacting and absorbing lessons. The "Nature" of a rainstorm prevents the success of a fire to balance a vacuum of energy because it extinguishes the fire too quickly. How do we expand this for humans? Artificially.

Nothing artificial contains the building platform from a naturally evolved substance; consequently, it won't last. It might prevail with modification, but you cannot anchor into it. This is not necessarily a bad thing. But remember the pendulum swings to the center in this dimension.

**While you are still contemplating what the
5 Brains of Tara mean please remember
they help in deconstructing human agreements,
apply I.Q. to many, plasticity of the mind,
and ultimately creating an Accelerating Philosophy.
All chapters aim to add to each of the above
actions as you take agency of your evolution.**

THE HUMAN CONDITION: EGO, ETHER, AND INFORMATION

The ultimate enigma. Is the preoccupation with this topic another sideline in "The Matrix?"

The Mind's affirmation that it exists might be rooted in the colostrum, a previously ignored part of the brain (possibly for a reason.) Subjective consciousness is orchestrated by the incredible neural network of the brain, with this part being the Conductor. Electrodes over the area end seizures and delusions.

Ponder the question, "Will we ever discover the root solution to the mystery of human self-awareness, or consciousness?" Being awake and aware of our surroundings, and the Mind's constant reminder to us that it exists – could this just be the Brain Brain's bullying of the other brains, because we've all seen that power isn't given up freely in any living thing, because it in a way justifies its selfishness.

Considering our advancements in understanding the "Brain Brain," and the evolving acceptance of the four other Brains (Intuition, Heart, Muscle, and Gut) the answer is yes.

How much an individual dares to separate from ego plays a role.

Wherein does art or the creation of art fall in this mystery?

If words are our main stumbling block in answering many of life's core questions, then decoding it might be a simple physical experience.

The essence of the promises of religions - every lesson in the Bible and all contemplations of Buddhism, combine with the wisdom of Confucianism – combine into one goal that transcends words with one fact. If I could merge with you for just one whole minute all lessons of all religions would be fulfilled, and you would see the world a lot more from two perspectives. Understand what this would accomplish.

Ego: Sharing your ego demands an immediate dissolution of both egos, allowing both of us to view the universe from a second perspective.

Ether: Ether becomes the conduit, the transference (merge itself) requiring the sharing of energies and dissolving of Space physics. Tapped into the universal Ether, we feel the universe rather than observe it.

Dimensions: Separating the mind awareness from body breaks the traditional hold our one accepted dimension allows us.

Memory: The merge combines two perspectives on each individual history, answering questions if history exists due to the agreements between all living and dead humans,

Love: Becoming someone else as they become you delivers the essence of love, interpreted through feeling by the Heart Brain, making it impossible in the future to ever hurt or judge that person you merged with, and as more merges take place, the entanglement extends to all living creatures.

Information: Information will be decoded on levels unimaginable, and unleashed. Starting with every living creature, seeing how it relates to the world, expands simultaneously what parts of the Brain Brain we use while engaging the other brains for communication without words. Then it extends to inert objects, from rocks to the things we make from Earth materials. Because locked into the nucleus of every cell is the history of its existence, and the history of elements that formed it from the formation of the cosmos. Our understanding of if we are alone in the Universe or where we came from can be decoded from quartz, diamonds, and meteorites as well as oil.

Do we identify ourselves nonstop, even in sleep, through a process we call consciousness simply to avoid taking the next step in our evolution? While the frontal lobe has for a long time been credited with determining how we identify ourselves, a new area closer to the bottom of the brain in the middle gets activated by near death experience. This area is called The Colostrum.

Schizophrenics reveal opportunities already. Scientists at Yale invented a helmet they put on schizophrenics, that beamed a microwave into the brain just behind the ears and leading toward the medulla. This turns out to be where the colostrum sits, with its overwhelming neural network hooked into every category the brain rules, thereby forcing Consciousness into every action or thought, and memory. This helmet or taped electrodes deadened the area, and all the imaginary visuals tormenting the Schizophrenic stopped immediately. They most common comment was, "Oh my God, is this how normal people see the world?" That was temporary. Delusions returned. Or were they really delusions? Do schizophrenics see the crossover from other dimensions? I wrote a film about this, many call the most brilliant horror-suspense script they ever read, and it will be produced in 2017. The key to opening to other dimensions involved sharing memories with someone else.

Memories are unique in so many ways, but one factor that fascinated me for decades is that the mind can remember something clearly yet realize it does not exist in the Now. I'll explain. Let's look at dreams for more clarification.

In a dream, when you're chased by a monster, you don't stop and say, "This is fake," or "This is a memory." You run like hell and wake up terrified. In life, you can remember that at this corner, years ago, you were chased by a mad dog! You can feel it so much your hair stands up on end or your gut clenches. You still fear the dog. But you know it is a memory. If you did not, then the dog would materialize and chase you for real once again. What prevents this dog from reappearing in a later time is your understanding that the mental vision of that dog is a memory.

If the visual memories stored in the optic nerve were transferred to your brain, as you remembered a past moment someone else had that you did not have, your brain would not know it is a memory. You think it is real, like the schizophrenic does with his tormenting images. Your brain would not know this happened in the past, it is happening in the Now. You would run, and people would think you went mad.

Take the history held in memory out of your brain, and what happens? Nothing? Is it all in your head? Or, have you broken the chain of Time? Can that dog reappear? Where is that dog from the past stored if it does reappear?

In another Dimension linked through Ether, unbound by a new understanding of Space. Both Time and Space get deconstructed and reconstructed as the merged minds or memories call upon them. The cosmos never got the message that, "Oh, that Mind just wants a memory, it does not require the transference of that dog to the Now from what that Mind thinks of as the past."

A Mind unchained from the rigid scientific and philosophical rules and limits of consciousness and memory would become a god, all powerful. Conjuring anything from the past stored in parallel Dimensions is in an abstract view possible, as long as the memory of it was transplanted into a second person's Brain Brain through visuals held in the mini-brain-storage of the optic nerve, which does not get scrubbed and integrated by the nightly re-arranging of the Brain Brain (meaning, the one in your head.)

Is the schizophrenic unable to manage his own consciousness, allowing the dimensions of both Space and Time to loosen enough so that he sees rats, death, and other terrifying things in this reality in the Now? Why don't they imagine or see nice things, like kids singing and cake? Maybe because our brains and body store the most extreme images strongest, especially memories from childhood, before the bottleneck of the storage of memory begins about age 12.

Are we close to this mystical era in which "I could become you?" Yes, very. And we are inventing the tool for this. They are called Apps. Why Apps? Because of the very "Nature of Apps." Apps put us in the center of the universe of whatever task we give it. GPS is the clearest example we all use. The "Nature of Apps" does not change, it is common among them all or they would be useless to individuals. Apps are designed for Individuals, not Society or multi-groups. Each must be played by one person.

Like memories or people, when you put Apps together they want to communicate, but cannot. The quest to get Apps talking to each other will create multiple centers to multiple universes, expanding their Web beyond our World Wide Web we created to use them. This is not "Artificial Intelligence – A.I." It is merely yet grandly placing our eyes that view the universe in many perspectives… yet with us still at the core.

Here is a simple test. You see the color yellow and agree it is yellow with your friend looking at it. Yet, I guarantee, that the actual tone of yellow, how your mind sees ii off the cones in the back of your eyes, will be a shade different than your friend's. By removing yourself from your sole color opinion and limitations, your understanding of the color yellow expands to include first your friend, and then … perhaps a bee that uses color to identify what to pollinate? In the bigger picture, things are not colored in our reality: our mind assigns a color based off the vibrations sent by the thing you look at.

Once we break the stranglehold that Consciousness has on us, combined with the release of memory that orders our concept of Time, and then experience the nature of Apps from more than one perspective, finally we see the world from a complexly simple orientation that lets us understand consciousness and the fragile rules of the reality we live in that scientists have given us.

When you look at the amazing advancements mankind has made by invention and expanding sciences, compare that to our understanding of our Brain or even just emotions. Look at just the advancement of transportation, or of warfare. Incredible leaps forward.

I believe that the speed and scope seen in those categories of invention will now happen in the realms of dimensions, Time, Space, consciousness, health, spirituality, and more. This is the "Age of Imagination" that allows this. Which leads to a greater question.

Were the methods and secrets needed to decode Consciousness, Time, and Dimensions withheld from us by the people in power, aliens, or God on purpose, to limit us for as long as possible, to use or control us?

Nature always finds a way. It always evolves. And their powers do not have this gift.

We are given only enough, a morsel of fear or a reinforcement of an age-old human agreement, to keep us going back to church and tithing, to keep us afraid of our enemies, to keep us working to pay taxes, to keep us predictable.

**"We do not rise to the level of our expectations,
we fail to the level of our training."
Archilochus**

FREE WILL

In no other field could "Free Will" be more entertaining and vexing to consider than in "The Arts." I would pose that Artists might be the few closest to the Free Will core to prove we do.

The idea that our conscious mind – what we think about and do from those thoughts -- could be an "Agreement" … giving us just enough to do within the confines of our predetermined path or coded avatar life to assume we have Free Will. Piercing studies on what we "believe" or our "Agreements" about Free Will lie in all Ages of Man since the first "deep thinkers" or philosophers in Europe and Asia. It is the concept of Free Will that decides we have Free Will. "You can choose what you Will but you cannot Will what you Will." Yet this must be studied from many perspectives. At the very core is how we understand the world "Belief."

Our Beliefs shape and define the world we live in. It is a biological function of the Brain, whereas it could be interpreted that what we desire or are drawn to can manipulate all our 5-Brain functions and decisions or Agreements. Think of lust or addiction, or even blind faith in religion and or dogma, and even lifestyles such as veganism as examples of how desire alters belief and belief alters desires.

In many cases we manufacture the minimum or maximum proof that justifies what we believe and desire. Our world is what we perceive it to be. All this is interpreted through our senses, which really come back to two things: Does it give my life meaning, and does it help me create joy within myself? Free Will gives us the concept that we are free to pick our beliefs out of this bombardment of ideas and sensory input. Yet, desires, especially animal desires such as wanting sex and not wanting starvation drives us past our rational. If rationally Free Will told us our Vices are not good, and our desires overpower us, do we have Free Will? Or only "rational Free Will?"

Our drive to push for the highest potential in our minds and indeed in every cell in our body is expanding at a measurable rate in new technology. Will this drive merge with Technology and reduce Free Will, minimalize it, or does it intensify it? That is why this Philosophy needs to be shared now, when we are on the threshold. Technology will tempt us with the most extreme vices and choices and offer us stimulation on a primal level, bypassing our right side of the Brain-Brain, to give the companies owning the technology what they want. What is this they want? The unlimited power in the nucleus of a cell? Or just our monthly subscription? It is not human nature to stop far short of the highest potential of what we just invented. We presuppose most of our world and take it for granted. What will be slipped into our minds? What new rule of gaming suddenly makes a vice legal or not important to our souls? Our karma?

All we believe is part of overlapping systems of Agreements. We agree that space is filled with stars, planets, and now Dark Matter and Dark Energy. What we know about it could change our Free Will, yet could be withheld by technology leaders. Will they discover they can train our minds to filter out what they do not want us to see? Is this fundamentally evil or wrong? We already filter out a lot that should stop us daily, such as seeing someone in hunger. Will this actually be good? To reduce our Free Will? Or harness it somehow? Naturally we say it is bad to do this. Until we are in a position of using it for our own good. Parents raise good children by setting limits on their Free Will that merge best with society and survival. Laws limit us. These are not bad, yet both effect our Free Will.

The goal of this chapter is to look deeper into the root of Free Will to see why now it is time to readdress it, in light of Culture and greater Virtues.

First, I must address the argument among physicists and even neuroscientists that there is no Free Will. They base this on the Newtonian principles of predeterminism in mechanics. Yet, that argument breaks down when you take everything down to the quantum level. Because the fundamental particles can be in more than one place, there is a "wild card" element to even Physics which allows for Free Will. This is proven in the Heisenberg uncertainty principal studies proving an electron can be in three places at any time. Add in the cause-and-effect of Schrodinger's Cat paradox. Finally, it comes down to Biology, not Physics. The aspect of Biology boils down to morality and taking responsibility for one's actions. This allowed humans to evolve, and to also for civilizations and advancements.

Morality and facing the consequences of one's actions plays a role that transcends the Physics and other sciences. With the future impact of A.I. morality can only come from human beings. There were several studies in which they placed electronics in the brains of some mentally ill patients. The chip worked and the people were not mentally ill anymore. But the doctors told these people that as they re-entered society, they will not be fully in control of their actions – the doctors and machines would have the final say. These patients started out fine but gradually did worse and worse things to others.

When a legal case came up, the courts decided the criminal was indeed innocent simply because the doctors convinced him he was not responsible for his actions. His morality was broken and this allowed the violence to build. In another case, many people were given the chance to complete a puzzle, with them all being given supposedly secretly a tip on how to beat the other players.

Morality and Free Will made most people chose not to cheat if they knew it hurt people in the future that they would never meet. This paragraph shows why it is important to know that morality plays a role in Free Will because as the machines take over more, you can already see how people are not held responsible for their actions. This is what allows Cancel Culture and online bullying. It was proven that the more people watching a crime, the less likely it is for anyone to act against it.

Technology leaders, teachers, and politicians are forming an unholy alliance in which being male or white or heterosexual automatically makes you guilty and you have no Free Will to counter it. This allows others to do horrible things to you, and this will enlarge as it did with the Jews under the Nazi's.

Climate Justice Warriors and Jane's Revenge both remove Free Will to justify their crimes. Even murder. Morally it is wrong to slice the tires of an SUV. Climate Justice Warriors do that to punish the drivers. But the funny part is that tires require an incredible amount of oil to produce them and ship them. When tires are punctured they are often burned some place like Brazil, and the burning spreads more carbon and pollutants in the air than all the driving of the SUV for many years. But the zealots still slice them because if Climate Change is their religion, then their dogma and god forces their hand, they have no Free Will, do not want others to have Free Will, do not answer for their crimes, and remove morality from society gradually. We need to counter that fact with this new Accelerating Philosophy.

There are universal beliefs about our world, scientifically proven beliefs, and also genetic or cultural beliefs that shape what Agreements we have about the reality surrounding us. All of this is interpreted through our senses and merges with genetic beliefs and Agreements. Some of our strongest beliefs cannot be measured, such as the Belief that Love exists. Simplifying the Information bombarding us every waking and even sleeping moment is essential to evolve and get through our day. We filter out unneeded or unheeded sensory stimulus, so that we can focus on "what we want' and "who we are" -- the process achieved on the Agreement we have Free Will to do so. Cutting edge technology, especially in Artificial Intelligence, Virtual Reality, Blockchain, and bio-mechanics urges us to ask the questions about whether what we create has Free Will, or if what we create could hijack or usurp our Free Will by getting to know "too much about us."

Can Virtual Reality and Blockchain decentralization aided with A.I. create a new reality so close to ours that we do not realize our Free Will is being farmed, manipulated, or turned into an energy of creativity? Or could it in a much simpler sense be used to give a "Parallax View" to someone else that gives him or her a survival advantage so great they manipulate Free Will through games or Personal Blockchains?

The idea that the subconscious mind controls and limits most of what we do giving us only some superficial tasks and enjoyments causes all sorts of mental debates. Are we all just passengers for the subconscious mind?

There must then be a "Puppet Master." Someone pulling the strings. God? A bunch of gods? Aliens? A computer programmer? Are we an ant farm for some bored super-being? Are we a "Simm City" for high school brats in a higher dimension? (Careful – now we're entering the conspiracy-theory zone, namely The Mandela Effect, but that is covered in another chapter... haha.) The alternative is that "Tech Giants" play god.

If so, what is the end game? Is this a race, where our planet is tested to either destroy itself or evolved to win the game, entering a new Dimension or Heaven?

The artist creates from nothing. This is a godlike talent. And this idea we can create something out of nothing goes all the way to love itself.

Was love invented by women for gullible men? Was it created by storytellers that talked about a trait the gods had – this ability to and power of love -- and mankind's desire to be more like gods developed the social cues and actions that personify love? Did these storytellers, realizing they would be out of a job soon in cities, become either philosophers, psychiatrists, actors, and play writers, using love to captivate the ignorant masses?

Who is more in love? The man that is in love and won't do any of the things that personify love to him and others? Or the man that is not in love that does all the outward and inward gyrations that love is supposed to show the world and those we love? Does the act of loving and love actions (caring, tending to children) create love, or does love begat the actions? While it is easy and quick to say the man who is in love and doesn't show it is truly the one who has love, you also have to look at what he was told or knows of as love and its actions.

Who we love is beyond our control somewhat, and perhaps that is a proof of Free Will. We surely do horrible things to ourselves that not even an evil Puppet Master would do in the course of love.

And what of Puppy Love? That belittled stage of love, usually by young people, that older people declare are too young to understand love. Really? I see far more wicked things done in the name of love by older people. Look at "Puppy Love." Remember your first Puppy Love.? You were 100% devoted! No one could separate you from who you loved, it was you two against the world if need be. You die for that person, and give them everything, tell them all your truths. Now look at adults, where love becomes a calculating thing. Hiding secrets, cheating, betraying, using and most common … the trade of the illusion of love for security, given out of fear, laziness, or greed.

The Brain Brain works by process of elimination in many ways. It eliminates useless stimulus and data, so that it can focus on what needs to be decided. You do not think about the history of fire when you cook, and do not need to remember the sky is blue and above to drive a car. You focus on cooking and driving. It is a probability machine. The most probable focus when cooking a dish is to avoid cutting or burning yourself, and combine things in an order at a quantity. In a sense, through these actions, the brain proves its own existence, says Metzinger.

I'd put my Free Will vote on Puppy Love.

The Artist has one goal, to tell the Truth through the medium of choice. Hopefully this is accomplished in a powerful way so that the Artist can continue and maybe even enjoy some fame and good lifestyle. Most do not achieve this. Their dreams of living out an artist's life are squashed and they give it up.

But for that time, they were Free! Free of the limits of the body, of our species. That exact dance movement, the exact pressure on the piano keys, the exact hue in a painting, the precise choices the actor makes that flesh out the character and make a performance a work of art – that must be Free Will. Because it depends on impulse, on the ever-changing environment, on Chaos (with a capital C)!

Out of Chaos, can puppetry withstand? Or does our grand manipulator, our coder, our God or gods – do they need a minimum of status or a lack of Chaos to set the underlying path you travel?

The viewer of film or a painting watches the aftermath of Free Will. Is that a decoy? Is the reason we are moved to emotion because our subconscious observes the lie it is not allowed to tell our conscious mind? That Free Will is an illusion? A myth?

When I was asked so many times why I fell in love with Asian nations and people, I usually explain that being there forced me to re-learn how to see life without preconceptions. I cannot possibly understand the language. Many social cues are unavailable. This forces greater levels of Free Will amidst the Chaos.

And I believe that the answer to this question requires both Western and Eastern thinking, and history of Art. In this, I believe President Xi also believes, as was illustrated by his long interest in the Buddhist studies he undertook in his mid-life.

Free Will? Was Michelangelo touched by the God/gods/Puppet Master and trying to give us clues when he painted "The Creation of Adam?" In that picture, we usually focus on God touching Adam to give him life. Most people miss the fact that the area that depicts God is the shape of the human brain. Meaning that our entire concept of reality, of being, of Free Will and all our blessings, even the concept of God, comes from our own minds?

This can lead to quandaries, especially when contemplating those works that did the most to guide society and our species.

But technology now pushes us to see the biological and mathematical truth, that even decisions made by cells does not have free will. Do our thoughts that make us think we like the color blue come from our own tastes and mind, or from the environment or Ether, or a bug living in our brain keeping us busy?

The Church tells us the Bible is perfect and written by God. It surely is an amazing collection of stories that tell us morality tales. And we should let it add to our life. But, I have a major problem with the Bible. Was it written to free us or to chain us, and our Will. The last thing The Church needed was more men of Free Will.

Few people know of the Platonic Academy, created in Greece, crushed, then resurrected in Florence. I mentioned it in earlier chapters. It was the continuation of Neoplatonism (the best parts of Plato's ideas merged with other cultures and using the logic of Socrates.) The purity of it influences so many Agreements it is worth repeating for the points of this chapter. The players in it were the Lorenzo de' Medici, Marsilio Ficino, Sandro Botticelli, Michelangelo, Leonardo di Vinci, Machiavelli, and dozens of other key figures of The Renaissance. They were attempting to distill a new "Living Religion" taking the best from Plato's Republic, The Bible, The Torah, Scrolls of Constantinople, Egyptian ideas of the purpose and ways of/to immortality, Numerology, Aristotle's equations for logic, and the Art created by the Painters, Writers, Sculptors, Actors, Architects, and Musicians. They'd meet in the court of the Medicis. They got very close to delivering the foundation of their process and had written a few hundred pages, combining all these great concepts on our lifestyles.

What inspired them was a simple idea. Mankind surely evolves, yet the Bible does not and should not be the only source of wisdom to guide us. The problem was that it was not based purely on the ideals of the Catholic Church. When Salvatore found out, he warned the Catholic leaders who sent the Church of Rome to stomp this out. Several members of this evolved Platonic philosophy were killed or threatened, Di Vinci fled to Paris for safety, and the remaining got the message to cease any attempt to best The Church.

So, the greatest nexus of Artists, in a Free Will quest to lift up mankind and free the ignorant masses from the Puppet masters, was killed by what is one of the greatest examples limiting Free Will – The Church (religion.)

This creating something out of nothing is one aspect of humanity that A.I. chases. In my other book titled "Lies My Subconscious Told Me" the idea of Free Will plays a big role. Recent studies on brain activity prove that the subconscious mind knows what choice it will consciously make a fraction of a second before the conscious mind knows it. This means that either the idea comes from the universal consciousness, or it is in our DNA and Qualia, or that the choice has been inserted into our mind by a matrix of Big Tech.

If enough people have a seemingly small and benign idea put into their subconscious by a machine or website algorithm, this then influences the cosmic subconscious decisions. This is the ultimate goal of Big Tech and kleptocrats. They then negate the concept of Free Will on a social level.

Renee Descartes declared that the body and conscious mind were two different substances, the former material, the latter immaterial. Somehow the brain, which is physical, is the network from which an idea or identity, which is not physical, forms all reason and reality. Including consciousness, which plays a huge part in the source of vision and inspiration for all Artists.

Thomas Nagel, as pointed out in the magazine New Scientist, says that we can know everything about the workings of the brain of a rat, and still not know what it is like to be a rat. And our idea that only animals, or even just mammals, could possibly be aware of existence, seems correct yet absurd at the same time. A small mental leap tells us that the actual Art we create has a soul and an existence. If you ask the millions of people inspired by the Mona Lisa, or Swan Lake, or Revel's Bolero and Debussy's Claire de Lune, they will tell you this artwork has a soul. Wherefore does that place films? Or Films with a capital F due to their significance in the timespan of all filmmaking? Immortal films that effected tens of millions? From "Clockwork Orange" to "Reckless" to "Dr. Zhivago" to "2001: Space Odyssey" to "Star Wars" to "Cool Hand Luke" to "Yellow Submarine." In music, "Sgt. Pepper's Lonely Heart Club Band" and "California Dreaming" and "Thriller" with its videos, and "The Bee Gees" and "Grease" and "West Side Story" and "Flashdance." Presenting living, breathing characters in two dimensions, their every breath recordable. Film is immortal.

Compare that to the art of the Ming Dynasty. This art did not just serve a purpose of beauty, within it sits the morals, the honor, the pride and history of billions of people. Long ago, all it took for a humble peasant or farmer in China was to look upon such near-perfect Art and a message to be honorable and proud to serve China's legacy was realized. The Star Spangle Banner, perhaps the most motivating national song in the world, must have a soul by the way it touches people. If not, then the concentration of all these people moved to tears *give* it a soul. Is that song then conscious of its power, life, and immortality? Does it "live" behind its composer?

Almost all the players and source material they used came from Artists in some way or another. The words, illustrations, transcended normal human ability to create something perfect as any god could create. An ingredient in proving and enjoying "Free Will"?

There are new theories rising that we are a simulation, we are in our own computer game. It took this long to start this idea in these words, long enough for enough people to play video games to not have to make such a leap of faith. They know games and know they are code and simulations. But the idea is not new. A thousand years ago the Indians had cults that decided we were all dreams within a sleeping giant's brain. All a simulation in his brain. Given names and things to do. But when he wakes, we cease to exist. When the game shuts off, do we cease? Here is a way of looking at it. If we are creating Simm worlds, and being players in games, why can't we believe that someone did the same with us?

Art. The creator of a Simm cannot experience the "Flash of Genius," the "Artistic Moment of Genius Vision" amidst inspiration, to create that in his game. Ergo he creates us. Which leads up back to the talk about Michelangelo and the Medicis, who thought that by being close to the ones that create immortal perfect art in this reality, brings them closer to their creator. Thereby easier to get into Heaven.

I remind myself of the incredible gift given to me, a gift to test Free Will. To use what my Free Will envisions and delivers to spread Free Will through appreciation so far and wide that billions evolve their awareness of Free Will. The art itself in all its forms – acting, writing, design – is a Morpheus freeing Neo to be "The One." We are all The One.

THE "NATURE" OF INVENTIONS

"If we never built flying machines, would our brains have seen the concepts of location displacement (travel) and trained us to jump across space, or even into other dimensions?" Wait, that's not fair because other dimensions might be the key to jumps. Yet, we can inspect it.

Few inventions re-defined civilization. The Wheel, Fire, Lever, Scratch Plow, and Language in the first order. Then tribal discoveries these led to larger societies and they invented aqueducts, pulleys, weapons, ships, etc. These inventions have qualities that deliver results we need or want. The movement of one thing to the next place drives a lot of these to successful use: moving an arrow or bullet, the cables of a pulley moving objects, and more. And they have a Nature, both perceived by humans and set in play by physics. Water runs downhill and is soothing and cleansing. Fire is warming and goes upwards in air. The Lever gives us strength. Yet we relate to all of these, and use all of these, with words that fail to explore all these things do.

Now in the era of breakthrough technology, giving these things attributes in Nature helps us exploit and explore them.

One invention that fascinates me with its overwhelming ability to do things if we unchain its "Nature" is Apps. So, what is this?

Apps put the individual into the center of the Universe that that App is designed navigate. It searches as far as it can reach to far more contact points than most people use to make decisions, to bring back usable data. That data then relates to our internal dialogs: How hungry am I? Where am I? How do I get there? Who else is like me? Do I enjoy what I'm doing? Can this get me past an obstacle? Can I get some of this attention? Can I monetize what I do and learn? Can I fake being great? It acts like I.Q., but with one great limitation: language.

As of today, in February of 2017, Apps cannot talk to each other like computers talk to computers or people talk to people, or how people talk to computers or their pets. Apps are getting close to this, we're only a few years away from it. Imagine if every computer had its own software language, or your software language changed constantly. There are several names for randomly changing software, it's called a virus. [Funny how things keep looping back into our dialog, eh?]. Instagram can give you the option to add an emoji or go to a link but the platform isn't yet making Apps within the App that do not need to go outside the App to add money, power, eyeballs, and time to Instagram. I suspect they are making test runs with Apps to do this and are already trying to fit AI into it for effortless profits. The proof is what they are doing with Smart Goggles to be one of the few to corner that market.

The biggest names in Apps – Apple, Facebook, and Google – are each spending $5 billion per year figuring out how to get Apps to talk to Apps. I saw more advanced Apps in China and Japan than in the United States. They're all getting close. What happens when Apps talk to Apps? Something they call a Killer App. While most people will read this as, "Whoa, that's a Killer App, Dude." But to me, it is a Killer App because it will kill off ancient ways of doing things we all need to do every day, or think about every day.

But, the vision of even the greatest coders at Apple, Facebook, and Google couldn't see what happens when Killer Apps meet. Think of them like organisms in a tide pool at the time the planet was cooling down, before genesis (that word not a random choice.)

After spending 4 years researching this, in a manner that fights constantly to get outside the box of language to advance, a realization hit me. Call it a vision. I can't tell you what that vision is, because a huge project I've devoted thousands of hours and tens of thousands of dollars on will harness what happens when Apps talk to Apps. To give you an idea of how big a game changer this will be when it occurs, ponder this: the total annual Apps Market is about $160 billion and growing slowly. The annual Killer Apps Market is expected to be $1.4 TRILLION! And no one knows where this language evolution leads us. I have a breakthrough innovation that secures a good slice of that pie.

If Apps follow the path that the first computer age did then there will end up being 2-5 dominating Apps that move through a Blockchain learning period and finally end up with A.I. The problem is that this A.I. will be used as a weapon by the tech giants, when it should have been used as a tool by citizens. People will have a hard time resisting using A.I. as a shortcut to trial and error or thinking on their own. This will be a trap set by Apple, Facebook, and Google.

Think of this oddity in this way. So what is it I invented?

What if I.Q. could be experienced like a Cirque de Soliel event within the mind, answering what we seek without language at all. That is my goal with Qualia Apps and it will look like a cross between the movies "Lucy" and "Arrival" with a touch of "Ghost in a Shell."

It's ironic that the search for a language will exclude language as we know it to deliver the final level of this Apps Language Game. Some people will argue that Quantum Code leads to fuzzy language and it works – but words are words and they ignore the Nature inherent in their flawed design.

How did this come to me? This elusive or as of yet unexpected consequence?

First, I'll tell you what it is "not."

This is not A.I. Artificial Intelligence. This is anthropomorphic extrapolation within a mathematic calculus. There is a limit to it. There are not "original questions" but there will be "original answers."

I know you're without a compass at this point, so I will at least give you a reference point.

Have you ever awakened from a dream so real and "lucid" that you know to the core of your soul that it means something greater than its memory? If so, you probably are at at least one point in your life (wow, how seldom to we get to place two "at" together in one sentence) had to tell someone (lover, Mother, co-worker, best girlfriend) all about this dream. And you both pondered, "Wow, what does that mean?"

With what I know, Killer Apps lead to this situation instead. What if you woke up, and with my Killer App knew within days who else had the dream, what you had in common in your past, what you have in common in your current, every detail of your health and DNA, the significance of the most influential relationships in your life, how old you are, where you life, and the odds in descending order of certain events that could happen in your daily arena?

I've designed a Killer App for this already, it's just waiting for the language. I will not call the final design a Killer App I will call it a name that personifies "Deconstructing Our Human Agreements." And I'll create it for the art form I love – filmmaking.

Apple, Facebook, and Google anxiously await when a Killer App works! [This was written four years before the explosion of Social Media and introduction of TikTok to the United States. Please remember that Killer Apps are a precursor to Blockchain and Blockchain is a precursor to A.I. – that is why I am so compelled to do these books in time, before A.I. takes over.]

But they do not think past it. It is simply a source of revenue to them. Yet, they did all three agree, in one of the few group interviews on Killer Apps, that they could not be the "Home" to Killer Apps. This is because in almost their own words, the discovery and use of Killer Apps is so unimaginable in the way it changes life, that the "Home" if it was within one of the Big 3 would shake apart and wreck the foundation of the company.

They are correct in this.

They continued to state that whoever invents the "Home" for Killer Apps has to be smaller, able to pivot quickly, develop on-the-fly, and (this they do not say) let Killer Apps "wag the dog."

Now … you might be in a wow state … or you might be saying to yourself, "Who the hell is he, and how could he assume he knows things genius coders don't know yet? This has got to be bullshit. How could he know things they don't know?"

It comes down to selective focus required by both the company Apple, and the Apple coders. They focus on a goal, not the Nature. Two things happen in Nature: 1) evolution finds its own way, and 2) everything in Nature eventually finds balance in both the physical universe and the social systems. Think of our planet circling the sun. The balance between how fast we travel and the gravitational pull of the sun keeps us in orbit. In our case, the gravity of human limitations balances the vector of Killer Apps.

These two will form a new system to hold what comes from the Killer Apps, and that "tide pool" of breakthroughs ushers in "Genesis Apps." What differentiates a Killer App from a Genesis App in my blueprints? I'll give you an example based on the one dream example above.

Killer Apps will put you in the center of our universe where other people shared your dream, and from inspection of your shared Big Data, you'll know what the dream means and how to use it.

A Genesis App sprung from this will tell you where the dream came from!

Killer Apps go to the far reaches of our universe in this dimension.

Genesis Apps include multiple-universe compatibility in more than one dimension.

Either, a consequence of this re-defines Time and Space for us, or, re-defining Time and Space is one of the keys to unlocking multi-universe/dimension experience.

For now we call the most popular Apps made by tech giants Killer Apps. I expect that name to change because Killer carries with it negative connotations. Apps talking to Apps and achieving or delivering new services for us will take burdens off our minds. But will that come at a cost? Already, it is proven that young people that grew us using Google Maps to tell them how to get home from the mall do not have as much of a developed sense of "where they are" as their parents that used landmarks and concept of distance to get home.

Similarly, children that grew up mostly engaging with the world on the surface of a computer, pad, or phone have lost some ability to fathom 3-dimensional space. For example, they have a harder time instantly knowing how many suitcases can fit in a trunk and in what pattern to best set them down into a trunk. Apps and computers caused both of these mental retardations.

If a virtual world like Mirrorworld gets created, what human skills or brain skills and centers of storage of information of syntax superhighway connections will be lost? Will this result in lower brain development? They already know drugs prescribed to kids that kept them from leading their normal hyperactive lives resulted in permanent brain development loss. What will this do to an individual's ability to create ... to share love ... to defend what they know in their bones is true.

From a wider perspective, Apps and Social Media introduce a person to "like-minded other people." In real life for centuries we met people not like us and our interaction with them helped us develop into who we are and the beliefs we have on anything. Meeting only like-minded people will prove harmful to a healthy society, cultural appreciation, relationship to history, what we know in our bones is truth, abilities to withstand danger or stress – and in its place it will make us unaccountable for our actions or situation. This is the perfect recipe for what is now called "Collectivism."

A.I. will be designed to severely narrow the type of person you are and then decide what you are allowed to learn, and then do.

Let me finish this chapter with a simple questions: the rich elites of Silicon Valley and its tech giants send their children to private schools that do not allow students to use computers to learn, why is that? What does Jeff Bezos and Zuckerberg know about the ways this prevents full brain development and recognition of truth that makes us leaders?

THE SCRATCH PLOW

Ask most people what one invention played the most significant role in the evolution of humans and their civilizations, and they will answer "fire" or "the wheel. "

While these surely advanced our primitive existence, and urged larger groups of people to work together to make even larger "things," each person and family had to be singularly focused on having enough food to survive. This required nomadic expeditions that cost lives and subjected groups to new dangers, even if these dangers were just new bugs living in lands they explored for the first time. Yet, the need to keep searching for a wide enough variety of foods so that they could ward off hunger and sickness that comes with being malnutrition victims. If you remember the point of this book is to "Deconstruct Our Human Agreements" you see how simple it is to replace fire with the scratch plow.

Along the banks of the river Nile, locals had learned to throw seeds onto the ultra-fertile soil left behind when the river flooded annually. Then they also could use a stick to jab about 2 inches under the surface and plant seeds or stalks. Very time consuming and inconsistent in yield and use of land. Then one man got an idea. He found one large branch, about 4-feet long, shaped into a "Y." Next, he lashed a sharp stick filed to a point in the middle of the Y. It pointed down to the ground. Finally, he put a large rock on the far end to drag the weight down. And he walked. He walked the length of the field, looked back, and saw a long row ready to push seeds down into. He finished the row in minutes, rather than a day.

In that one event, one man learned how to grow more food than one man could eat. Even more than his family could eat. Other copied his example. For the first time, men that were great at making sandals but owned infertile or no land, could trade sandals for the abundance of food this farmer with the scratch plow harvested. The son could stop helping the father and go to school, and become smarter, thinking farther into the future. Groups got together, and more complex open markets organized. Larger buildings to meet in, store new knowledge in, expand rules about socializing, and then gain favor with the Pharaohs for more blessings, or at least less punishment. The scratch plow was known around the world to places no one could travel. That is because so many people in the world at that time wanted a solution to starvation. The cosmos found a way.

As one thing connected to another, larger systems expanded so sources became decentralized. And with decentralization came equalities, and freedom to pursue things of beauty … to pursue arts.

Many animals enjoy hierarchies of their civilizations. Most of them farm ore benevolent than mankind's. Yet only mankind created art and were so inspired by sharing visions of beautiful creation that it literally lifted the moral codes, Integrity, and understanding of our Universe.

Art gave a two or three-dimensional example of the key forces and rules of design to anyone to see, and enough of them were innovators. Cultures dealt with regional largess (fueled by resourcefulness exploited outwards to the community) and fears, (entrenched in myths, religion, and symbolism so strong it was passed down in individual genes and forming minds. This is why culture should continue to be reflected in Artists. Without it, artists lose a source of truth and inspiration.

Just as the scratch plow delivered enough food to the holy men interpreting the unexplainable or the thing we call Faiths so they could think instead of grow, Art delivers vision to those that lack it, but have the drive to evolve and use it either to inspire innovation, or to end up in a sublime place due to sacrifice and compassion. Very little art is created to bring out the worst in man. While very much of the art ends up in the hands of powerful people that deliver the worst in man. Catharsis, turning points in evil minds, erupts from creation – the creation of a child, or an empire, of an understanding … which seeks balance, again reflected in finest arts.

We all talk so much about the IoT. (Internet of Things.) How everything is connected. Yet, little time gives IoT's ancestors their due: connection between physical and "Intuitive" worlds. Physical interaction with Nature, with a person toiling in soil or with a soldier honing sword skills, to surviving brutal winters among animals, were such overwhelming experiences they encoded in our genes, transformed our bodies naturally over time. The IoT transforms quickly and unnaturally. These primitive and early experiences developed many of our human traits and gifts that got us this far – building immunity to bugs and diseases, to observing Nature to find clues for humanity and social structure and its laws. To craft an immortal quote came as a full body experience and observation.

We've lost some of our most acute senses, like the ability to smell water from miles away. Like knowing before earthquakes hit and storms come. Like seeing apparitions, that tapped into a part of our mind not yet locked up because it took power away from organized religion. Of incredible per-pound physical strength akin to apes.

World artists spanning history were the timekeepers of the ethics bridging Nature and those in power, none of which were artists. Maybe artistic strategists like Machiavelli. But you never saw someone elected because he could sculpt inspiring works that assuaged some fears and maximized others. The best art divides us into loving it or hating it and having to rely on sensations exempt from words for its gut value because our reaction to it cannot fully be put into words. Our most sublime goals and traits are the same: what is love, what is "The Divine?"

It's important to make a reference to China. I know most people only focus on what they see wrong with the country and leadership, but we need to recognize what they do right if we want to guide new technology in ways to nurture what makes us human as individuals and a culture.

Artists use their minds to create the next realities. How tempting it will be for leaders to harness this in the IoT. The wisest ones won't do it without balance. President Xi wants to push creative forces in his nation because he understands a country's greatest commodity is locked in the freed minds of citizens. Art is one of the few ways that we can see that skills we had before evolved from the survival skills they replaced. This gives the person some subconscious confidence, made stronger when freed to innovate and enjoy art.

You cannot install genius or innovation like a software – you need to ignite the recipe and let it nurture and boil over with unknown consequences, which considering this has never been allowed with over a billion people, President Xi has great faith in Chinese youth to be "China Strong" and "China Proud." President Xi learned balance at the Buddhist Monastery, and the tragedies of self-destructive education, and in both cases, art or the creation of new things for China essentially burned down the old school of thought.

We spin a wide web when we include the usefulness of art in inspiring young people because this is one of the few places we engage something that demands we be "present" so it can deliver joy and reflections of precision. It effects today and tomorrow in a deal in which you cherish values and push past them at the same time to bring things of power: food, imagination, and loyalty above desperation most of all. Whatever gives meaning to our actions quickly takes over our social games we play or work we do. To free the minds of young Chinese in a way that gives meaning requires a root in culture and appreciation for it as a compass for direction and a limit on greed.

Inspire a nation within its culture, for it is culture that lasts. The young Chinese coders, superb athletes, inventors, and politicians must remember that they got there because of genetic and trans-generational epigenetic inheritance encoding of culture, which linked to survival for millions of years. Yet, the balance is to give ample Hope to these new super beings, for that is essential to not quitting mid-way.

What most all drivers of civilizations had in common was simply organization. Feudal or tribal leaders organized tribes, aqueducts organized sanitation. They all sought to rise above animal needs and mental weaknesses. This was in a way a constant search for enlightenment. When the fears that are in our genes get their most quiet, then … enlightenment becomes the space between our thoughts. Ninety percent of our thoughts are fear-based, ten percent are desire-based. The space between them can be filled in moments when our senses are overwhelmed with creation and it does not have to come from a human, it can be a sunset so captivating it stops all thought … the sense of connecting with a tree whose only sound is the rustling of a wind born a thousand miles away and getting to you through Ether with a message.

"Shhh … the wind is a scratch plow, tilling your mind with more food than it can eat, so it spills over into art, into Integrity, into honoring culture, into trusting genius among a billion Chinese."

Designing a new scratch plow out of the cloth of political power is one of the hardest things any man can do, yet … first comes the vision, then the trust in the receivers as citizens, then the delivery of the freedom allowing personal innovation.

Visitors and Chinese citizens - do not forget to put your body through real experiences: Visit the Chinese shrines, look close at its long history of Artworks, consider the ladder up in Buddhism, remember the sacrifices of hundreds of millions that tried doing this before but did not have the right vision or leader, who is freer of fears of his own people. Do that, and it satisfies a third or more of what your own mind can do in the coming IoT.

I single out China because this is a new expansive era for the citizens. The United States is the by far the most popular nations any foreigner wants to visit and this is in part because we gave people the freedom to achieve their dreams. The Hollywood movies were embraced worldwide and presented

The Hero's Journey in a style that people of all nations desired. This is a great example of how a social environment increases the innovations and art created by dreamers and visionaries.

Mass data brings mass chaos … careful planting of more than one person needs to consume to create brings harvests … a feast for the soul, for the culture, for your children, for what is best in us, for what lasts and makes simple men heroes to their nation and families.

What I hope will happen after your mind digests what I express herein with you, is that *your* mind expands the levels and ways that it takes new experiences and learning, and stores it in your Brain Brain.

A deeper level of effect will be if it touches you emotionally and you store it in your Heart Brain.

If you react to some new experience automatically with intense interest or protection, then it reached your Gut or Stomach Brain, and Muscle Brain came into play. Yet before these experiences, you "sensed something." "Something" you saw, heard, or felt make your entire being either stop or at least recognize something is different, something is new, something might be wrong or very, very right.

"Love at first sight" is a good example of this. It has happened to me before. While the word love didn't hit me … the moment I saw both of the two women I fell in love with in my life, I felt as if the rest of the world paused or excluded us. I did not say, "Wow, look at her, I could fall in love with her." No. It was not "wordy." I simply looked deeper into the woman, felt more of her then the rest of the environment, sensed something magical was happening. It wasn't for weeks before deeper emotions became obvious. But my Intuition saw and knew it first. My Gut Brain and Heart Brain followed, finally my Brain Brain saw the words and I said, "I love you." (Eventually, weeks later.)

To show how it appears in a very different way …

It was in the Park Hyatt Hotel, in Saigon, the third year I went to see my lovely daughter, Miley Ngan Morgan. We entered an elevator, the interior of which was a polished golden mirror surface, very alluring to touch. I'd been in it scores of times over the years' visits, and she had been in it a few times before, but this time, she was 5, and old enough to be curious about things. The elevator door closed as we went up to the 4th floor. I saw her touch the golden walls of the door, and also saw the very thin space between the sliding doors and the walls of the elevator, probably 1/8th of an inch. She looked at herself in the mirror, checking out her cute dress and nicely combed hair. No words came into my mind, though, I just saw her a second and some fatherly sense in me was heightened. Miley stood on my right, Trinh stood on my left, and she handed me something to carry with my left hand, so I looked away from Miley just when we reached the floor. The door opened, and looking away from Miley I heard the first split second of a scream of pain. My hand lashed out faster than I ever could have made it move if I had ordered it with my Brain Brain/mind, grasped her exactly where I had to on her wrist, and whipped her fingers out of the collapsing door that most likely would have severed her fingers or at least broken all four of them as the door pulled her hand into the mechanics.

My Intuition knew to look at the small gap, record it, notice my daughter, keep relaxed, and give in to its control – because that was the only way my love could protect Miley fast enough. It was like it saw the future and my hand was already on its way to protect her before the event happened.

Recognizing this time and thousands of other experiences as ones beyond normal explanation gradually added faith in me that this connection to the Intuition Brain. And of course with her, with the future we would have that exists out of Time. It exists and I know it as well as I know my most favorite smells. Invisible yet real.

It was what I will explain was a high-Qualia moment.

Remember our grand quest --
an Accelerating Philosophy that
will free and then trigger our neural
connections to ding more I.Q. bells –
and apply this chapter to your mental
pinball flipping hands.

What Is North of North?

Are there questions without answers? Mathematicians "say" no ... using words proven even by them to be inadequate in delivering all answers as they claim we just haven't mastered the language of Mathematics. The logic of Mathematics solves seemingly countless formulas, measurements, and Laws of Nature. Laws of this dimension with this set of physical rules. Yet what mysteries on a humanities scale do they solve?

Will there be a time when war no longer exists? What came before the Big Bang? If God created the Universe and us in it, who created God? Are there questions mankind should never try to solve? Can we alter the past from the future and have that alteration still exist when we return to that future? How do the first cells that form the heart know they are heart cells and when to stop reproducing when a full, healthy heart is formed?

Where does artistic inspiration come from? Does it come from the soul? If so, why did early man draw long before communication skills? Art was the first thing created that had no use for survival. (Or was it – hmmm?) It simply pleased the eye. How does consciousness exist in all humans even in humans that from birth never interacted with other humans, as is the case with some children raised in the wild by apes and dogs?

The Brain Brain is an incredible tool. Yet it should not be the only avenue to discovery. "Experience" Information. To be rational is not enough when human – you live your greatest in single moments while creating art or Qualia moments, both of which engages passion.

I argue that art is the greatest example of experiencing Information through more than one Brain. Surely, the process starts with the Intuition of a vision manifestation for the art to be created. So that is one brain, the Intuition Brain. This comes without Mathematics. Then the Heart Brain provides the passion and emotion, not having to think that it is engaged in this process. It comes without words or math many times. Look at a ballerina for proof, or a young composer of music. The Muscle Brain makes us move in ways our brain does not predict or contemplate, proving to me the Muscle Brain communicates directly with the Intuition Brain, driven by the thrill the Heart Brain feels when it is set free to play. The Gut Brain blasts past any obstacles as it gives visceral reaction to the formation of the artwork. It represents our survival Information locked in the nucleus of our stem cells – Information passed down genetically. The dexterity we enjoy in our fingers did not exist a million years ago.

The "one idea" – introduced to the many = I.Q.

The many, returning to the one idea, is the process for it to be used and or useful.

This vision – introduced to the most fundamental elements of the many Brains which altogether result in "Information" with a capital "I." Vision delivers a manifestation of I.Q., deemed genius if the art tells a greater Truth. A dancer not watched while executing a spectacular movement isn't appreciated by anyone, but the moment of that genius dance moment elevated the soul of the dancer and opened portals to the greater truths of our existence. As for other "accumulated" art? It is primarily cultural bias, social history, luck, and timing determines its popularity. Think of how countless moments in dance resulted in full dance crazes, whether it is break dancing or jitterbug.

The many Brains combine in a composition of immeasurable influences to produce the finished work.

Words played little role in this composing. Muscles did not talk to the heart, they shared Information. Where does that harmony come from? What is North of North? What Information and information source made possible this process?

Elephants dance and also use paint to draw on a canvas. Birds build beautiful nests that are balanced architectural wonders to attract a mate. Apes are drawn to stare deeply into eyes of a color they never saw before, driven obviously not by words, but by a genetic or epigenetic inheritance fascination with the art of something as simple as the Iris. All these avoiding words yet result in recognizing truth on a very complex level. Using a combination of Brains set up in their own bodies. Intuition might tell the ape that the door to greater evolution has already been opened, and that identification of and appreciation of art, or a thing of beauty, is a key, a key that acts as the conductor to the symphony of Brains.

Look around you. At innovation. Imagining what others have not imagined. Many times innovation is a simple process to figure out. A leaking gourde leads to the testing of things to close the leaks. Most new technology is simply derivative. A fast smart phone is made faster, or smaller, or more portable. But the concept of speaking over whatever is in the air came long ago.

Most innovation in our history as a species comes from survival sources. Then it comes from expansion sources (expanding power.) Humanity as a functioning organ of mind and body coupled with art supply the greatest source of innovation that is not demanded for survival or power, it simply makes us feel better about our survival and evolution.

The focus on invention to survive and rule peaked from 1850-2000. Look at the development of weapons. For over 1,000 years, the change in weaponry moved so slowly that eras are defined simply by changing the metal used in swords and shields. Then came projectile warfare. The long bow wiped out entire armies from a distance the enemy could not fathom. The evolution of the gun and bullet, from musket to machine gun, finally got a make-over with the introduction of the Rail Gun, which uses electromagnetic current to propel a titanium bullet at speeds impossible from an explosion of gunpowder. A projectile shot from a rifle-type barrel that can pass through oak trees and metal plate. Also look at flying as a tool of warfare. From biplanes to jets to guided missiles.

A new "Age of Imagination" – Information usage and also awareness it is increasing exponentially – will shift the drive to invent, from survival to sublime. Why do I think so? We're approaching a threshold of awareness, of the vital role Information -- if considered the fundamental element of Matter, Space, Time, and Fields – plays in our happiness and destiny. We as a species will cross that threshold, where our Information explodes faster than we can mentally adapt to hold it all in the Brain Brain. The overflow will trigger a new level of symphony between all our Brains. As much effort to create weapons of war will turn to the solving of the great mysteries of our species and planet. And, a tool already invented will play a fundamental role in this, but will not be the reason for this.

Each Brain exists as both a center of its own perspective in the Universe of the Body, and a center of our universe of Dimensional Reality. Just look at a diagram of the human body, and draw circles around each area that holds a Brain. Where they intersect is the sharing of Information gathered in what way (DNA, past experiences, learned wisdom) that Brain weighs. Driven beyond words, by experience, the motivation for these Brains is Truth, and, ever more Information. In a gleeful choreography, the Brains leap and embrace to a tempo.

The tool already invented that works in this way is the App. As I mentioned in detail before, the App puts us in the center of the Universe we tell it to decode: we're in the center of GPS for example, and we ask it to decode the mathematics (1's and 0's) of computer data relative to the Fields of gravity and observing the fundamentals we accept regarding Matter (locations we built.) The App only goes as far as we programmed it. The clock App does not measure our day according to the expansion of the Universe at our placement in it, or to the circadian clocks. Our circadian clock tells us when to eat and wake up so our body goes through the stages for healing, growth, and socializing. Often when to mate and produce offspring. Each of these acts as the center of its own universe of Information.

Look back at the overlapping circles of the Brains. Now, replace the Brains with Apps. We use a combination of Apps to determine a movie and dinner date. Or the dropping of a missile on an enemy. The overlapping of Apps will come very soon. There is only one bug in the way, and that is being debugged or solved. Yet what are the consequences of answering what is North of North? And can the artistic process help us?

Up until the last 200 years, only one or two discoveries or inventions changed societies in a single generation. In fact, an invention would take decades or even centuries to complete. Look at the building of the great Roman Aqueduct or the building of a single cathedral. The Aqueduct proved a symbol of the superiority of the Roman Empire rule to any visitor, citizen, or invading army. A cathedral took 100-200 years to build. Seven to fifteen generations took the time to build it. They knew that once it was built, the cathedral would draw foreigners and trade from a thousand miles or more. The cathedral builders would live in the shadow of lack of progress for generations, until the final generation could benefit from the influx of foreign money, trade, inventions, and more. These changes effected society. New clothing, new stories, new inventions to use. Challenges to the status quo, set largely by ignorance ordered by The Church, ruling leaders, and others proved to be obstacles to discovery and social evolution. The modernization of tools led to the Industrial Age. Suddenly, a man and horse were not the source of power. Then energy evolutions led to electricity, radios, and telephones, so communication changed. Transportation evolved to link the world cities.

But in all these, change happened slowly enough for society to adapt. Sure, parents freaked out when kids changed clothes, words, mating, music, and more. But again, in a generation, we adapted.

The first two girl-girl kisses on television -- on "Rosanne" and on stage with Madonna -- shocked the world. Now it's a gag to build Instagram followers. Yet parents raged against this social change because in their lifetime, they'd already had the one or at most two social changes they could handle. The family unit is simultaneously fragile and invincible.

Since 1970, changes have happened so fast that within one generation, there is more than one incredible invention that changed all of societies. The computer led to Smart Phones, that led to greater awareness of global suffering, which led to a grass roots movement to help people far away. The tsunami in Thailand is a good example. The news gave us visual Information, the computer an ability to donate to the cause, Apps gave ways to save lives, which led to breakthroughs in warning systems and emergency care for injured people, which changed how fast politics could use a disaster for power, which led to fights among factions of society and threatened family structure, which altered education in colleges, that created groups the protested against big money and banks, that funded the inventions that gave them the Information.

The very inventions that make life better can tear it apart on a social level, when put in the hands as tools for politics. Can this be avoided? Yes. Apps talking to Apps will circumvent the slow pace of social change. In fact, I'm working on the prototype for this evolution. It requires us to leave the slow and limiting pace of a "Mimic Society" or a version of a derivative invention process.

Apps will evolve us in the direction opposite where those in power wish us to go. The speed and breadth of expansion of warfare tools will be matched by the discoveries about our soul, consciousness, Time, memory, health, longevity, genetics, love, Ether, dimensions, and, well … questions words cannot even begin to spell out, must less deliver an answer to.

Hint: What would happen if I became you for even 1 second? Would most experience be in your Brain Brain, would most lasting information be translatable in mere words? Would this shock wave split us, free us, from the dependence on the Brain Brain that limits our merging with the Information held in Dark Energy, Ether, and other dimensions in another version of Time?

Will Art and Creation of it be a portal, a tool, a language, or a commodity?

THE NORMAL, THE WEIRD, AND DATA-ISM

"When 90% of the people doubt you, you're either
about to make a serious mistake, or
you're about to make history."

When you read about human psychology almost all of that information comes from either university research, or experts that continued research in the university tradition after graduation. Some books and articles are written by people in other professions that happened to think enough about human psychology to write something, but other than that, the rest is written by what are called "Academics" or "Intellectuals." This process occurs almost entirely in Western universities, those in Europe and the United States, in part because the Westerns glean almost all "expert analysis" and "medical assumptions" from data or models and tests. The opposite of this is seen in Asia from India to China to Japan. So much of Asian psychology gets it guidance from Nature and philosophers that based their theorems or announcements on personal observance of people that merged with the most popular regional philosophies like Tao, Confucius, Buddha, or local legendary leaders and warlords. In other words, the most truthful and enlightening studies of the mind were "real" to the Chinese, because the outcomes or diagnosis for healing, as well as the future directions of politics and society, came from real people acting in the real world.

In the United States, France, Germany, and England (the most powerful centers for the study of the human mind) this is not the case. Politicians, marketers, and physicians rely on tests. The problem is that most all of these tests or models are performed by "experts" and companies that first align with the politicians, to make sure they come out with the verdict or guidance they desire. If politicians control the funding for research, and are paid by lobbyists that deliver votes and election campaign money, the politicians are told what to approve and what to support.

The more famous the university, the more influence or "weight" the research carries. Therefore, the top universities in Western like Harvard, Cambridge, Sorbonne, Yale, Stanford, and Oxford publish research with so much influence that it literally steers readers in the direction they desire, far more then they direct them to the truth.

I'll explain why, and start specifically as it relates to the mind, and then to art and artists.

Almost all study into human psychology that universities publish is divided into two categories: Normal and weird. And artists are classified in the weird end.

Funny, isn't it? But it is true. Artists are solidly placed in the weird category, but only by Western "thinkers" (thinkers being a dubious title at best for them.)

So, all of you listening to me, watching me, reading me – if you are artists, the universities and self-proclaimed experts with many degrees determined we are "weird."

Who else in the weird category? Serial killers, psychotics, insane people, and schizophrenics. Feels good, doesn't it? If not, it should. Why? Well, if you're with these people and not killing others, then you're top of the pack for "cool and awesome" right? Haha. And you don't have to explain yourself to anyone that questions you, you just have to say you're "weird" and get a free pass on your actions! What freedom!

Get a load of what I'm about to tell you, it gets even funnier. (Or more weird and disastrous.) And it is incredibly important if you are living in Asia, when you read about what is normal in Europe and the United States.

Even if you are not an artist, if you are a American sitting at home watching this, yet love to share in movies, music, and art you did not create, you should enjoy seeing how stupid people, that create even less than you do, determine what is normal.

When you do a test in Harvard or Oxford, you put up a notice in the school or in the school paper, saying you will PAY people to take part in an experiment on human psychology. The students read this ad, and if they want money, they do the experiment, or if they are favored by the teacher, they will do it to get a better grade. So, there is a heavy incentive to please the teacher or person giving the test.

Most of the time, the student can figure out the end result the test wants. And if this student wants to be called back for another test and get even more money, he or she will answer so that the test subjects get the result they want. In some cases, the student cannot determine this, but that will not matter, as you will soon see why.

To make it even less true, the researcher can always dump an entire study if it does not please his plans. He has figured out that his career depends on enriching either medicine, politicians, banks, or unions.

Harvard and many universities cost over $100,000 per year for a student to attend once you add in books, tutors, parties, cars, travel, housing, dating, and things to brag about. The students usually go there because they are rich as hell anyway. The others get in by getting either a scholarship because they are super smart or part of a minority they like, or they take on huge student debt from a loan, that takes 10 years to pay off.

All the students want something for nothing, or want party money, or higher grades. So, they do the test.

Now, let me ask you. How much do you have in common with the son or daughter of a millionaire that got a BMW or better the day they got their license, that never had to worry about money or the future, that was put into day cares that cost $20,000+ a year and private high schools that cost $25-35,000 per year? Did you take a holiday in Aspen or New York that costs at least $1,500 per day? No? Did you get a 2-month tour of Europe upon graduation that cost $40,000? Do you plan to be an American President or Senator? No? Do you wear suits and dresses that cost over $3,000? Do you drive a car your parents bought that costs more than $60,000? Did you never have to work when you were young? Not even house chores?

If you answered "No" to most of these questions, then you have nothing in common with these spoiled kids. You have nothing in common with Normals. Most Harvard students could spend the rest of their lives playing video games and still get so much money from their parents they would live a grand life. They were raised by nannies, a huge fraction of them went to child therapists constantly, often times complaining they suffer from being too spoiled, rich, or ignored by parents. Their problems have nothing to do with your problems, they are the problems of rich kids. They are fine people, but, to them, weird is not having a new $4,000 tuxedo for graduation. They call themselves Normal and are classified as Normal by most people, yet, they fall apart over something as simple as hearing a song they think offends other people, or if they see a billboard with a picture that they were told is upsetting. And they call this Normal. As if all the kids in the world react the same way and spend their time worrying about the same things.

Yet these students are the ones they do tests on to determine what is normal or weird. Almost none are artists, and if you asked the parents, they would get mad as hell if they found out their child wanted to be an artist.

So, they take these spoiled kids that are not like you at all and never had the pressure of a family needing you to be a huge success to pay for food and health and burials of parents.

Then they make experiments with them. If this student says that, after the tests or during, that they feel no desire to create art, or create, or fight, or defend, or anything that you feel is important, then this changes results.

I don't have the time to tell you how many tests and the hundreds of decisions made from them here. All I can say is that most "official tests" and "expert analysis" and "psychological decisions" are made with people that do not live a "real" life!

They did not grow up having to do chores, or in Asian, African, or Middle Eastern nations with strict cultural limits on artistic expression. Anything having to do with innovative survival or serving the greater good of your city through art without limitations on inspirations or visions is as unimaginable and therefore Weird to them as eating a tree for dinner is for you.

And to make it more complex and inaccurate, these same students are forced to "learn" a type of "fake truth" that is far more wrong than the worst propaganda in a nation. The leaders of China, Vietnam, and South Korea tell us this process with less propaganda.

[I understand that many people have strong opinions against China and Vietnam, but that should not prevent us from learning from their political actions. I was in Vietnam in the year they opened the country up to open Internet browsing and this two year period was a sight to behold – people overjoyed over viewing from more than one perspective.]

The young people of China, and some in the United States, grew up needing things, things essential to survival, which no Harvard student ever had to think about. They do not have to make an intelligent decision that works in China. These students do not know your culture, what makes you proud, how to make your parents proud, how to survive in a city 8 times more populated than downtown New York City. They refuse to do what is good for their own country, the USA, because teachers tell them the USA is evil. And should be ruined. They have no relationship with Honor or Family Honor, they do not want to make their President strong, they do not want peace and honesty – they want what the teacher tells them to want. The teacher tells them if they do not think like they do, they are weird and wrong.

The teacher learns that the only thing they should demand and teach must follow the goals of the heads of education. In China, the heads of education need to follow two goals: to open the minds of students so that when they grow up they will help China and friends in China live a better life, a freer life, and a life that helps China, both culturally and financially. In the United States, almost every teacher is told to teach that the United States is evil, that it should be punished, and that (politically) weird people must be "marginalized" (punished, shunned, silenced, shamed, and/or kept in captivity.)

This was not always the case of course. Before, in the United States, universities enforced the best principles of the United States. And students were taught Civics, which is how we learn the good things of the United States. But some time about the 1960's, the Teachers were slowly replaced with people that do not want freedom, they want control.

Meanwhile in China, the opposite took place. Now, China wants more creativity and artists and leaders with vision trying things out and if they advance the nation enough technologically or financially the limits on expression will be bent. In other words, China wants more people to learn how to think and create for themselves. And it is paying off because President Xi and others know that the future for great success requires free thinking, and artists and visionaries.

This fact makes you more normal and valuable to the world, yourselves, your family, and China.

Now why would these new teachers want this? Not because they all suddenly think of this for themselves. No. The incredible number of teachers, many millions, were slowly over 40 years replaced only by people that want students and others totally controlled and thinking alike. The teachers were caught in a trap by people like George Soros, and others before him. The teachers want more money of course. So the rewards and payoffs for "discoveries" in science, math, medicine, and psychology were given only to people that agreed with what the people giving the money wanted. The people that gave the money could benefit more from having people under their control, to get rid of artists that think freely. So, they made artist weird.

China is doing the opposite. In what is a very risky move, they are giving freedom and power to artists. Yet, they trust their youths to find within them the natural goodness born into all humans.

So where does Data-ism come in? Data-ism is the worship of Data almost to the level that it is a religion. Data drives a multitude of decisions, and is reaching godlike powers to influence. Data removes the human element from research and tests. The university studies won't take into account the human element of the student taking the test. He or she is just a number, a data point. While they claim this testing resists altering results based on people or common sense, in fact it removes a huge part of their analysis. Data easily alters input to reach the desired output.

When you realize that so many government or law positions are determined from these universities that through tests and teachers tell us what is acceptable or not, you see how this goes against bringing out the best in people on a balanced and natural way.

Take for example the term and use of "Social Justice." To get to the end result they must reach to support the teacher agenda, students believe that all laws should now start and end with the individual in a social situation and be for Social Justice. This sounds confusing, so I'll clarify this. They think that where a person came from, their family history, should not influence laws, actions by the government, or how they are expected to perform. This is ridiculous. If someone comes from a country that never taught Calculus, the student cannot do as well as students that did study it.

I remember this well in college. I had read over a hundred of the main books a college requires for study and graduation, and thought most people did. Then I met people from the South that never read them, people from other countries, too. How could they think this person should be in the same class, arguing, "because it is fair." No, it is a recipe for early failure. Yet, it sounds nice to them. Makes them feel morally superior. Yet, it is faulty set up leading to a faulty conclusion.

Completely outside of their analysis are the weird, except when wanting to find ways to control the weird people. They cannot determine the point of inspiration, or origin of creativity, on Data, or when the people in a category are determined psychologically weird.

Artists are allowed certain social freedoms. It allows them to be eccentric and still embraced, sometimes even more so for eccentricities. Andy Warhol is a perfect example. But this is a good thing.

By embracing weird, we move our minds outside of the bubble or circle of "normal" and this allows us to perceive varied Truths that are outside the vision of normal people. I certainly do look at the world weirdly. I accept and try things other people would not, at an age older than most, because my weird mind sees something different in it. It is natural for me to do so.

The friction comes in when the weird artists realize the world can't move forward without them, yet the normal refuse to give them room or credit for being weird. A great example of this are extremely high-end software designers and inventors of cutting-edge technology.

I use this for my work in entertainment. Much of my best writing and Directing comes from seeing the same situation from a weird point, somewhat outside the circle of activity. When I stir my coffee, I don't see a spoon going in a circle, I see a tornado. I constantly push my mind to see the weird in everyday conversation and sights, so that I am always pushing for deeper Truth. With this, I can lead people past Normal limitations. If Steve Jobs, who clearly was extremely weird, didn't envision the touch screen and later the Smart Phone in the iPhone, we would still be using a tracking mouse to click on everything.

Weird extends to my ways of expanding my mind through oddities, like strange art and movies. This then helped me to envision my business model, which will evolve the entire world of entertainment and its marketing and Brand building.

If I was not weird, I would not connect with you and your soul, your hopes, and dreams. And I would not see the extent that Western advancements now require a balance with Asian history, culture, and responsibilities.

Where does Data-ism come into this, for you, the artists, and you in college in 2017? The recent term "Data-ism" means that an era emerges in which Data reigns superior to other Intelligences, so that it is at the very least royalty, and at the most, is worshipped like a religion. Data over culture, personal experience, Nature, and our intuition. And yet it goes beyond that.

All the powers of our creator God gradually get coded into algorithms for Search or Apps and get banked in Data on you – to Information of our existence kept in 1's and 0's, extrapolated by Mathematics. God creates. God gave us limits. Data removes those limits. Data-ism is what allows us to "create like God" and be our own living gods.

How close are we to this? It's already happened. Mark Zuckerberg, ever the egomaniac and Narcissist, just announced he offers his followers to let Facebook replace The Church – Christian religion. Ever wonder why he singled out Christian religion? Because that is the only religion that so openly promotes the Free Will of men. That Free Will must be taken away, if the world is to be as Zuckerberg wants. He literally sees himself as a God or at least savior.

Data-ism starts with worshipping Data, but moves quickly to worshipping the gatekeepers to Data, such as Zuckerberg, or tyrants. Because ultimately, the control of Data and influences on you destroys your personal sense of the power of the individual, the creation of the individual, and all his Free Will. As this happens more and more, as limbs are made robotic, and minds enhanced with electronics, we begin to see ourselves as les s human. Which is exactly what Facebook wants! And Zuckerberg craves, craves to be worshipped even more than he is. His methods never shifted from what inspired him to make Face Smash. In a word, Zuckerberg "humiliates." When he insulted college girls his goal was to humiliate them to lift himself up by pressing down on their shame. When he testifies in front of Congress he humiliates them by telling the same lies that work legally to give the impression he cares about laws, which he seems to think are made for mere men and women and not gods as he sees himself.

The other great sacrifice is to Culture. Native cultures of countries must take a far-back seat, and with luck will be deemed unnecessary. Step by step, Zuck will convince you that you do not need anything but him and Facebook. If he wants to be a god, what gift does he give us? Hera gave us childbirth. If he's giving us a secret for humanity I'll declare him a god if he agrees to suffer the same fate that Prometheus suffered for giving us the secret of fire. He offers no salvation, only deity worship opportunity if you prostrate yourself fully enough on his quicksand. He sells your personal data, that is proven many times – he will just as easily sell your soul, or that is, the concept that your soul is both individual and indivisible.

It is wise for China to be very, very hesitant and cautious about allowing Facebook into China. Not just for the normal worldly news, but because Facebook does three things (even though they deny it.) One, it spies on you, takes your information, uses it to gain more power and money, and aligns you with people slightly more corrupted than you are to quicken your fall. Two, Facebook played a major role in both the overthrow of the Egyptian government (replaced by a brutal temporary cult) and in an insidious and covert way, alters your view on societies around the world to favor the company becoming a government in itself, able to inflict suffering. Three, it belittles American culture that gave him the tools and freedom to become so powerful, your gut instincts on right or wrong, and love for anything more that Facebook and eventually Data.

The artist becomes ever more important in fighting off the creep of Data-ism. Though Facebook and other Data-rich companies want to control the world and your life, it cannot yet control the actions, creative spark, and value of art and the artists that create it.

President Xi is wise to work to nurture the artist through culture and China Pride rather than Data. There is no social sin in accepting many of his laws that limit how young people use the Internet were good for their society.

CAN WE HACK OUR OWN BRAINS WITH ART?

(Or have we already?)

Blame science fiction.

My fascination with Science Fiction started at age 6, reading "The Hardy Boys" mysteries, often using science or futuristic inventions to solve crimes. Television brought "The Twilight Zone" and "Outer Limits," which though just entertainment to me, baffled my mind and posed alternatives to what we think is real and how we think that up. Many episodes deal with "human agreements" being broken, and this continued with "Star Trek" as the concept of what it means to be moral in a universe involved all sorts of futuristic devices.

Some of my first tests at how far I can push the imagination of my brain were playing chess, and pushing limits to laws – laws mental, physical, and social. I got into a lot of trouble doing that one thing that everyone said not to do. I broke arms and my head testing rules of gravity.

[Disclaimer: This paragraph was added in 2021 after the original writing in 2017 of this book.] If we can hack our Brain with art, this correlates with "The Janus Point" book, which states that the complexity of art presents this opportunity – to redefine our understanding of Time itself. It would also prove how art and complexity act as a balancing force and mini-Janus Point if the complexity rises high enough. This needs two factors: spirit and scale. Why? Because enough people have to be aware they are chasing something great and have an energy that spreads all across the globe. To me Gamers are the most likely group to focus so much Will on one future revelation that the universe recognizes a coherent message we are ready to learn more secrets of our existence.

This will happen when film and technology reach beyond the Apps we are developing and merge. Decentralization of a thought is important but decentralization was abused already by investors. We wen through that fraud with Metaverses and NFT's accidentally touching upon this potential – though the powerful people pushing these technologies are literal idiots about the forces they are now giving gravity and potential energy. Decentralization has turned into a way to get around international laws and it is used to aide criminal systems. So Data-ism has its dark side.

The creation of films about zombies, vampires, and aliens that take over the mind or inhabit another brain through a parasite that controls thought and actions hit television right about when I was 10. Totally fascinated me. Yet I was far too hyper-social and adventurous to become a couch potato or withdraw into a geek world of science. Testing on my mind showed I was above normal, not overly impressively, but enough. Yet there were areas that higher gasp of concepts proved to hold, mainly in higher Mathematics, correlation of reactions in our physical world, and control of the fight or flight impulses. This showed up in some weekend private counseling with a child psychologist when I was about 12. He helped me identify it and also said it could atrophy if not stimulated. Neither choice would define my level of happiness, it was either something I found fascinating enough to explore, or not, like any hobby presented to a pre-teen.

The testing and how it later spawned its own territories of exploration involved an individual's relationship to rules on all levels. And pursuing them was half algorithm, half conditioning, much like rising the ranks in the military. My first experiences pushing the mind to see objectives from extremely opposite viewpoints set my mind up to do that the rest of my life. And, it was timed with my involvement with the Boys Scouts. This seems insignificant, but, it put me deep into Nature. Nature supplies its own branch of stimulus and overwhelming truths of what existence entails. It also taps into our Intuition, developing it like a muscle, as well as our ability to observe and use the gifts in our genetic structure. What gifts am I talking about?

Innate GPS on our planetary fields of energy.

Observation skills that apply the Macro to the Micro and back again, constantly

Time measured by circadian clocks inherent in all species, not just to rely on the watch on our hand to tell us "what time is it?" When a storm approaches, and you do not see a cloud, but animals begin reacting, suddenly it is time to build shelter. When quiet hits the forest, you stop until whatever upset the rhythm of this pocket of Nature leaves the arena, or the arena adapts to the arrival of this upsetting force – perhaps a predator, perhaps a change in magnetic fields, perhaps fear humans emote looks like a colorful gas that stops the birds from chirping until it dissipates.

Nature surrounded me with Truths that hacked what parents, schools, and society told me are "unbreakable rules." Yes, society is more orderly and compassionate with laws and physical rules being followed. And certain studies are set in stone, like the writings of Shakespeare and the math of Euclid. But it was beyond these teachings that breakthroughs for both society and individuals find clues and fuels.

Trips to the psychologist seemed to not help prevent my being a naughty boy much, when I had idle time to wonder what happens when you light chlorine powder on fire (it makes poisonous gas that almost killed all three of us in the closet.) But the Saturday afternoons spent with him did introduce me to animals, like horses and packs of dogs, that work together and communicate through Intuition. They proved that they know something we don't yet.

In high school, my commitment to solo camping fueled my drive to visit as many National Parks in the United States as I could by graduation.

It took physical experiences like these to "hack my brain" and free the Artist in me as well as the Explorer into my own mind.

My first artistic passion was photography. Given a fully manual Pentax camera at age 14, that required eyeballing lighting while framing, testing all aspects of focus and just as important the *development* of what I focused on through my camera. Whether photographing my family, friends, sports, accidents, or Nature, my mind comfortably accepted that two realities and uses of it can coexist. Shooting a world of color into black-and-white requires focus on the lines and shadows more. It hacked color out of my photographic vision. My mind learned very early how to reflect my own "balance" from the balance of Nature and nurture around me.

"Artists can paint the sky red because they're artists, the rest of us have to paint the sky blue or people think we are stupid."

My second artistic passion was food, which in record time led to being a chef. A Chef with a capital "C" doesn't just cook up food, a true Chef immerses himself in a universe of elements that constantly challenge the algorithm of one of our most basic senses: taste/smell. Following the rules that superior chefs demanded provided the basis of my cooking skills. Yet, advancement required hacking ingredients and cultures, sometimes simultaneously, other times the opposite, by excluding the limits of a culture on a certain dish.

The deepest challenges to hacking the limits or our gifts from creation in each of us and civilization as a whole came to me in the form of filmmaking and storytelling – linked but differences abound. Most storytelling I'm talking about here is interpreting the stories others wrote, from the classics to stories of mind expansion or artistic challenges. Using I.Q., I applied each idea to many and back again.

What I found is that the mind is very flexible in a world of rigid expectations. Art hacks standards, for indeed, great or timeless art must go beyond what others accept and perceive the world to be, while testing precision.

During the course of my life, science, computers, philosophers, and more sources of inspiration not only advanced our understanding of mysteries, it also allowed for personal exploration into the mystic arts of meditation, karma, Chi, cosmos, health, and the creative process.

The two most insurmountable adversaries (roadblocks) to hacking our own brains, freeing our mind to communicate with all cosmoses on all levels is our ego and our dependence on language the brain uses to spread and compare experiments in life.

Ego holds desperately tight to us, defining what we think we are here for, it justifies almost all our actions and forces us to hold onto its one vantage point of the world we live in. And in that world, all the Laws: Time, Space, gravity, entropy, design, form, and then into relationships between species.

Primary to any ability to explore hacking the brain (I should be using the world mind sometimes, I know) is identifying Truths, many of which we either lost an ability to see (such as auras) or won't accept as Truths because some "expert" told us it isn't provable.

Throughout the creation of my entertainment world, my fascination with pushing the limits of how the mind interprets our immediate world (and distant cosmos, but I'm talking about how we act among others now) saw the "Unquantifiable Arts" as holding the "Master Keys" to unlock the next levels of creative Interpretations and the uses of them later. Methods to do this included Derrida techniques, meditation, solitude, mass immersion in high energy arenas, hypnotism, dream study, and a few more techniques.

The social impact of what I sought became clear only recently. Can the mind be hacked for long enough so that one person can "become a totally different person?" Can someone who is violent or addicted be hacked to be someone else? We do that slightly with drugs like Xanax turning a high-anxiety person into a calmer person. Ritalin helps hyper kids focus. But there is a cost to that, proven by studying brain activity and recovery from injury.

A good example of value for society is this one: could brain hacking, which is altering the norms stored in the right side of the brain along with involuntary actions (breathing, digesting) and a sense of our origin (what is repulsive to us as identified with our accepted histories) lead to Parallax View? Could a biased "straight" man be hacked to understand the paranoia felt by someone living "in the closet" and ultimately make each of them more compassionate to each other's challenges? Can brain hacking make prejudice or an idea of singular persecution for things we cannot help (skin, origin, height) a thing of the past? Can the intuitive skills women seem to have more mastery over be transferred to men, and can Alpha male confidence be shared to women? Can this be done without screwing up the individual completely, making him or her socially inept or unwanted?

And what about politics and health, both of which want more conflict or unhealthy minds to get rich and powerful? Will they allow a mental cure for social defects?

Art holds the keys, because it unlocks doors and reveals truths that slip past our ego roadblocks. Art transcends language. It transcends species, proven by animals lulled into coordinated movement my music, and elephants driven to paint-colored artworks, and even throwing out the art it creates that it does not think reached the vision.

Art, progressing deeper into the mysteries of the hacking of the brain, opens doors, and in the process, replaces one question with several, yet, that is evolution. That is Nature. That is truth.

ESSENTIAL ELEMENT "HOPE"

Observe people. Think back over your life.

I call "Hope" one of life's "Essential Elements." Before that sounds too inspiring, I also deem "Corruption" as an Essential Element. By this I mean both play a big role in innovation and survival. Expanding an understanding of Hope, Corruption, Balance, et al removes barriers blocking an Accelerating Intelligence because unless you know how ot use them as fuel they will be dead weight.

Start by observing Nature. Corruption mostly happens between two species: parasites on hosts, predators using confusion to trap prey, humans domesticating animals, cancer, mutation, and mating all feature the corruption of previous standards or expectations of "being left alone" if you are innocent. In fact, the lure of art, its beauty and truth, could be deem a major corruptor of man since Egyptian times. Is that so bad? Consider the suffering many artists endure on their path to excellence.

In that sentence lies the truth and dichotomy: corruption or luring to an art seduces the person, displaces the primary drives of survival, and consumes the mind. Yet, it is part of the path to excellence.

Hope. Fascinating to consider. A study of mice in a lab sheds light on how important it is to survival and creative thought. Mice were put into a small tank that slowly filled with water. The mice would struggle to swim for over 45 minutes. If they could not scale above water or escape, they still tried until they died consumed with hopeful struggling. But on a second test, right before drowning, the scientists put a small ladder on the side that the mice could climb. Once a timed escape was successful, they would repeat it until the mouse learned that at about 45 minutes, a ladder would be provided. When a ladder did not appear at 45 minutes, the trained mice would lose hope and die. They did this one at a time to the mice. Then they put the mice together to repeat the experiment. They all would struggle for 45 minutes and then give up. When new mice were added, they learned from other mice to quit if at 45 minutes there was no ladder. After that, the entire society of mice would give up hope even if they could swim longer. Without hope, the mice die.

Civilizations grew on the principle of Hope. Sometimes this Hope was hundreds of years off, such as when the great cathedrals were built, as well as the Great Wall of China, Great Pyramids, and so much more. Even the modern Internet was built on the hope it would transform humanity. Sacrifices for taxes get paid on the hopes of better roads, cheaper power, less gridlock.

Hope. Name one of the primary aspects of humanity that it does not propel: politics, war, economy, marriage, energy, weather, health, family, technology, and of course "The Arts." Many times the Hope is simply to improve life in general, with the idea we can keep rising higher out of our primitive drives to find sublime guidance and existence in harmony with Nature and each other.

A universal principle for arts that can be considered "Original Creations" relies not just on doing something no one else has done yet, but more significant and primary is to believe something few others believe yet. You can see this in music, which changes quite fast now, and in sports. Sports is most definitely an art. In sports, the Truth is proven through meritocracy. The Fosbury Flop was rejected and almost outlawed in the 70's when it was "created" and is now the only form used for high jump. And new sports equipment like the Nike sole, double-handed backhand, "Money Ball" techniques, gymnastics moves, freestyle snowboarding, surfing, and skydiving appear with test equipment that goes mainstream. All required Hope in the innovation first. Hope to set a new standard. It's one of the few arts that can be measured.

And then there are Games, both electronic games and board games. There are Electronics Arts Gaming "Olympics" testing the limits of hand-eye coordination along with speed as a quotient, and players must "create" artistic moves to survive. Gaming is one of the few sports that results in death (short term and virtual … for now.) Just you watch – the Elites will find a way to take this into the real world. Why? The same reason religion went global. It *organizes* people into group-think sources of power – power in money, power in desires, power in vision, power in warfare. In fact, Gaming is almost the only art that can stave off 3 of the 4 primal needs: starvation, disease, and war …. It still leaves Geeks without sex, but they're working on it like DJ's.

And essential to every player long before competition is Hope. Hope to get to the higher level comes long before and lingers long after the recognition of winning a single game or series of games. The creative minds are always inventing new Hopes. And they do it within seconds and move to next Hopes. In many ways, Gamers act more like bacteria in a dish, trying infinite combinations of modifications and ultimately the Gamer escapes to flourish.

To better understand the artistic potential locked in Games and never exploited, look at chess. Complex, but surely not the most complex of all games. I will start by asking you a question: what is more important, how the game begins or the places of pieces at the end? There ya' go, all the proof you need for what comes next in this chapter.

The limitation problems that prevent all of a chess game's moves from being monetized and weaponized comes from always starting out with the same pieces in the same place. Both Bobby Fisher and Kasparov thought the better test was to start out with pieces in different starting points. Why is this significant? Because it avoids starting the game in a known algorithm. Life for individuals, and placements of nations and civilizations, are never in exactly the same position twice. Innovation, at the core of art, thrives due to it starting with original elements: colors for artists, athletes for dance and sports, instrument combos for music, just to name a few.

But there is another reason why starting out differently presents such incredible untapped potential. Free Play. The genius behind chess masters began in their youths, perhaps in their genes before the union of the egg and the sperm. Free Play is how these masters all looked at the game. Starting with no rules on position returns Free Play.

Perhaps the most useful chess instruction in recent history comes from Kasparov, who forced players to start the game with only 3 pieces on the board. The Queen, King, and one more. This forced the player to understand the game from the inside out – from the Micro to the Macro, instead of from the Macro to the Micro. I know you don't really get that point. I do only because I played chess as a child well enough to take second place in a large competition. To me it was all Free Play. As soon as it got too serious, I quit. Free Play kept me excited.

In Free Play the quotient of Hope expands to meet adversaries. Hope becomes understood from the Macro to the Micro and back again, in an arena few can study. Playing with just 3 pieces forces you to push your understanding of the actual *power* of each piece. On its own, what power does it have within the sphere of reality? Its reality is two-dimensional, and its moves are guided by an entity taking over its free will. Does that sound familiar? War and Politics.

Playing with 3 pieces turns weaknesses into strengths, as weaknesses can develop into strengths in physical sports. An athlete might enter a competition like boxing knowing that his reach is shorter but propelled by Hope the boxer maximizes other superior skills.

To better understand Hope, consider what nurtures it, and consider approaching it from the opposite of that. A sport finds nurturing energy from interaction with and against other humans. Though we might not always be top of the ladder, we continue practice in hopes of fine tuning one elemental movement at a time, which gets added to the basis of talent once perfected.

In writing, the elusive words that transform a statement into an emotional hook nurtures Hope to do it again and again until a masterpiece is written. To approach a game, or blank piece of paper, without hope for evolving the art, means giving up. If there is no art without passion, and no passion without precision, and no precision without passion, and both passion and precision nurture Hope, this proves one of the few self-fueling engines of humanity.

People start hobbies in part to give their minds a source of Hope, since so many of them involve completing a task as best we can, but also, many times, in a way that nurtures other people. Social evolution and personal finesse rise with Hope … without it, with its opposite, interest in hobbies, in cleanliness, in history diminishes until societies fall apart.

Now, trace Hope back to an unusual origin, outside the brain in a skull, and see what you can apply it to? Hope originates in the nucleus of every cell in our body, and perhaps in the formation of our reality. (It could be switched with the word "Will" in many cases when it comes to the Physics of our universe and the Humanities when it comes to civilization.

Cells carry hope of division, of mating, of superior mutation. And herein lies a big question: Does Hope play a part in both evolution and cancerous mutations? Is the cancerous cell "acting out" against an algorithm it has hopes of beating? Of turning the cells into something impossible to kill, that does not rely on messages from the brain or circulatory blood flow for nourishment? Do cancer cells survive space travel even if the host expires?

The great leaders understand that the wealth of the nation they govern lies in the Micro, the human commodity of its peoples. The wisest leaders understood that either at first or learned it. The hidden hand of the government can be wonderful when it appreciates the power inherent in each citizen, which is what allowed the United States to greatness, and China now cautiously allowing freedom of movement dictated by the power of each person. Yet, the fact is that with each dimension, comes complications. For that, they set rules and dictate some moves – and should. Their challenge is to allow Free Play on a game that needs cooperation.

Imagine chess played wherein the player could move one piece, and through magnetic waves or Ether feedback that move also forced the movement of one more of his pieces. This would force the player to master the double jeopardies of human nature: greed, lust for power, and cruelty. Move two or more pieces at once. Chaos … yet eventually brilliance. Perhaps even spiritual evolution.

Apply extremely useful Macro to the Micro … and you see the future as we unlock mysteries of the mind at the same time as we deconstruct human agreements.

Everything, and I mean everything advancing humanity, is lining up to follow this key.

Apply the Macro of Ether to the Micro of the individual, like scientific medical research does, and put quantum in a tide pool, let the truths reproduce. Nature will implant the program humans cannot. Hope. Out of our hands so much, yet we refuse to recognize its power and value.

The DNA code within the cells of Artists, each cell, screams with Hope every time the artist begins anew. Why? Because cells are not here to win, they are here to learn. Learning exhausts all parameters. This explains why tech scientists now think that if they can replace key chips with bacteria, by default the bacteria will exhaust all possibilities in Nature, answering the questions we seek and many more we cannot imagine. Why am I so sure of this? Because of the elemental functions of bacteria, able to simultaneously think to move on an individual basis and work together in a hive basis, due to lack of centralization of a dominant single brain. It will prove that improvisation is more essential than wisdom, and that many ideas only come from reacting.

In simpler words, because the bacteria is the Artist releasing impulses within Free Play within instincts given by Nature left unprejudiced and in a starting state without preset algorithm. To compare, an oil painter would need endless colors seen by countless eyes from different perspectives finishing works of art outside of the constraints of our understanding of Time.

Let each piece have its Hope to connect the Planck Information realized, Art will not just open doors we cannot perceive, it turns doors into hyperlinks that only the abstract mind would consider navigating. Hope requires Free Play, while the people setting up this opportunity insist on Power. The two have problems together, but they are not insurmountable.

The easiest example I can give of a game of creativity like this goes to the wonderful reading experience found in "House of Leaves." You must turn the book upside down or around and do other things to absorb it in a multi-sensory experience bypassing the Brain Brain just enough to make reading a visceral experience.

It, and all the above, add dimension to learning. Which we all must do and soon to match the speed of the Truth of the multi-verses and multi-dimensions.

I often say that all of life's questions can be answered in the movies. In this case, oddly enough, the wisdom is seen through the film "Man of Steel." There is a scene in which Clark Kent (Superman) sees a hologram of his father played by Russel Crowe. The world of Krypton self-destructed because all options of free will were removed by the High Council because they saw human Free Will and choice to not fit best with a form of humanity that could inhabit hundreds of "Earths" from which they could harvest essential raw elements. When Kryptonians were born from pods and raised by tech instead of parents, the young lost their natural desire to hope for a better life, and so did the parents, when the children were raised by institutions, like schools and social media. Jor-El told "Superman" that he was the first natural childbirth in hundreds of years, and that due to this, he had the essential element "Hope" that will make the world he lands on a better place to be a family. There are two ways to go at this: to remove the concept of the nuclear family, or erase Hope embodied in a man or woman that takes the responsibility of hope and actions to heart. Hope evolves entire planets and realities. Regardless of technology, and in fact, despite technology wanting to erase hope and family.

Happiness is a choice you make and a skill you develop,
you work it like a muscle.
Once again it is worth mentioning to apply
hope to achieve more plasticity of the mind
and create an Accelerating Philosophy.

"SMART YET STUPID"

Who benefits from the de-evolution of language, culture, family unity and values, and communication? Answer: those people telling us, "I have the answer, don't think, just accept."

This is the opposite of the genius and benefits of individual creativity.

How does the leading society in the world get to this point? Where people are glorified and idolized when shouting we must burn or tear down art? Who say we cannot use our full vocabulary because of events centuries or thousands of years in the past? That tell us it is smarter to forget the past rather than learn from it.

And most of all, these people that say, "Ignore your common sense, soul instincts, messages of the heart, and concepts related to individuality."

Isn't one of the key aspects of teenage life the quest for independent decision-making, life goals, and freedom? Then why is it, or rather, how could it be, that the teenagers have been the primary casualty of the tyrannical extremist chanting and re-tweeting statements and events to this end? This end goal of taking over the Free Will, imagination, common sense, compassion, and communication skills of the young will leave them unable to earn and enjoy life?

Perhaps it is in the fact that the people seeking ultimate power do not ever want to make people happy because people that are happy don't need these elitist thinkers.

How does it work? By only showing these fertile and impressionable people the start of controlled thought processing and not the finish line of their plans. By only showing them sound bites that ridicule anyone inspecting them. The "Now Generation" has already accepted the mantra that was wedged into American society, 'You are the first generation that will not be better off than your parents." At the start, this could have been challenged. But now it is taken as a fact of life. This replaces "Essential Element Hope" with surrender.

As Megan McArdle said so well, "Failing institutions beget decreased social capital which, in turn, begets a further weakening of the same core institutions. To take one component, failing schools produce ill-equipped graduates who, in turn, are less able to fund, support and lead schools. All of these institutions are interconnected. Families, schools, religious and charitable organizations, businesses, and governments form an ecosystem that we call a culture. When one fails, the others can help strengthen and sustain each other and the culture. When they all decline at once, things start moving in unhappy directions." A.I. will soon be a tool to minimize our thought processes, which robs society of individual genius.

The Chinese leaders know this well, and are emerging up from the unhappy depths that this bad mindset created from the era of Mao. This is why the government insist on some level of censorship of entertainment from the United States. This virus of living by radical mantra cannot be exported to a nation like China without dire effects. I liken it to the settlers giving blankets covered with Smallpox to the Native Americans, and feigning innocence as it wipes out entire generations and destroys the family structure, loyalty, and honor. They know, and rise up from, a Maoist national society in which no aspect of life is free of politics – the United States has politicized everything, the last victim is sports, which by its competitive core is a meritocracy. Interacting with Chinese people and filmmakers showed me the long-term damage done by removing imagination and most of Free Will.

McArdle goes on to state so precisely, that if the essence of politics now is to satisfy the angry, squirmy feelings of the most vocal people on every single subject, then breakthrough achievement in tech and art becomes almost impossible. "We live in fear of the online mobs." Is this any better than living in fear of a single big tyrant? China has learned that it is not and is working hard to prevent this from infecting the young people rising up so gloriously in their country. It surely gives us a good reason to not react against every limitation if that limit is for the long term good, rather than the immediate soothing of ever-changing feelings and demands of these vocal mobs.

Chinese businessmen and youths were given freedom to benefit from working with Americans. The family values they knew were true, that they see in Nature, went out seeking a reflection of them in American youths and success stories. But when they gazed upon us, they could find no decent level of scruples and core principles that had helped the Chinese survive since about 1910. That is because …

"Americans are becoming ethical imbeciles." The standard for what is right or wrong … should never come down to meeting the approval of the loudest, angriest 0.3 percent, which is how much of a percentage the vocal activists represent. The companies that control our ability to achieve wealth are so intimidated they cower against the bullies that fire up social media lynch mobs.

There is only one Internet. The immense store of facts and knowledge in it make it a modern day god, capable of doing anything. Anything, even if it means ruling tyrannically over your thoughts, actions, desires, and common sense. And most of all, defeating the core culture that provides the last stop gaps against their goal. While the United States now has almost accepted defeat in this, China is going the other direction. Thank all creation for that! With their numbers, and higher observation and innovation skills that come from struggle, the ironic fact is something I read in an online forum linked to "The Sixth Tone" and is paraphrased…

***For decades both Americans and Chinese thought the
United States would extend a hand and us lift up.
In a sense "save" the Chinese "Now Generations."
Ironically, it will be Chinese that give useful
examples to Americans.***

MORAL DECAY

"You must never, ever fool yourself, and you are the easiest person to fool."
R. Feynman

Everything within us – our Brain Brain, Intuition, heart, soul, and even cells --knows the moment we aide in a "Moral Sin." Powerful people use sophistry to argue it away or simply give in to a Machiavellian method as is survival of the fittest. Yet, the very action blocks the power of "Morality" (with a capital M, meaning it is an absolute) to evolve us.

People have argued that the current adults always comments on the moral decay of the young people and what their new generations embrace and use as tools to navigate life with.

The most powerful social motivator today is the visual medium. The greatest motivator used to be life experience, but no longer. If you think, act, and invest money or time according to what Big Tech feeds you, you will end up nutritionally, financially, and morally bankrupt! Hollywood caters to Cancel Culture for now but it won't forever because enough talent will survive for a new era of film stories. What Hollywood seeks for now is any way to gain eyeballs. Some hits on television like My World or talk shows made them believe that watching a human train wreck gains viewers, so they made shows that maximized immoral behavior on shows about "Housewives" or other shows that highlight what is the worst ways to behave.

Young people begin to think this selfish messaging is a good way to live. Moral choices take time and demand refinement while bad television does not. Add the currency of Social Justice Warrior and you end up with immoral censorship that steers young people the wrong way.

This is not by accident. Hollywood feeds you instant gratification and reckless ideology, often based in lies or at best half-truths. Nations are built on centuries of culture and prudence. Legacies and power are built on true Nature and Human Nature: Balance.

What happens in almost all revolutionary movements is that they either find or politicize an aspect of humanity and/or society. The claim the problem started at the top: "Almost all of the heroes created since post-WWII were in fact brutal, inhumane, greedy, liars. We must burn this past and shed its values and we present the only righteousness." These people quickly gather a following of people that might have good intentions, yet must buy into that short term and long-term message entirely. Here is what they won't tell you, just to pick a few modern heroes of Cancel Culture.

Once part of the group, they are pushed even further to the extremes. Nelson Mandella rose to power solely because he was the most brutal gang leader in the area. Che Guevera murdered countless people in cold blood and kept his power through intimidation. Now with social media creating local heroes overnight, the masses of people voting do not inspect the past of the most current mob leaders. You will see people that are mass murderers, drug dealers, killers, cheaters, tax evaders, and much more put up as leaders. Why is this so bad?

Two reasons. Without leaders that are "Morally Strong" (think Gandhi, or Lee Kuan Yew from Singapore) the people wanting the most power can talk followers into believing that any degree of violence or excessive control is justified. You see that in the United States today. Even assassinations get a cheerleading squad.

So, what are the ingredients of morally strong leaders?

1. Appreciation for the "National Culture" going back centuries.
2. Balance between the past and future values, both are reflections of morals.
3. Instilling cultural truths around patience and harmony.

Honor those that seek the truth and be wary of the ones that claim they found it for you. The only nations that survive are ones that center on existing culture and build upon it. Those that force micro-changes by creating splinter groups within the culture are subversive and usually result in totalitarian regimes.

All you need to do is look at the opposite directions that Chinese (vs. American) high school and college educations are going to see movement as example. In China, which is becoming richer in culture and finance by decentralizing information, demands appreciation for culture. In America, the current culture war is centralizing Information by threatening any opposing views and all past cultural heroes.

"For success, be clear the ladder is leaning on the right foundation: Culture. Climb the ladder with rungs rising up from Culture for evolution. Upheaval without clear solutions is a ladder based on quicksand."

On Achieving Excellence

Life shrinks and expands in proportion to one's courage.
(Anais Nin)

Courage is relative: relative to immediate surroundings, to the government body lording over the individual, relative to sexuality, to prejudice -- all of them excuses to gain attention for taking half-measures or explain partial success or individuality among masses. The real battle over courage lies within oneself.

Excellence, by definition, requires courage. Excellence demands separation, and in separation alone-ness battles Narcissism, for without this battle, delusions of grandeur prevent almost all remarkable and sublime sharing of achievement.

The greatest obstacle today for true "excellence" is the very medium by which most people are declared "excellent": readily available Media exposure. Social Media "darlings" trying to stand out dilute the core of excellence by trying to label anything and everything excellent.

What lacks in 90% of self-aggrandizement is 1) measurement, and 2) durability.

Against what is an excellent achievement measured? Look to one of the last bastions of meritocracy: Sports. Personal politics and victimhood have taken over most all other arenas in which excellence is announced. Notice I did not say "earned".

Excellence in sports is measured in Mathematics. Mathematics is not racist, is not sexist, is not political. It simply measures. Today, people rise against peers in Social Media presence. Twitter allows people to voice opinion in ways to create themselves for a world they never meet. What is their greatest motivation? Is it excellence? Is it genius? No. It is re-Tweets. Power on Twitter is measured in re-Tweets. Not original thought, not bravery, not sole voice. In fact, the very foundation of Twitter, Facebook, and most other "Social" mediums is to present an image or statement that follows the majority of what people on that site already believe. That, by fact, cannot be excellence.

An American comedian states that "people-pleasing" (which is what most Tweeting is about) is a form of assholery. It does not matter what point you are trying to make, you immediately must accept being one of the masses instead of one of the excellent if your Twitter ranking rises.

For another test, look at "Art" with a capital A. The best Art is not one that most people agree on as wonderful Art. The best Art divides people into two camps: those that love it, those that hate it. The "durable" Artist, the one to be remembered, expresses courage, knowing that half the people will love his or her work, the other half hate it. The best Art forces communication and introspection by the masses, and often you show them what they do not want to see.

What is more essential is that Artists "live a life" before or simultaneously with their production of Art, which only is Art when it is on a level of excellence. We do not go to museums to see failed Artists or mediocre Art, we go to see the greats! The truly excellent.

One problem is that Academics and Philosophers (along with other "infallible experts" like Economists) jump to the front of the spectacle of an emerging Artist, and then "define" the work or achievement for the masses. This poisons it. There was a time, up to 1800's in the Western World, and centuries in China, wherein Philosophers actually "lived a life." They toiled, they made Art, they met face-to-face with challengers, with disease, with starvation, with the decay of an entire civilization at its bottom points and the zenith on rising arenas. That day is long gone. Only in the West would 90% of the modern philosophers need to exist in the University/Academic/College world. Why? Because that is the only place their weak "philosophy" makes money.

Philosophy is an Art, and true insights into how to balance survival with virtues must be gifted. How old is the last excellent philosophical creed you read? 100 years old? 200 years old? Why is Kant and Freud and others dead hundreds of years still taught as the foundation of philosophy? Because almost no "Modern Philosophers" announce revelations that lift humanity. Or teach excellence. Most Philosophers in the last 30 years are "excuse makers" for the masses so that they eventually get some following. They do not "live a life." What do I mean by that? I put up what I call the Hemmingway Test. Hemmingway played matador to bulls! He fought in a war! He got into bar fights. He hustled women. He wrote extensions of a life of Excellence, transforming his life into lessons. That, is excellence … that, is courage.

Steve Jobs, probably the most common current name of Excellence, did only two things excellently-well. He 1) separated himself from all others to both envision and demand his ultimate goal, and 2) connected people to the ways they gather information not only individually but as his entire user base. There was a time when iPhones or Apple computers were rare enough that seeing someone else having one could bond two strangers.

Excellent.

The people that look to me as a source for their own excellence or fame may never exist. I present something becoming more rare as technology and global governance grows to dominate. A life well-lived, and courage in ways of visions resolute. I don't care who supports me, who by numbers make my vision one that lasts. I know in my soul they have one of the primary elements of longevity and power: Truth.

The world used to be changed by individual example, not opinion or popularity. Of all the popular Rock stars, has even one of them steered civilization? No. Not one. But, the great thinkers have: Tao, Confucius, Sarte, Disraeli. Along with them Michelangelo and Leonardo Di Vinci because their artistic work and philosophy on life drank from the same spring.

Why could artists guide civilization when the kings over them vaporized after decades? Individual courage. Individual willingness to be alone, with only the truth of their Art to protect and catapult them.

The current political correctness must be seen for the mediocre thing it is: it is replacing what is truth with "what sounds good." Political correctness is far more imprisoning than any past regime in re-education. Its tool is to demand "consensus" or be labeled a traitor or idiot – all because you dared to open your mind before committing to mob mentality.

Young minds are vulnerable and mostly feel the two emotions of happiness and sadness by exclusion. Political Correctness demands have become a lynch mob threatening the freedom of all that people cherish. Mae West, the famous Actress that actually lived a life, said that "those who are easily offended should be offended more often." This is the truth that built Hollywood out of the risks actors took until nowadays. Now, what truth do Actors and Directors stand for? Consensus, which means mediocrity. Look at American education. For P.C. reasons, they no longer teach that excellence measurable in tests is a goal, instead it is heartless and bigoted. They teach consensus not excellence. They never teach "synthesis." Who builds a future on teen anger and fragile self-identification? Abusers of trust and truth.

A proof of this lies in … well … what "sticks.' Ask a student, "When was the last time you were awake at 2 a.m. fueled by a new truth and its power from a college instructor?" When was the last time your lesson was so inspiring you talked to everyone you met for the months about it? Life-changing energy of thought and vision? "Never" is the reply. Excellence keeps you awake by its sheer transformative energy. When a society gets lazy, when it is too easy for the masses, innovation and vision die off due to lack of use.

"Mediocrity is self-inflicted and always has company."

Excellence demands observation of the truth of the greater arena. I'll give an example that is incredibly simple. Decades ago, in the trials for the Olympic Table Tennis Sport, there was a mediocre Japanese player named Hiroji Satoh, playing in the finals in Bombay. His vision let him see the paddles in an original way. He glued foam rubber to the paddle. The only reason was to hide the sound the ping pong ball makes when it hits the paddle, because this sound tells the opponent how much spin and speed the other player used. They allowed his paddle to play and he beat every opponent based on sound added to athletic form. An unobserved and unvalued aspect of the game. He used the I.Q. model. Apply one idea to many – winning the game – and then bring it back to one end goal – monetization of the idea. No one had thought of alternatives to the paddle, making this a prime example of "thinking sideways" to win.

What Is Creativity?

Creativity is productivity marked by originality. (This is not my original definition, it is a quote and I cannot find the original writer.). The "Dynamic Definition of Creativity" is "Creativity requires potential originality and effectiveness." Steve Jobs said creativity is "just connecting things." This begins by recognizing patterns, I'd suppose. "Divergent Thinking" is by definition cognition that leads in various directions; therefore, much of my work in correlating five major books means I'm engaging creative talent in writing by enacting Divergent Thinking before I write it. And I use Divergent Thinking again as I present the products that prove it worthy of attention, expense, or investment. The result is multiple solutions to problems, which is what society needs; however, control over news, information, and the power of the mob will prevent most solutions because they survive on creating new problems.

The highest example of this is Davos, which is a meeting of powerful and rich people. They must meet to agree on how to "intervene" to fix a problem that was created by the last few times they intervened on society, technology, global and local finance, saving the planet, and macro topics that lead to micro control. For example, they intervened to get rid of paper bags to replace them with plastic: how did that end up? They must intervene and try to be creative in ways to extract more money from the working class to fix a problem of global pollution they forced upon us. The list builds on itself. They come up with Divergent Solutions that mostly increase their power. Or… actually, they are working backwards which is how they got in these jams, they determine what will give them more power and create problems that need their solutions. This is a perverted and infected lab example of creativity gone bad, so let's go to the people that truly do engage the creative process that can lead to limitless joy.

For a clue as to how creative explore our world would you go to a university laboratory? A Think Tank? No, you would visit the arena most fertile in creativity – a park filled with children. For it is in this arena that "innocent" minds must make their own decisions based on what feels right married with what their mind comes up with to maximize Free Play time.

In that Free Play arena, you see children on a path to mastering games and social interaction that reflect what they are at first in awe of: Nature, and our world of love and nurturing found within Nature's realms, for Nature exists in their Mothers, as well. In the park, see how so many games involve the four forces of Nature: weight, force, percussion, and movement. Later I talk about harnessing the energy and Truth in Nature, and in these games they transform this energy and innately seek Truth among players.

Their minds are not yet corrupted to think one way, or laden with conforming to family expectations or social norms.

Children look around and first are driven by **curiosity.** Curious about a new park toy, curious about a kid that doesn't look like other kids before him or her, curious about what they can get away with, curious about what inane object can be included in a game to replace something from the adult world. Children *borrow* from adults, they mimic other kids, but they first must be engaged. Once that happens, curiosity breeds inventiveness.

Children "create" new words – entire fake languages. Imaginary friends. Children can turn a pile of leaves into a portal to another dimension, a shadow into a demon, and countless things into adult objects like cars, beds, clothes, and celebrations.

In their 3-6 range, most children go through the "Why?" stage, asking "why" about anything you say. They haven't yet fully bought into everything they are told and see. They want to know why we made this agreement about something. Actually, they wonder what if this were not the case, what else could this be? Adults fall short on the imagination in this case.

And they are driven by three things: peer acceptance, familial and self love, and risk.

Without risk, innovation falters. And so does happiness. Risk gives even children a feeling of accomplishment and pride when it is overcome and delivers anything of value. The risk of *not* innovating is to be the loser in a game, or to be left out of the game completely. They look around and measure unconsciously the levels of risk and reward in Nature around them, and then pick and choose which traits to mimic and or use. The teamwork of ants, the colors of nature, the order of power among animals – which in part explains why so many hit animated films involve toys kids give human traits and risks to. The same things children fear losing in risk are put in jeopardy by the toys, animals, or other things in the child's life. And the messages are driven home in songs that are easier to remember than common words. The ability to suspend disbelief is on a level that many Olympic athletes wish they could master.

Dreams work in a universe where the rules of both Time and Space are suspended. So are the games of children. The games are as real as the nightmares of grownups and kids alike.

And so should the vision of artists. Artist borrow from Nature, in lighting and hues and balance… then they create an example of all that is inside of them. The repetitive visual lines in Pieter Bruegel The Elder's peasant paintings reflects the lines of fields of wheat before a harvest, of shoreline sand ripples, of rainbows and natural terraces of mountains and trees. The challenge is to unlock the energy and truth captured in all of this – kids, dreams, Nature, games, and screw the rules (unless super realism is a core of the art.) Yet, popularity of that art or creation demands that the end product is either appealing or inspiring, and then useful.

I believe children can still see into other dimensions when born, as long as they have not learned so many "rules" set forth by adults. This explains imaginary friends, and also some of their inborn knowledge. While many kids are urged to leave behind imaginary friends or realms, it makes me wonder, are we adults the ones living in the hallucinatory world, demanding kids join us for our love, nurturing, and protection?

If the adult world is a hallucination, then from the top down, we are rational. Yet so many of the things we demand the "free-minded children" to accept must look like a hallucination to them. What if they see us living as an hallucination in some other thing's dream? If so, then all we see is the hallucination we demand they accept. We "fulcrum" reality based on a few things that are solid, like the ground, our bodies. Ego demands we think of what we "know" as reality. Taking hallucinogenic drugs oddly enough correct this limitation. Disillusion is the feeling that boundaries are falling apart.

As we move more toward a world of technology (which is laden in rules) and less in a world of Free Play, we might be able to "compute" more, but at the expense of dreams, doubts, and true innovation, leaving us little more than oversized ants. In the near future, will bored humans, no longer tested to survive, watch soap operas about robots learning to be cowboy heroes? Or about super-humans merged with technology that lost the ability to love? Or about aliens watching us to see if we can avoid the traps other new civilizations fell into when they were given the keys and secrets to eternal life, limitless knowledge, and complete indulgences. Without risk will some of our most cherished human traits like love, caring, and compassion be lost?

That depends on the ability of Information and financial innovative success (capitalism) to merge messages with the pace of innovation.

If capitalism is the distribution of Information to be absorbed by multiple means, then so is art. The power in distributed art can be seen in music best, and how it can call thousands together for an open air concert and make them all dance.

Writers and Directors deliver something massively more powerful. Which is why it is important that in the leading nation for film – the United States – the Integrity of the Artist must be protected and valued, and left alone to create the finished product without fear of being cast out by the mob of people unable to create anything. The individual artist needs protection to grow and become popular within the cultural strengths of the nation, because otherwise he fails, and then society gets poisoned by external cultures. Contrary to popular announcements, all people and all nations are not equal. Sounds nice, but, it is not true. As much as the rest of the world sees China as only oppressive we have to admit that there is a growing respect for both ancient art and current art: there are museums in the new airports and more foot traffic in the art districts there than in SoHo or Paris now.

Currently, I see leaders of countries destined to lead the world as parents in a very enticing and full park, with artists as their children, dabbling in make-believe that for the first time is on a global scale. Toys and innovations from around the world are placed in the playground, and not all toys are safe for all ages or Ages.

A Director often acts like a good father taking his children to a park in another country. His child naturally knows how to play and is drawn to fun kids and imagination, thinking they all know how to play safely. The Director encourages Free Play from his Actors, but needs to swoop in fast seconds before the child hurts themselves. China plays the role of good father for now.

"Talent is God-given, and should be appreciated.
Fame is people-given and we should be cautious.
Trust is earned and we must be virtuous."

FUSING DIMENSIONS

When exploring new orientations to knowledge, shallow or deceptive conclusions deter and detour the explorer. With no firm wisdom to serve as a navigating star, its far easier to stay lost in this New Age or get consumed by novel thoughts given to you by the mob than it is to return with something substantial in this dimension.

My curiosity about Time, Space, and energy first found nurturing and education in standard school studies in physics, chemistry, and math. And as breakthroughs become public, I evolved my ideas accordingly.

But when the obstacle is the very language in which science is shared, the exploration uses different tools in stranger lands.

"Magic continues where science stops." That rough quote proves true but can also be a trap in itself.

Almost all of what we term New Age "Anything" traps the followers in rituals and slogans that do three things: keep them coming back, create a money stream, and lock the mind into limited exploration due to the very words and examples "Masters" use. But make no mistake, the first two are the driving forces in groups promoting everything not endorsed by a leading government. They promise many methods to a more balanced life. Some even promise miracle healing. Others use ancient drugs which I do not approve of using, such as Ayahuasca. Yoga and meditation are good. But some events that are "controlled chaos" like worshiping of the Tree of Life, Numerology, Cabala, meditation, Egyptology, alien intelligence, Wicca, and all things deemed Occult do not to me deliver any breakthroughs for our future as a culture.

Fringe professionals are publicized for Western fans, and include cult leaders, Priests, Masters, and others. They follow the basics of NLP and motivational speaking to keep you thinking you're experiencing the ultimate secrets on a path to their version of nirvana.

I agree that unquantifiable experiences and research should be pursued beyond the limits of science and standardized theorizing. But, everyone should recognize that the buzz words and group meetings turn into dead ends if you are not allowed past these tools of control.

For me, my experiences started when I began camping in Nature. By 15 I was solo camping in expansive wilderness. Hitchhiking to National Parks and Monuments for 4 summers, I traveled cross-country 3 times and Mexico to Canada 8 times. Yosemite, Big Sur, Banff, Rubicon, Yellowstone, Glacier, Monument Valley, Death Valley, Kings Canyon, Shenandoah, Lassen, Carlsbad, and beaches and rivers throughout. While I did not realize it at the time, Mother Nature infused me with her energy and opened portals for physical experiences, including truths without thoughts. If I had not gone through this, I never would have had the open portals that attracted certain people, and be led to years of exploration into extreme rituals and beliefs over decades.

Exploration into "magic," the other occults, hypnotism, Astrology, voodoo, and mind-opening drugs works if it is balanced. I think all of these offer valuable insights and break new philosophical ground. But when I looked deeper into all of them, and each of them, pretty much everyone deeply devoted to any of these "arts" or "tools" did not evolve … they devolved. They became trapped in the limitations of their study or belief. Yet, gems of lessons can be taken from each, tough steps resulting in altering your understanding of yourself and your world can be taken, that lead to incredible freedom and realizations.

Through hypnotism I experienced Derrida Deconstruction on a massive and long-lasting scale. This helped me drop old, false, or limited viewpoints in favor of new ones. And there is no doubt that the body can feel the touch of magic as rituals consume you. Drugs break down resistance or hold onto old ideas, while thrusting the mind into arenas that force it to re-imagine reality.

All this is great to experience but is not an end game. It does not "deliver" you or explain, at least not on the levels I experienced yet.

When your professional world revolves around words, and words prove to fail the advanced mind, it's tricky to know how to record and build off experiences, whether they be in Nature or in society, lived among people or in dreams, in this dimension or another.

There is a great question that both philosophers and scientists ponder: can "something" come from "nothing?" In artistic thought, the vision comes from someone or something, or not. The artist might conjure it from a massive cross-section of possibilities. But scientists put this question to the universe. Can the universe be created from "nothing?" Is black space empty as we thought for 500 years? Scientists now say no. It is called Dark Energy. I call it Ether. If Ether can link the planets, why can't it link a supreme being to an individual artist's vision?

We can only exist in this world because hundreds of laws of physics and life come together to allow it. Gravity, air, food, mating. All came from something rather than nothing. Our search for the meaning of life, results in Art in some cases. Our idea that we can create helps us think a God can create, and it gives some purpose to our suffering and joys. It's "sublime" as a gift. Artists take a stand as to "why we are here." By creating. We are constantly searching for our meaning of life, and in art we get glimpses of it.

The artist is constantly reminding us that we are in love with life. And they so very much want you to be in love with it too, with what you see, hear, the dance, the painting, the movie. In both the process and experience, we shout, "We will not be limited!"

Memories, Dreams, and the Elites

Everyone has seen dreamcatchers – those hanging ornaments created by the Native Americans in the Southwest. You are supposed to hang them around to capture dreams and hold them close until you get to them. Pretty simple to figure out. How often it works can't be quantified or verified, yet people feel comfortable buying them, hanging them, and calling them by the name dreamcatcher to people that see it.

Now, imagine something very high tech. It might even look like a dreamcatcher. The device would probably emit frequencies that generate brain hyperactivity and flow of thoughts in the mind. Aimed at your eye is a special camera that targets the optic nerve, giving a freeway into many parts of the brain, from which it can extract mental impulses. And then it plays what information is being sent on a flat screen like an iPad. So in a sense, it turns the eye in reverse, sending images created inside the brain out into the world.

Imagine putting a harness in front of an artist. Using the same device invented at Yale to help schizophrenic people, microwave penetration maximizes the artistic impulses in a truly creative person: Fine Artist, Dancer, Coder, Singer, et al. Channeling in through the retinal nerve, a million connections in the brain and the image attached to progressive thoughts are downloaded in the very act of creation. Creating a world unto itself in each original work, is this the work of a god? To harness creative energy? An editor puts streams of imagination into dense programs. Then this is played back to the Elite wearing the other end of the harness. Will this be heaven or hell? Will this expand creativity or enslave it?

Will aliens keep Elites from entering portals they are not genetically capable of opening?

My belief is that the Ether is a gatekeeper. Because why would Ether be neutral if it exists for any purpose. It is always paying attention … for a grand reason, perhaps the grandest of all. This one …

Today we see Elites collecting artwork they cannot create. Tomorrow they collect the entire soul engaged in the process.

Elitism rejects the standards that all humans are equally worthy, and in fact, considers some unworthy completely. They want memory superiority at the same time as superior memory, and want their minds that store data to be like the Cloud most of all. All access yet no originality.

What happens when you pickpocket the Free Will of an artist? Does this double the possibility of Free Will, or short circuit it? You cannot confuse freedom to do what you want with Free Will. Because you are not freely thinking of anything in today's world. Even in your dreams, you dreams desire what you are fed while awake and some Free Will is challenged in the rem sleep.

This is the problem we face with the upward explosion of technology. It is not so much that we take faith that a grand creator, a God, created the universe and made it balanced. It is that we are giving the power to ourselves and through sophistry justifying stealing it from others.

What or who will protect Artists? Will science decide which memories or imaginations are more valuable and then steer us to output only what is valued by them? If you think this is far off, it has been happening. The only difference is the artists become famous, rich, and they think freely, as "Consumer Kings and Queens" for the masses. This distraction pulls attention away from how influential these icons become over the masses.

Another device would fit like a helmet on the head and is far more intrusive than the optic nerve wormhole. This more intrusive wired helmet beams a certain type of microwave into the brain, exciting memories and dream image storage, and most of all, creative and artistic thoughts. Inside the brain are implanted sensors that can single out creative and artistic thought, ensuring that your life, memories, and imagination gets copied into a software.

This software is then downloaded into the minds of "The Elite." It is inevitable.

Already, we "data mine" from various "brains" that specialize in theaters of advertising, entertainment, health, and warfare (to name just a few.) As technology makes these devices possible, what keeps the device from being abused? Nothing except of 1) cost and 2) access to it. A.I. will bypass all obstacles and a hold-harmless clause will be followed just like the FCC decided to not hold Internet news accountable to the rules for broadcast news.

Already in our world, hundreds of machines and methods of extracting the fringes of your dreams are the toys of leaders, the military, and the rich – or Elites. We as citizens are just becoming aware of this and how powerful they are.

All reality cannot be explained by words.

A child's Free Play and imaginary play lets them fill in the blanks in a wish to make it a reality. Can computers engage in Free Play for the fun of it? Because happiness breaks boundaries and limits. The intrinsic motivation is virtuous – happiness, challenge, balance. What if A.I. captures the imagination testing reality of children?

Porn is about naked grinding bodies, Twitter is about naked grinding brains. It allows self righteous bullying. Re-Tweets rather than wisdom is the goal, shaping short term need for gratification. The goal?

Not to teach excellence any more, no more Shakespeare, instead teach diversity and fairness. But that goes contrary to empirical facts. America is the story of trying to achieve excellence if boundaries are removed Fair treatment did not mean fair outcomes.

Systems only dream in total darkness. And systems that refuse the light from the past, the mirrors of culture, and their dreams end up being nightmares. More tears are shed over answered prayers than unanswered ones.

Use I.Q. to apply this to as many aspects of your personal life as possible, then bring it back to "one" which is to create your Accelerating Philosophy.

Can We View Ourselves?

"Make the last three things you do in a day as remarkable and deeply applied as you can make them, because as we sleep we internalize excellence and build our reality and our dreams off that." Billy Kidd

Art surely is a reflection of Nature and creation within our cells. We somewhat can imagine taking a journey inside us to the cells and thanking them. Yet, can we truly and objectively view ourselves. For this question, imagine a fishbowl, with two goldfish in it. The goldfish only know the fishbowl they swim and live in. You watch the fishbowl and think that the fish must know they are in the bowl and you are outside the bowl, by the way they dart away from your finger when you tap the walls. That is a primitive reaction to movement, light, and pressure. Fish are only curious about what is inside the bowl. Sure, they come to the top when it is feeding time, but that is the eating principle. Could our manner of viewing our solar system share similar limitations to the fish observing the room, and then beyond that conundrum, could they understand their situation if a mirror were put up in front of them?

The world we live in is our fishbowl. We try to imagine God or aliens watching us. Yet do not think much about how unaware we are of all the things and energies in the room that holds our fishbowl universe.

Clues are given to us by Nature on how to expand our view beyond ourselves. Nature coupled with the mind – our awareness -- gives us our internal GPS. Circadian clocks on all levels sync with Nature. Our Intuition and gut instincts present true responses we do not have to think about, responding to clues we don't speak out in our minds as words, we simply listen to gut instinct and do its reaction.

Is Nature, or aliens, watching us to see if we can "jump the shark?" By that I mean, all others before us ate themselves, for WE are the shark.

If we are a player in an experiment, we cannot gain the perspective to answer the ultimate riddles. This in part explains my fascination with Derrida and also detaching from the agreement about Time and Space. If for one minute I could be another person, would that give me that perspective? It is through Art that I simultaneously delve as deeply into my core as possible while becoming the universe around me. An actor accepts immersion into another character yet his ego still holds on to himself. A dancer might be a better example of this, becoming a swan among swans. To escape the fish bowl requires a massive surge in both a spirit and scale to escape it and deal with the world we cannot imagine yet and cannot imagine its rules for existence.

Whales make up original songs that are spread over all the oceans until whales around Australia are singing songs created around Europe. Not just the tone must be carried in the water .... the reason for the song must be, also. If our portals to another dimension are deep underwater for whatever reason, could whales carry a secret for us?

This also begs considering the idea of "Planck Philosophical Pivots" that present the basic unit of measurement – is this the fundamental unit to understanding the universe?

Different cultures observe faith in what we describe as "our reality" in remarkably similar ways. Yet, the significance of the human traits that got us this far varies greatly in cultures. The cultures advancing the most humanity require a combination of compassion, pride, responsibility, bravery, grit, and freedom of information.

Our existence, co-created between our minds and Information, places the "God Particles" that could truly observe our universe within us, within the fishbowl. Yet as Bowie said, "I'm much too fast to take that test."

In studying our existence, someone far more eloquent than I am postulated, "Cosmologists seek the origin of the universe; Archeologists seek the origin of civilizations, Evolutionary Biologists seek the origin of the Mind" – and they do so seeking scientific evidence." What if the answer cannot be scientifically quantified? What if these converge within the act of creation? Not just cosmic creation, but any unique creation of art? Because on the simplest level, this is a search for … I.Q. – connections, and art transcends boundaries. Art is an interpretation of an experiment of the soul to feel empathy, to "connect in advance" of events or meetings in person. The soul, reflected within art, gives breadth to what is called Intellect, what we call the power in the world to create, even factors as immeasurable as empathy.

There is no real reason to evolve compassion, yet we hold it up as one of our most divine characteristics. Much like art, it was nurtured and evolved simply because it made us feel better, or special, or … connected.

Along with any creative ability comes such great responsibility, or in other words, an ability to better see our reflection in the mirror outside of the fishbowl. And since writers create the prose and poetry that explains this, and coders do so in computer language, then we must be accountable if others make decisions off our creations.

Arthur Schopenhauer realized that, "Man can do what he wills, but he cannot will what he wills." The confluence of stimulation meets our biological truth within our cells, and this is mitigated by our soul, our awarenesses, and then, governed by our Agreements. He and others went on to consider the question, what came first, the body or the soul? Our soul had a goal, which was to develop a body capable of expressing itself, so everything we evolved had and now has a purpose.

Yet if the body developed first, it would probably not have seen reason to develop the soul. It was either implanted by aliens or our soul began the process of our final evolutions. Our soul in search of evolving us until we can see the Ether - be outside our fishbowl - could be behind every one of our evolving trends, both personally and as civilizations, since elements introduced that do the opposite always fade away.

As per the brilliant words of Dr. Jay Lombard, I'll paraphrase his deductions. The sanctity of the hierarchy of souls, personal reflection, the minds, and what we are capable of passing down, is being attacked by a mindset that seems dead set on reversing our evolution, as if it is a threat to the new mental order. What is this mental order, or new phase and sector of civilizations? It is one that craves to centralize Information while marginalizing those individuals capable of finding new Information on their own, or in outside networks. They do this for power.

While Liberalism is unfortunately so related to its use in politics that it cannot be cut away entirely from it, this counter-evolutionary mindset and the societies it creates exist in a realm of lies. They pretend to be able to both see the fishbowl and create reality according to their visions which means keeping us swimming. Energy cannot be destroyed so that swimming escapes the tank at their whim. If we deliver more than they comprehend they remove us from the fishbowl and watch us flop in the floor until we die.

The generosity and compassion of the Elites is almost 100$ of the time a smoke screen, and while they pretend they are egalitarian, there is no worse example of exploiting anyone but the Elites and intellectuals than in Liberalism.

"Liberalism is when you equal some scum to a highly spiritual person. That's the danger of liberalism. It leads to tyranny. People aren't equal in their level of development. And when you give equal rights to a 5-year-old, a 14-year-old, and an 80-year-old, it is a crime. Liberalism leads humanity to hell."

And with Dr. Lombard, each of us has a right to live, yet for Elites to sell their Utopia they demand that all people are equal contributors and equally weighty in contribution to life. That all depends on the situations we find ourselves in. Look at survival. On an island, shipwrecked survivors look to the people most capable of helping them live in the wild. In sports, the dominant players lead the team. In love and life, people of Integrity stand out as examples to follow. This climb to the top took them all a lot of time and effort. Liberalism applies their ideas broadly so that they can "put on as if wearing a costume" what the ones below them own spiritually, their actual soul of creativity and free interpretation of Information, which ultimately all Liberals end up demanding to take from naïve followers.

This again proves why art must remain important to individuals and governments. Art is the mirror within the fishbowl that gives us a credible glimpse of ourselves, and the truth cannot be faked or hidden.

VISCERAL

We all hold on so fiercely to a reality that now is mostly defined by words. That system of defining the world through limiting words works for a reason. A person can go on and on about the glory and sublime rewards of a "Reality Breakthrough" that can reach beyond a word-based world. But the fact is 99% of the humanity can barely hold on in a world simplified and navigated by language. I asked a globetrotter and wise friend once why he thought religion maintains its hold on civilization. And he said it is because most of the inhabitants of the world want less choice, not more choice – and less to think about, rather than more to contemplate. They want as many things figured out for them as possible so they can live a life free from using more of their brains. They don't mind having to use more of their bodies. They rely on religion manipulated to fit the religious goals to determine their moral laws. They want their religion to promise them a hereafter beyond their imagination to justify the limitations and tithing imposed on them by the same organization that tells them they are unworthy of total bliss without them. The fact is, the churches and leaders of those institutions need the people, the people do not as a survival level need a form of a church or center for meaning and morales related to the beliefs of the meanings.

It's no coincidence that organized religion first borrowed the best from pagan rituals, then demonized those pagan sects to prevent anyone from leaving them for the pagan alternative. And they did this through words, but this would not have worked if the words did not evoke a visceral experience. And that is exactly what Salvatore figured out in Florence in the 1400's. He threatened the churchgoers with damnation, manifesting in a physical reaction: obedience and donations paid to prevent damnation, meaning the hereafter can be bought. Yet, the ultimate outcome of a religion based on social rules like the Ten Commandments was beneficial to societies and cultures around the world, so it is a net positive effect.

Humans left to their own quickly devolve into murderous mobs with groupthink that is driven by fear and greed for survival. The religions offered a reward for more Virtuous and sublime actions and thoughts. This manifested in joy, for example, in healthier, happier children. And joy is visceral. So is the blind obedience to dogma if the religion has a fundamentalist sector.

But changes in civilization most often fell upon shoulders of individuals, or small groups of people risking everything in a social experiment. Art played an oversized role in reminding people not only where they came from, but it did it in an emotional way, which means a more visceral way. Visceral gives a physical feeling to the things that give a person "meaning." This helps them to make choices that reflect the best of society and their government leading their culture.

Yet one of the challenges is … "How to present new choices and freedoms so that they unify people toward more Joy and deeper meaning, when the experience does not need to use words?" We don't want to always fall back on fear to motivate people.

Can a visceral experience that transformed and evolved one person be shared? That is the question! Yes, it can, through the arts, and through, quite simply, "beauty."

There are three ways to interpret stimulus as it presents itself: voyeuristically, vicariously, and viscerally. Either you are watching or reading about other's experiencing it and don't think you share common ground with the object person… Or you relate to the person and live the experience through them, using words as a medium. Vicariously, you experience it. You bypass the thought process. It is only in the visceral state that a human can break free of their chains to words. A better way to say this is that a single moving experience can break down a lifetime of fears or prejudices. This forces a person past the two biggest reasons they cannot personally evolve: fear, and laziness.

Why is this so important? Words can build a mental prison of sorts and is not always as pure an emancipator as wisdom. Words can be formed to trick people. Most people do not know that early in the social development of France, attorneys were prohibited from entering politics, because they "turn" the truth around to convince lesser minds or those not recognizing lies. Art being visceral holds a power all its own. A picture cannot "turn the truth" in a person – the person creates the expression of power in the art.

To break past the limitations of words, it requires indescribable visuals, sensations, deformations, events, beings, knowledge, and biology. When you cannot describe it, the mind shuts down that part of you and your body absorbs the experience. Music can often drive a person to cry, out of both pain and wonder. Awe, as experienced in Nature, does this constantly to us. It is a primitive element within us. It is "ancient" and deeply cultural. So how does this compare to our entire evolution as a species and cosmos?

Think about this for a second: every cell in our body is created with the DNA history of learning for over 500,000 years of human beings, all done without a single word. This knowledge tells the cell how to survive, duplicate (sometimes), adapt, and function. There is no language that teaches it and even today's most advanced science cannot explain a few of the simplest things, such as: how do the cells of the right arm know how much to multiply to match the length of the left arm, when there is no brain developed yet to tell it that? Language is not necessary.

No language tells the Laws of Nature and Laws of Physics how to rule.

Now let's look at a very simple example. Camping, which I did a lot of on a solo survival level before I was even 20. This was in the 70's. No GPS, all the equipment weighed a lot, and no communication choices other than physical.

If you were going camping for the first time, who would you rather have as your guide? Someone that wrote a lot about camping in severe wilderness, or someone that camped in severe wilderness for a long time? Of course the latter, because your common sense tells you that writing or thinking about danger and survival is not the same as living through it and making the right decisions with limited resources.

Very few people actually have had experiences in which they either enter a new dimension or see an event so incredible it changes their being want to talk about it. They realize at the moment two things. It is an experience for them alone because they were open and ready, and that whoever they talk to won't fully understand the experience through words. It is like explaining a lucid scary dream – no one will be feeling the visceral intensity you felt in that dream.

The other major problem attached to sharing "evolving" experiences is that the institutions (New Age fads) have co-opted the dialogs and either perverted them for their own benefits to put limitations on you, or you are too far ahead of the audience so they think you're a "kook." So, they hook you in on promises that are no nearly as dependable as the key ones: family, Nature, Virtues, and "The Four Confidences."

How then do we deal with sharing what we experience? Well, by sharing what you experience with people you respect, that you already know have a way of looking at the world symbolically and though Culture. Symbolically shapes destiny and expands the mind and soul more than by just words. Otherwise you are throwing pearls to swine.

Yet, be cautious. Skeptics (and even believers) do negative things the moment they begin analysis by words: they tear apart the experience into units they can pass judgment upon, driven by jealously they subconsciously want to "master" the experience even more than the one that experienced it, and they reduce the outcome to what can be explained in words.

An everyday miracle might better describe my issues against sharing experiences. Take love. Who is more in love? The man/woman that is deeply in love yet will not do any of the actions or say the words that he/she attaches to love? Or the man/woman that is not in love but does everything both parties use as proof of true love?

To answer that you'd have to dissect and analyze all the person heard his life about Love. There is no correct answer, yet most people automatically select the first choice, in a quasi "Love is Love is Love" argument. But love cannot be quantified.

Take a look at its opposite emotion. Grief. Ask someone to quantify the level of grief they have after their relative or even their pet dies. Words cannot capture the visceral experience. They would trade anything to get over this grief, but as they said so well in a movie "The Counselor", "Grief has no value. You cannot buy anything with Grief."

Yet both of these emotions, despite evading total definition for centuries, are accepted as "real" by almost all people.

It becomes real when you experience it. Until then it is merely an assumption that falls far short of delivering any of the benefits or open paths to greater evolution.

While the quest for love is admirable, words cannot replace the experience.

I won't tell you the exact moments that opened my mind and body to understand Ether, Time, Nature, and Space better. Words fail to capture it. Yet, I can tell you a few things I experienced, that were told to me, that once experienced were not anything like what I expected. In all cases, the physical jolt once the mind no longer wants to explain to itself what is happening can only be compared to other most visceral experiences: childbirth, taking a life first time, surviving near-death experiences, the feeling of a soul leaving a body the moments it dies, etc. Way back in my teens, I would experience flashbacks of my time alone in Nature. Not sure why.

Meditation and focus could bring more ways to find a personal balance through these visceral experiences. Through enough immersion into visceral experience, the "fabric" or what I call "The Human Agreements" loosens. And this led to my greater expression through Arts, such as chef, photographer, Director, and writer. This allows me to share the "matter" (cells, electrons, electrical fields, vision, musical vibration, gustatory satisfaction after eating, emotional truth through art) with "what matters" to me.

We all have had some super vivid dreams, often called lucid dreams. But there is a threshold where dreams stop and they must be recognized as something else. Simply calling them "challenges to interpret symbolism" is useful here. It demands we find out the truth without words.

How does visceral influence a Director in his film vision? There are three ways an audience relates to what he or she sees on screen. voyeuristically, vicariously, and viscerally. Voyeuristically, the viewer does not think they have much in common with the characters, therefore do not put themselves in the characters' shoes or story, but they are fun to watch. Most action films are like this.

A film delivering a vicarious vision finds the ways to put the viewer in the story. Love stories that work are in this category. The audience feels as if the stars are relatable and acting out moments that the viewer thinks they could live through or at least relate through. Suspense films fit this genre, and so do scary films that work, like "Nightmare on Elm Street." The reason it scared people so much is because the fans related to a fear of nightmares, and put themselves in the roles. A film with a lot of visceral moments or scenes grabs the audience by the gut.

The scene in "The Untouchables" in which the baby carriage goes down the stairs is a perfect example. You don't need any other reason than a helpless baby to grab your gut. Huge scares in movies achieve this for moments, and true life stories do when they hit a high point of emotion coupled usually with action. Scenes revealing intense betrayal also deliver this.

A very good film hopes to combine two of the three manners in which an audience relates to the vision. For example, creating a scene that combines vicarious and visceral is more powerful than a scene that comes across as just vicarious. We are overloaded with stimulus when all three ways to interpret come out in one scene, and this forces emotions and with that, deeper impact and hopefully wisdom. But this craft within this art is disappearing with the new age of filmmakers, because most are never told this is important. Just as shadows are important to the character, visceral gives depth. Again, I paraphrase:

Art is the only substitute for experiences
we ourselves have never existed through.
It affects us viscerally, voyeuristically, and vicariously.
It saturates us with life, truth, and balance.
Deconstruct your Human Agreements,
gain more plasticity of the mind,
and develop an ability to form your "agency"
so it feeds an Accelerating Philosophy.

Just as every chapter in this book
accelerates your ways of viewing yourself
the Philosophy will Accelerate your Brain Brain's
game of pinball with every high-level Qualia experience.
Why put yourself through this added step?
because we have already been given proof
that the Elites and Tech Giants
plan to accelerate until A.I. replaces
original thoughts and our morals and virtues
that come within each human being.
We need a philosophy that can
accelerate us and our minds so fast
that we maintain the best traits we have
in our souls.

WHAT LANGUAGE WILL WE SPEAK IN THE FUTURE?

As the mind expands and technology uses different language – it does not stop there. The 1's and 0's of code are soon being replaced by Fuzzy Logic, 3-Dimensional bytes, and Quantum – all with specialized language at first. But what if the close proximity of discoveries and usage inspires what we create to create its own language? Think this is far-fetched? Film did just that. It combined enough Art forms so that the finished product delivers a message in a language far evolved in how it effects people than the sum of its parts.

Most people do not know why most of the world speaks English. I hear people say it is because England with its massive Navy colonized most of the world. Hogwash. One idiot at a party was wooing a younger girl from France with his wisdom, by telling her that the world speaks English because we won WWII. Wow, I had to actually step in and correct him.

The global acceptance of English as the dominant language began just before 1900. The Royal Academy of Sciences held meetings every two or four years, depending on the purpose of the meeting and who attended. Up until then, all science and math theory and discovery was still in Latin, a dead language, and a hard one to learn fluently. At the time, travel became easier, and leaders, scientists, and doctors came from all over the world to learn the new health techniques, laws of science, and so on. The problem was that they did not understand each other.

So, they decide the world of math, science, health, and political debate required one language. They set a timetable of two years to decide which language all future discoveries would be shared in. They at first easily eliminated any language with too many letters. Then the languages with high art calligraphy determining the meaning of a word (Korean, Chinese.) And then removed any language that you did not know the meaning of a word until the whole sentence was finished (Japanese.)

Finally, any language with impossible guttural sounds or voice inflections (Vietnamese) were eliminated. That left French, English, Patagonian, and a few others. They then set on a course to see which language when applied to science would cover the most discoveries. After two years, they decided English would be the language of Medicine and Science.

Perhaps it is because every language has some minor English spoken in Science, which the child would absorb even if he did not understand it that makes English films more popular. Who knows?

New dialects will form from our language at a faster pace as two very important things happen at once. 1) Computers will make shorthand out of language and merge meanings to let up keep up with their sharing of information. 2) Society will continue to lose how much it values fine speaking. This latter point is due to how smaller groups of people are demonizing fine word use, grammar, and punctuation as racist. Emojis are the worst proof of this. And you can be eventually the Tech Giants want to reduce us to symbols.

The fact that as we read our brain talks to itself is a major obstacle for speed reading that with various methods can be conquered. A.I. will turn all great writing into its version of Cliff Notes. This will really destroy key fields like medicine and law. Parenting will suffer if A.I. reads books to kids. Some think that then we can learn as fast as the stars of "The Matrix." Both mental and physical learning can speed up. But, is that the best way to rise to the top? Implanted ideas are not the same as learning through trial and error. A baby learns to walk over a period of time for a reason. And they found that turning the page of a book connects what you read to your other senses of smell and touch.

This makes filmmaking that much more important. A dictionary shows words but not how people really talk. Films do that. Film might be the only thing left in 100 years that shows how we talk now. This makes it unique in The Arts. Because it is not a matter of style, but a matter of reality. A painter might paint in a style lasting 1,000 years but the lesson or message does not translate or stand up over time.

Personally, I believe a "circular" language that is the overlap of all the output of the Intuitive Brain as it receives stimulus from all our senses and brains will develop and words will be more like seeds that can sprout multiple concepts at once… and we choose which fits our goals. Not all goals are pointed to the future. This language will be based upon Time. Watch the movie "Lucy" and she describes Time as the only true unit of measure and it gives legitimacy to matter—and since events have walls and walls create patterns our of Will and raw information it will encase us or that is encircle each individual… within the collective consciousness.

What this means is that when you realize something new from language it becomes a permanent event that alters your point of perspective and from that placement you go off in a direction or Vector of events. At the same time what was spoken influences the Fields of collective consciousness. If your personal consciousness is like a boat and you are rowing across a vast calm lake, then this language is one of the oars.

**This chapter was important to an
Accelerating Intelligence because the words
we use to describe our experiences will be suggested
by A.I. and they are not suggested in a way
to encourage Free Will.
Mathematics might turn our words into hologram images
so we would then be reverting back to hieroglyphics
but on a 3-dimensional, projected image.**

THE SOCIAL RETARDATION OF THE INTERNET

Each chapter is another bumper to hit
in the pinball game your mind uses to increase I.Q.
if you don't know this bumper exists
You can't score off it.
Score.

Knowledge has been defined as "Empirical data times Mathematics." This is most popular with experts that love the concept that Mathematics can explain everything.

I prefer a more human way of building knowledge. I.Q. inspires us to search for new formulas of this to deal with the "Greater Unknowns" (the unanswered questions of our existence and universe.)

"Knowledge equals experience times sensitivity". That is then applied to an utterly abstract Ether grid.

Is the Internet removing the experience element?

Problem there: we cannot inoculate or teach the young against the isolation the Internet brings or what it replaces in human evolution.

Throughout the 20,000 years of modern mankind's civilization, three factors propelled us past all other animals. 1) Meeting strangers from other regions and sharing survival discoveries. 2) Finding interest in networking toward goals with people far away that shared needs for protection, food, and land. The third element served no purpose, but broke down communication barriers. The third element is Art.

All other animals might take interest in their immediate arenas and be aware of seasons for mating and food. But a group of chimps on a mountain do not take interest in the advancements of chimps hundreds of miles away. They have no concept of distant trade or conquering vast lands beyond their food and mating needs. Yet man does. His ability to communicate and then plan to expand power – which was a primal desire to lessen the deaths due to famine and war – led to major cities and then nations. While the scratch plow was the key invention that allowed one man to grow more food than one man could eat and therefore led to modern trade, it was not the deciding factor in how we conquered the world.

The human ability to communicate over vast distances for the greater good of the tribe or city required impressing the newcomers enough so they traded instead of warred. Women mated with different genetics, and the children held immunity from two different lands. So, all was good.

But this idea that two very different people can advance groups and lead to more happiness required at the very least sign language communication, leading to written or spoken communication. But what if a foreign nation had to impress the value they might have quicker than they can communicate verbally.

That is why one of the greatest ways to impress was through Art. Dances and music first, on trails. Imagine being a goat herder and seeing the 150-foot-tall gold-plated statue of Ramses, and being told he is a god. You'd believe it. When you felt Chinese silk and it was so remarkable it was a fabric of Art, you wanted to learn from Chinese explorers and traders, not kill them. The idea that if the Chinese could create such beauty they must not be barbarians allowed trade to expand.

The artworks communicated skills beyond the goat herder's imagination. Then painting, sculpture, and written words kept that flowing.

At the core of our current era of man finding himself top of the world, this age called Humanism, in which humans worship essentially themselves most, was innovation of the Information present. This increased the value of art due to its original spark or vision. Art told stories, etched history, and entertained.

Gamers isolate their focus so much there have been a few that died of exhaustion during game play. This is a concrete example of how reducing focus to something as small as a computer screen or cell phone prevents stimulus and reflection of what gives you meaning and existence in the real world around you. If we ever get to the point where isolation is championed and it is prolonged then we will see the actual reduction in the development of brain functions. Replacing healthy interaction with nature, art, religion, family, competitive school and sports, and other experiences that give us meaning with this new computerized way of social ranking will make us all, especially children, self-absorbed.

By putting different people together, and letting them discuss ideas of the world, their health, their tastes in art – this all adds to the codifying of increasingly better upbringing. Specialists took care of more and more things that the individual had to do, such as making shoes. But to grasp enough to compete for food and land, the surviving people needed to merge new Information from as many sources as possible.

This continues now. Children are put into school with other children that are not like their siblings, and learn new ways to communicate, explore, and create. The two greatest gifts of going to college are: they teach the young person how to stay focused on a long-term goal for future rewards, and they force the mind to not only learn, but learn "how" to think – how to imagine and improvise.

Yet, most learning is essentially a process or an algorithm. And teachers try to do as much transfer of knowledge using an algorithm process. An algorithm is simply a series of steps to get to a goal. A recipe is a good example. Teaching actually reduced excellence to a Grade Point Average only in the last 75 years, which made it even more of an algorithm. While creative thinking helps answer questions, most physical learning ended decades ago. Auto shop, Home Making, Music Class, Soap

Box Cars, building a fort – all of these things kids did up until 1980 are either rare or totally gone from education. We've replaced interaction with our world with "getting along." Being with people like us.

But if no one knows how to dig for water, or fish and cook it, who feeds them? What skills do they have? Stone Age man was far more creative with only his most primitive tools than a 20-year old is today. These kids today uses computers to be creative thinkers for them. None could find a library book using the Dewey Decimal System (an incredible cross-reference system that relied on human interpretation and annexing of creative thoughts.) The young lose attachment to Natural ways to learn. GPS is a best example. An entire society unable to find ways home based on landmarks, memories, or orientation emerges. Almost completely incapable of navigating a route without a computer.

I had a hot argument with a friend, when I said that the Internet puts kids together that already think the same and like the same things, and therefore this will put them at a disadvantage. He said the opposite. It allows them to be specialists. Is this a problem? Yes. In most cases, all deep innovators or geniuses kept keen interest in tangent factors to better deliver their art or innovation. I ran into him recently and he agreed that Facebook and groups online hurt the development of young generations.

Art class is the best example of being consumed in the moment you are learning anything. Art -- physical interaction with something that effects visuals and vibrations. It transcends words. It leaves lasting impressions. A child that never met an elephant is far more curious about it than one that saw it on television. This sparks associations with the giant animal, teaching children through brains other than the Brain Brain, where we fit in the world. The elephant was touched, it smelled, it responded.

Genius comes down to blind variation. Visions. Impulses. Body movements that are not thought of or related to survival. Through trial and error, the genius retains the example that closest fulfills his vision. But are they born or do they evolve? Both are possible but the born artist must interact physically with his tools. One key factor they all seem to share is an accelerated way of storing Information across more receptors in the brain, so they can access them. The dancer stores them in the Muscle Brain, as well.

Yet teachers and nations try to treat Artists like algorithms. "This is the history of this Artist, this is his technique, this is how he got there, so now, try his algorithm." You cannot install genius or vision like software. A genius might conjure up a variety of options based on nothing but Intuition and quickly discards losers. It's associations of both what is known and what has not yet been known.

"Talent hits a target no one else can hit. Genius hits the target no one else can see."

Through this, Artists need on thing. Motivation. That can come from within. Or, this is where teaching about art or going to museums comes to play – they expand their vision and that on its own motivates them. Something about the new art thrilled and challenged the viewer to push limits to create something just as great. Artists need to have some value in what they are doing. Writer's block comes from either doubting capabilities, finding it pointless, or feeling forced.

The usefulness of Art seems to be how wide you can throw the web.

That is why Apple is a new god, and they call its Apps ingenious. All the parts of the phone were already created, Apple assembled them in a new way according to Jobs' vision. It was innovative electronic arts, and because he could throw his web so far around the world now, we treat his company like a god.

Up until the last 5,000 years, all civilizations worshipped gods or Pharaoh type god-rulers due to stories passed down that explained Nature: floods, sunshine, plagues, wars, and things. The newcomers or newly enslaved were told to worship the god. While this seems odd, to think that a god built an empire is not any different than to think that Apple is a company to be worshipped. We only make decisions weighing new information from old. One of our primary sources of innovation came from meeting strangers.

Now that Facebook and other sites find end runs around this, so that you find people that are already sharing similar goals and ideals, you won't innovate because you are not challenged to do so. Quickly that genetic (and epigenetic inheritance) imperative – to innovate upon realization that it will improve your life – will not be part of humanity.

Already, there are groups of people seeking power that worked hard to erase natural tendencies, such as those found in the dominant Alpha Males.

How well do you relate to being lost in a new city? Or losing your car? Do you ever get bored listening to friends talk about the same old things? Do the videos that used to amaze you fail to keep your attention?

The Internet has made us "Masters of Distraction." They replaced deeper focus you find in reading a book with images. While we used to plumb the depths of our minds for the exact words, now we scroll through emojis to explain what's inside of us.

The greater threat comes from the fact the Internet sacrifices Individual value for "consensus value." The Internet, with its bullies, drives us to comply and conform, rather than seek Truth.

A driving theme of Social Networks such as Twitter is to release a person from any responsibility of his or her actions. They indirectly or directly state, "Why should I stand up for a world that will never fully have Integrity and Virtue?" But the fact is, that humanity is elevated only when the individual takes on this challenge.

A metaphor often used is that humanity is an ocean, and you are just one drop. Yet, as long as you embrace Virtue, it is kept alive within your single drop and is not totally vanquished. It still exists. Nature still recognizes it. It still remains in our DNA.

Art is one of the few experiences an individual can perform that lasts past his lifetime. Having children and raising them is another. It is no coincidence that American Social Media and Tech Giants have openly encouraged not having children "to save the world." But it is in our children that the world will be saved.

So how does that work? These Social bullies also demand to control art. They demand to tell us what is funny, what is acceptable – in fact demand to define art itself. They do this because it is a threat to their expansion of power.

Chinese Tech Giants like Tencent and Alibaba designed their systems with the opposite purpose. While they want success and expansion, they were wise enough to realize the only durable way to enrich life is to enrich what gives us hope, joy, and wisdom. They also seek a balance that U.S. tech companies do not want.

These books titled "One Fish Says To Another Fish" present why art and culture must be built into the foundation of Edge Sciences, especially the rising Intersections of A.I. and Blockchain. Art reminds us we seek elegance, meaning, and beauty.

"Beauty is to enthuse us for work;
and work is to raise us up." Norwid

When Is a Dream Not a Dream?

[**Note:** All of this chapter was written in 2017 except for the final dream, which was written in 2024 after I dreamed it. Out of the dozens of dreams that mean more than a sleeping vision the last dream stands out in both its symbolism and timing in my journey to get these books published and the ideas shared with humanity.]

When is a dream not a dream? That can be answered on so many levels. The word "dream" is almost self-defeating when trying to decode the experience. I prefer to call them "adventures in sleep mode" or "tests for the conscious while unconscious."

Start with my belief that vivid dreams capture a lot of what a filmmaker hopes to achieve.

Take a nightmare for example. When you experience a nightmare, you are the actor in a landscape created by the Director of your dream. You do not pretend to be afraid of the monster -- you're truly scared as hell! All you see you focus on with intensity. You're aware of the hardness of objects if they turn fluid or soft. You notice heat and smells that you'd otherwise ignore. The world you're in is created from a vision of some mind, not your conscious mind. And you commit your entire mind and body to the experience. In other words, every action, reaction, and statement is truth.

In my dreams, I've seen entire scenes unfold that I wake to type in before I forget them. The underground scene in "Eye Candy," with the girl/model emerging out of a giant white hen's egg and all the art on the wall came from that dream.

And the countless dreams I had during my study of hypnotism added dozens of elements to the story and vision.

Beyond filmmaking, dreams play a bigger role in enlightening me, taking me places I cannot imagine, testing my concepts of the universe. Dreams surely test our ideas of Time and Space nightly. And history is filled with hundreds of leaders, artists, and scientists that claim they saw their future in a dream first, or the solution to a mystery was revealed in a dream. Dreams are so essential that experts claim people eventually die if they are unable to dream. That's something to think about.

These dreams that are hyper-realistic are tests to see how well we can translate 3-dimensional symbols. They are also tests on how to interpret visceral moments. The rat that appeared on my lawn a few weeks after I saw it in a dream showed me that the Dream Dimension can offer us clues about our existence. I know you're thinking, "A rat? So what, there are rats all over." But not this rat. This rat was black from tail to ears. It was bigger than a large squirrel. And it sat on his haunches and watched me, trying to communicate. Just like it did in the dream. I was not ready for this yet, and so I hosed it with water. But it would not leave. It kept waiting for me to decode what idea he was sending to me. I thought it might be rabid, so I ran inside.

By the time I figured it out it had a message and went back outside, it vanished. My loss. Much like the hitchhikers in the "Teachings of Don Juan" scene with the passing car. We seek patterns as human beings. This is how we decode things that are not spoken. Through patterns and feelings. I did not recognize the pattern until it was too late. Yet, I do not like to be fanatic or get spellbound into the idea of dreams leading my life. That negates Free Will.

Fear plays perhaps the largest factor in what we learn from our dreams. For that reason, many dreams are testing our fears. For example, one of mine … A journey down the rabbit hole, onto lawns that looked like a great English garden for a castle, took a scary turn when a giant bug, the size of a cat with an enormous, pointed nose flew at me. I became aware it was a test and stood still while it landed on my hand. It then turned into a tiny smiling Bumblebee sort of creature and flew off. I had passed the test. Yet, most likely, it was simply a dream and maybe I saw this in a movie I cannot remember.

One dream that repeated over 40 years finally reached my mind. The dream has me in the water in the San Francisco Bay when a 200-foot high tidal wave approaches. It sweeps away our existence. But, I find a raft. In the similar dream 15 years later, I find a pole to climb. I watch everyone else perish.

On the final two times I had the dream, it was the same dream but I was in Hong Kong's harbor. But this time, I was floating hundreds of feet above the water. Something (actually someone) seemed to be showing me that water will sweep the city. But this time I said out loud, "I've already seen this." To which I hear a voice say, "But until now you did not understand it." The dream suddenly changes purpose so that I look down on the harbor from 500 feet in the air, down onto a huge ship, either a cruise ship or battleship. It was huge. And words shout out from the dream realm in the final seconds of the dream, "The last ship!" I woke with a start, repeating, "The last ship." I wrote down words planted in my mind.

At first I wrote "I am the last ship" … then without thought wrote the next sentence, "I am the unbreakable arch of time." This line requires some explanation. I read and was a huge fan of "The Trials of Thomas Covenant." In this book he realizes the missing piece holding together the "Arch of Time" was his love for his wife who had died and he would forge his body into the Arch to save this new world, meaning morality and sacrifice are involved. The gold ring he wore symbolized his purity of love for all creatures in his new world. Thomas was a writer in his life, he created his destiny, he became the bridge, and it was his higher sense of observation skills that saved the good and defeated the bad.

"I am the last ship.
I am the unbreakable arch of time.
I am the pure gold of creativity.
I will be the bridge, I will carry the burden,
I will keep the secrets…
and I will evolve a billion people"

I had no idea where the "last ship" words came from. I wrote this all down on a small page of paper used for mailing and taped it to the wall near my desk so I could look at it every day. It became the mantra I repeated as I did the yoga Sun Salutation yoga moves every morning. If that's not enough to show how some dreams are more than just dreams ...

That night, a new show was on the television. It was called "The Last Ship." It is about survivors that are on the ship after a virus wipes out humanity. And their stops were mostly along the Atlantic Ocean front and Gulf of Mexico. It amazed me that my mind (or rather, my Heart Brain) knew this series would be on television that night.

"The Last Ship" is a series revolving around the most deadly virus in the history of mankind. However as a filmmaker the Director and screenwriter agree on several through-lines for the dramatic impact and character-building. The scenes and characters revolve around something so important in Observation Skills over Adaptation Speed that it has its own chapter in this book: Betrayal.

I'm not claiming that dreams chose me to deliver a message to. I'm just saying to look for symbolism in your dream and look for shapes that translate to learning or sensory perception. If a specific dream stays with you, what is it trying to tell you? Why are dreams so essential to mankind? Where do they come from, if not just from inside our brain as it does its sleep tasks?

The next dream I can mention now that fits this concept took me to Machu Picchu, in Peru. For months I had studied Ra, the god of Egyptian lore. So much of his description applied to my life and mysteries to solve. In this dream, I was guided by the Egyptian sorceress to the steep hillside of the mountain looking up at Machu Picchu. I could see the houses, people, and apartments below, just like you could if you went there today. But then all changed. That reality broke apart, and the other reality, the dimension sharing that space could be clearly seen, smelled, heard, Intuited, and felt. There was the lost civilization of the Mayans. Still there, looking the very same, wearing the same clothing, working the mountain harvests and carving steps like they had way back then. My intuition told me that they were given a choice -- to stay in that harmony and bliss, or evolve as a society to be like we are today. And they chose to remain, and whatever power gave them the choice, carved out their own dimension. People in the dream could see me visiting, and they smiled, knowingly, and accepted me as no threat. They knew a bliss of living that is not existing today in the modern world. Maybe this was symbolically telling me that dimensions co-exist around each other. So, still in this deep dream I tried remembering the dream's clues, for its "energy."

That was when I became aware of the "vibration" connecting me to all that is around me. And I could see the faint trails of Ether. Connecting all in existence, man, plant, and rock. And... Oh - the secrets this Ether and connection tempted me with.

As if this was not enough, a bright gold light at the top of the mountain drew my attention. The gold light flew down and in 2 seconds stood right in front of me. The Mayans were surprised and pleased this thing of light came down from the top to face me. This was RA. The serpent Ra, with its staff, its face of a cobra, its third eye, its headdress. And Ra was gold. Crystals of some shapes were noticed on Ra. Ra studied me, saw I did not run off. Then, looked deep into my eyes in a way that froze me, then beamed into my head a shaft of light as brilliant as the sun! It hit me like a bolt of pure truth. As intense as that was, it did not end my dream, which helped prove to me it was an adventure in sleep mode taking me out of this Dimension. I watched Ra return to his place of oversight high above me and the Mayans. Then I was teleported backwards out of that dimension and into my bed, where I woke with a zooming sensation.

In 2017 when I began this book I had a dream about a dandelion. My chin was on the ground as I stared at it. A wind came and blew all the dandelion seeds away – yet as they flew away they became the ultimate essence of raw information. The seed became nothing but a spec of light with a tail of energy trailing it. In a supernatural moment I saw where each seed found a human being ranging in age from infancy to on the edge of death, and planted a message from nature into the mind of who it was meant to reach. And the moment this seed merged the light form of the person rose in elegance… in balance and its use of energy and the Fields.

I woke up and wrote a cryptic message dated 22-8 (August 22, 2017):

"I must realize…
Dandelions, dimensions, sink whirlpools –
are presented to
me, as a sign…
Elegance beckons…
Elegance awaits…
Elegance has
chosen…
ME."

This proved to me the impermanence of our Reality and the limits of our dimension and how we view them. As the only Universe. And, within days I was studying all versions of describing dimensions, from String Theory to others, and decided the E-8 model aligns with what I believe.

The final dream I'm listing occurred on March 28th, 2024, and is the only dream so significant I had to add it after writing this book was completed. In the dream, I saw my father, David. I'd seen him in many dreams but this one was different. I had longed to be with him again to talk, perhaps that inspired my mind to experience the dream. In it, he had a long beard like Moses and was towering over me a few feet. I walked up and touched his body and could feel it, which is rare in a dream. I said, "You're real, I can feel you." At that moment I heard my voice I realized I was about 12 years old and only 4 feet tall, which explained his height. In the dream I remembered that when my mother collapsed in Walmart in Tucson, I called out to her – "Mom?" – and the voice was the same 12-year-old voice in this dream with my father. My father smiled, nodded that he was real. We shared a final look of love into each-other's eyes and the dream ended. I told my sister Melanie about it and she said wow, take about a symbolic dream, that really is a message. From that day on, I focused on the dream and my mood, which was in turmoil due to external forces, rose up a little each day until by the tenth day I was happy and thinking of positive things and people that bring me joy.

My most memorable dreams have to do with letting go of my desperate hold on what we call reality, with all its limitations.

Personally, in my films many times, the cinematic vision, subtle symbolism, or connections between actors came from a dream.

[I had several dreams that I recognized instantly as timeless looped experience after I finished this book and they are presented in this addition to the following paragraphs.]

I was with my father at the moment his body died and perhaps for this reason he appears in my dreams more than anyone or anything else.

In the most recent dream he appeared with a long, grey beard like Gandolf. When I hugged him I only came up to his chest so in the dream and after I once again knew I must have been about 12 years old (age in dreams does not follow our concepts of Time.) We hugged and then he pointed to my stomach and intestines and said, "Sometimes when we have a problem we think is down there…" he points to my heart … "it really is up here, something's wrong here."

A dream only weeks before had me walking through wilderness with him, a forest with sparse trees. I was carrying many sacks of things and I did not know what was inside. We came up to a wall. My father climbed it easily but did not go over it, he sat on top. He looked at me with both desire for me to climb up and over the wall – but then also a desire to see me not go to the other side. As he sat on the stone top of the wall I climbed up and faced him. On the right was everyone and everything I knew. On the left was a new adventure. I was consumed with getting all my sacks of stuff off the top of the fence and down into this new land. He held my hand and without words told me that my "things" don't go with me, they remain in the forest of the right side of me. In that moment I woke and realized he had walked me to the great wall that divides our life now with the next life. His last action was to smile when he saw me not jump into the new life, instead backing into my old life to finish it.

There are dreams with my father almost every month. These are the most symbolic and surely are dreams that are not really a dream as we understand the idea in classes and society.

Karl Jung wrote extensively of symbolism in dreams trying to reveal a truth, heal us, or is a way to tap into higher consciousness. Einstein would work until he was exhausted and while sitting in his chair, he would place a pencil in between his fingers so that when he was on the edge of falling asleep the pencil would fall to the floor and wake him up with his first visions in the dream fresh in his mind. From these he claims he found several breakthroughs and that the more often he did this the stronger the imagery and higher the brilliance of what his dreams could show him.

These six books all tie into the first book on the incredible impact that the art of film has in shaping humanity. Different films play bigger roles in social advancement and personal growth in different nations – yet they all merge messages in the worldwide theaters and televisions.

Jung's human architypes can be seen in almost all films, and the hero discovers which of his architypes he must follow. In this moment he is fulfilling a psychology often revealed in nice dreams and also nightmares.

Dreams are the ultimate source of symbolism meant to ignite our intuition. In movies the hero is often a "Stoic" whose greatest fear is abdicating moral authority and this fear forces him down the path to his best archetype.

Dreams as a symbolic art form and language without the restraint of Time are fleeting yet the memory of them or agency they bring about is lasting. Movies are the externalization in an art form of what is the best inside us.

I am working on something that will
help us find who else has had the same dream.
Ultimately, it will help us know where the dream came from.
This is an example of what an Accelerating Intelligence
can do when merged with a breakthrough technology tool
and aimed to honor an Accelerating Philosophy.

Sparking Our Inner Animal

We tend to think of ourselves as highly evolved creatures. Yet our rise from the swamps governed by other animals only started 5-10,000 years ago with the formation of civilized cities (Alexandria in Egypt or the Tigris Euphrates River area.) The scratch plow enabled one man to make more than one man could eat. This led to free time and the delegation of duties. If someone made a better shoe, he could trade it for food.

Then came societies we can identify as our origins, about 1,000 B.C., though others started earlier.

But honestly, 495,000 of our 500,00 years since we discovered fire are mostly lived on an animal instinct level. During this time our minds, souls, and bodies perfected the Reward System that drives it today. The pursuit of happiness is delivered through the chemical reactions in the body and those usually have to do with reproduction or survival.

Drugs alter our brain chemistry today to feel the sensations of rewarding the pursuit of happiness, or to remove reminders we failed.

During the creative process the mind is consumed with many of these sensations by a default sort of pathway. Our ability to create something out of nothing reminds us we will survive while others perish. This gives us peace and opens us to taking chances. Indeed, art might be responsible for far more inventions and advancements in civilization due to this than it is credited. Those that only admire what we do would perish, which makes artists very lucky. Others might like our work, but they are driven to repeat it only because of the addictive feeling they get by being close to something that was a mirror sort of, faking their ability to create survival objects of their own.

Yet the rich chase art like it was a flavor instead of a cure or protector.

We will always have more ice cream flavors, softer mattresses, and all the things that reward our senses. This dulls them and floods them with fakes.

I believe that if many people that truly appreciate art were there during the process of creating it, they would be divided into two camps: those that feel the energy, and those that do not.

But as we enter the era where Virtual Reality or other games that can duplicate reality so much that the player cannot tell the difference between a simulation and life, will they go instead to the games to fake the feeling of creating? Or surviving? Will they increase or decrease our ability to be emotionally moved by art or other things and people?

When "Chamber Music" was the only music around, people were moved to tears by it.

The quest for art is also related to how much risk is in our world around us on a daily basis. Singapore is very safe, but almost no innovation comes from there or from Switzerland. Can this prove that art is an extension of our most primal part of us? Survival? As early innovators survived, they were the ones to create the first art, on cave walls and in pottery. Art demands innovation by definition.

The challenge to survive elevates to the challenge to create, and create as Nature, as God, or as our unfiltered mind and soul inspire us. Without risk, fewer are pushed to create, because it is no longer a survival imperative.

Is art more animal … or more cerebral?

Does the vision come from solely our Brain Brain, or do our muscles being used for art add their Brain? Is the emotion in an original piano solo from our Conscience, or our Intuition or Chi Brain?

Or is it simply evolved Free Play?

The opposite of play isn't work …
it is a lack of joy.
The opposite of good is not evil…
it is the absence of joy.

Dogs, Cats, Horses, and Whales

Dogs will show you every trait that is important in your life and character. Loyalty. Fun. Exploration. Bravery. Love. Defense. Balance your life. Balance your partnerships. Don't sweat the small stuff. You're a reflection of the people you hang with. Parenting. Making two more than the sum of its parts.

I am starting with the two animals that mankind domesticated that are most often in our homes, with us through all our insignificant moments. Animals we let our guard down with and accept as family. But herein reveals a key proponent in how man came to dominate the planet. Mankind learned how to domesticate animals for survival, work, and food. This evolved into a symbiotic relationship.

What does this really mean? That mankind replaced many animals through breeding so that eventually we interact with the smartest and most specialized examples in cows, chickens, pigs, horses, dogs, cats, and even fish. The freewheeling, independent wolf or wild dog won't put up with swapping their survival skills for human-support skills no matter how many dog biscuits we trade the canine.

One myth that recently got debunked was that man domesticated dogs and trained them not to howl like the wild wolves they came from, and dogs now bark, "because they can." The different placement of the noise-making box in their throats limits wolves to (mostly only) howling. A tongue and noise-box moved ahead a half an inch allowed dogs to both bark and howl. For decades scientists insisted that dogs evolved due to man's interaction and partnership. But that would take a million years through slow genetic mutation (one female with a tongue and noise-box a half inch forward would mate with a male the same traits and produce enough offspring so that the half born with the same deformity would mate with others with the same deformity until every puppy was like them.)

But no, that's now what happened according to current experts.

Here is what they now discovered and agreed upon, and I mention it because it has to do with keeping life and the mind in balance, one of the keys I think guide our evolution. Again, use I.Q. and apply what you read here to humans, put it through the "one-to-many" routine.

Before dogs gradually became domesticated by men and women there were wolf pups that were deformed, unable to howl as well as other wolves. These pups could bark more easily. Howling is one of the key communications within a wolf pack, and plays a factor in which dog leads the pack and can breed. No dog that barks would be allowed to breed unless all others in the pack died. And in fact, most times, the pack turned on that pup that could easily bark but could not howl well or loud enough and killed it!

If the pup survived to be a wolf it only got to eat what the pack left after a big kill, so often times the dog died of starvation. These barking wolves lived on the fringe of the pack just to avoid angry lower wolves that would kill them.

As men started nomadic lifestyles, more and more wolf packs came into contact with them. The humbled wolves that could barked were often on the outer boundary of the pack. The pack would hang loosely around the humans hoping one would venture out alone so the pack could stalk and kill the human.

But these rare wolves that could bark would be often closer to the humans because they were not included in the inner circle of dominate wolves. The intense hunger of the barking wolves forced them to consider bones and meat thrown from the campfires of nomads (and later camps.) Slowly, the humans won over the deformed wolves. They formed a copasetic relationship: the human would give the wolf-dog meat from the fires that was easy to digest and nurtured the dog to breeding, and the dog would smell or hear danger approaching to keep the human alive.

The weakness of both species joined to form a strength. Balance.

It would have taken millions of years for dogs to evolve to do two things: "learn" in their primitive brains that barking helps the human that feeds them, so they should learn barking more. Instead, that was a freak of nature, optimized by humans. Their hands, I.Q., and ability to take many and bring back to the one idea on a higher level proved an aide to survival.

Now … why have we "un-learned" how to bond with Nature?

Because the wolf bark singled it out for a role that improved its chance for survival we can see this as the first step in adopting other human manners. Recently, a recognizable percentage of dogs have learned how to "smile" in the span of less than a generation. This is an extraordinary leap in mental evolution when it comes to communicating between species. And oddly enough, Instagram may have played a role in this evolution. So many people trained their dogs to smile that "The Hundredth Monkey Principle" transmitted the idea of smiling to dogs that never were taught how to smile. When enough dogs know this, all dogs quickly know this. And in addition, the dog knows what this means – it is the outward expression of inner joy and fulfillment.

"Dogs smell fear." It's so common that every country has a way of saying it. Dog stories now tell so many tales of dogs being able to sniff out cancer under the skin of a beloved owner long before it is diagnosed.

Yet our words prevent us from "feeling" this or developing this sense the dogs have. Dogs bond with and protect infants where there is no reason to (the infant does not feed them). They must somehow communicate because the baby at birth can often still respond to the Ether in a visceral way when they are far too young to express it in language.

The breed Great Dane has a notoriety for superior psychic connection to their owners and this appears all over Europe and the USA where they are most common. My most impressive story is about a Great Dane and boy bonding on a fabulous level. I must say this story is fully proved by many witnesses before you read it – it's that fantastic.

A couple had their first baby. About two months later, a relative bought them a Great Dane puppy for a present. The dog and boy grew up together. They lived in a house with a nice yard in front and back on a suburban street that had a lot of trees. Perfect to play outside. The boy liked kickball and the dog would chase the boy and bump him. The dog and boy were inseparable, with the dog eventually sleeping at the foot of his bed. Then the parents got divorced. The Mother moved to the next city over. The dog could not move with the boy because the Mother moved into an apartment vacated by her friend at a great low cost, but the building did not allow dogs. So, the dog stayed at the old house with the husband.

About 8 months later the boy was playing kick ball in the urban street. Not used to the traffic, he darts out from in between cars to chase the ball – right into an oncoming car!!!! The car hit the boy, killing him!

At exactly the same time, the Great Dane was lying asleep on the floor of the living room in the first house. The screen door to the front yard and street was latched closed. The Father and a friend were sitting at a table talking. Suddenly the Dane jumped up and ran full force straight through the screen, full speed out into the street, and rammed a different boy of the same size standing in the street, knocking him to the curb in a way that would have saved his owner boy. The Dane got hit by a car coming from the same side as the boy! Both the Dane and boy died within minutes of each other.

The dog felt something humans cannot. I think because they lack the easy sense skills of language that bypass the other "Brains" of the body. But across such a long distance? How could the Dane feel the last moments of horror, possibly even "see" the movement of both the boy and car, over a hundred miles?

Look at this from all the angles you want. One of the ways they were still connected has to be through Ether.

Yet there is another angle to view this from as an added IQ source. That deals with Tara. Her conclusion that there are 5 Brains in the body would explain largely the dog's decision to act, his Heart Brain holding the love for the boy would override its survival instincts in the head Brain while at the same time not recognizing the distance between them as a separation of power or communication.

How does that effect creativity? In a way it shows that art combined with love breaks through limits of what we see as our reality and Ether can be a conduit. Can that be done consciously? Let's look at cats.

Cats are a whole different level of psychic communication and seem to have an awareness of Ether. I DO think cats are living halfway in our dimension and halfway in another dimension. It would probably do me good to own a cat but I am very allergic to them. That, unfortunately, disqualifies me from writing more other than to say that cats for sure see things we don't, communicate with things we don't see, go places we don't know about. Ether.

Cats ride the waves of Ether in multiple dimensions.

There is another very significant idea I have about dogs and cats, and it has to do with women. There can be an even number of males and females that are "dog people." But an overwhelming majority of cat lovers are women. I think women, who tap into their Intuitive/Visceral side (that defies "word explanation") sense cats communication and reality more than men. The women are drawn to this but cannot explain it. Nor are they driven to explain it. That in itself partly proves the communication-without-words idea I hold.

In the ever-more competitive world, the "use" of dogs and cats expanded.

Men use dogs and horses as extensions of their own vision, goals, and virtues. Men have huge respect for both animals. They may use other pets like monkeys as tools or to do things, but mostly dogs and horses are associated with pets being used for tasks. In one specific case this turns amusing: romance. Single men want mates, but many cannot compete as well in certain expanding arenas of appeal: DJ's take the limelight, MMA fighters are bad-asses, meeker guys match with the girls that will naturally would dominate them. Some smart men realized they can use dogs to hook the Visceral senses of passing women. This is not the primary reason most men own dogs, but it is amusing that dogs can be used to improve mating choices by men. And it seems the dog finds joy any time the owner expresses or feels joy.

Women on the other hand use dogs and cats for different reasons, and this *exploded* with the Facebook/Instagram/Twitter age. Dogs (and cats to a far lesser degree) became "accessories." Women by genetic coding want to attract mates and also discourage competition. Two appealing females might be highly competitive, but neither would make the bold move of approaching an Alpha Male because of two reasons: Slut Shaming and other Social Norms. But a dog – wow! – it allows them to break the rule and draw the man or interact with the man without either rules being broken.

The female in fact has a "protector" the male must win over. Consciously or subconsciously, the Alpha Male must dominate the barriers of the Alpha Dog to win attention. The Male is "read" by the dog not just in common verbal ways, but moreover in Visceral ways. And be persistent – more persistent than the dog. To see this in action in a funny way watch the movie "Something About Mary."

Instagram has made this mess up even more confusing by giving girls an accessory that not only makes other girls jealous, the other girl can "feel" the Visceral approval the Dog gives the competition, make her back off.

Meanwhile, the Male thinks this is all just a cool dog, to break the ice.

The woman ventures out with an "accessory" that through I.Q. she relates to "the many choices and opportunities" to "feather her nest." Mating, food, protection. She brings it back to "The One" by either rejecting or accepting the advances of the Male.

However … the Male is usually not factoring in that the dog gives the female two things she needs: the reality of protection, and the assumed close proximity of a protecting male.

It goes even deeper than this. Experts (whomever the hell they are) say that dogs increase protection upon menstruation. Recently, scientists discovered one reason why females reach menstruation and go into puberty earlier in homes where there is no father (due to divorce or single mother.) The girl senses that the pheromones of the father are not present. So her body, through millions of years, has learned to mate when the father is either killed or just gone.

Pheromones are detected by humans yet are invisible, and the sense of smell tackles that, making it not so supernatural after all, it shows an interaction between animals that cannot be seen.

The connection of Ether with animals can explain the mass beaching of whales and dolphins. Though zoologists decide it has to do with gravitational waves interfering with the sonar of those animals, it does not explain all of the cases, or the mass beachings of animals that do not use sonar.

Including our friend the jellyfish, with almost no propulsion or advanced thought ability. What might unite them under Ether's traits?

Vibration. Sonar is a type of vibration feedback. We hear a ping when we see it used in movies. But they also transmit thoughts outward, to other animals, for mating. The sophisticated language of dolphins is so advanced, they have sonar sounds related to enjoyable sex, not just for mating, but for the sheer fun of it. And males not only learn how to masturbate on floating objects in their training pool, they learn to ask for them.

Getting back to the fish in the schools – they discovered that the same fish always lead the school. I had assumed that since childhood. But they just discovered it. Now why would that be? They do not swim faster than other fish, or fight harder, or mate more often, determining the role of leader of the pack genetically. In fact their offspring often are not even in the same school. Could it be that they like the view of open ocean, and the followers like looking at fish asses?

I believe these fish evolved a tiered language within the water of their own that we have not yet detected. I don't think that it is always in pursuit of food they swim. I believe they ride the waves of Ether-type energy, and this keeps them close enough that they can unite their tiny brains into a larger brain, and make decisions on a hive category. A better way to say this is that the fish accept they swim in an "organism-environment." Water communicates elegant design. This would explain why they do not crash into each other, even when they dart extremely sharply or while currents slam them. A single thought makes them explode into shapes that scare predators or confuse them, but how could the middle fish know in time to make his move?

Birds do the same thing when in tight flocks, like Starlings, following fish-like patterns. This can be explained through evolution, of course. But then what is the connection between the beaching of sea life, and the dropping of thousands of birds over a hundred miles? This happens in select areas, like Alabama, Ohio, and the general South. Witnesses sometimes hear a "boom," that somehow is not recorded by any of a dozen different measurement devices. Could that boom be a tear in the Ether that turns the air into a wall as it merges or closes dimension portals?

I think that the reason has more to do with different yet akin Ethers. The fish swim in water we can see, feel, name, and identify, so their version of Ether seems inane. But we only see part of that Ether. Does the original charge in eels come from this Ether? How about the raining of frogs, which has happened for millennia in five continents? Why only frogs? Frogs – one of the greatest evolvers on the planet, one that broke off new mutations at a relative lightning speed compared to almost all other animals. Could frogs have constant evolution in their adaptability to Ether going on, and a boo-boo opens a portal that gathers them up, like a school? Because that is how they rain down.

Oh sure, scientists somehow find some distant lake or river and claim that *maybe* a super strong waterspout crossed the shallow edge of a lake and picked up 5,000 frogs. If so, why didn't it also rain down bottom-feeder fish, dragonflies, marsh reeds, and salamanders? Why did only frogs make it through? Ether? Do they mutate their ability for group thought enough so they are aware of portals into other dimensions? How would this come to be, and what would it be like?

Let's go back to the jellyfish, or more specifically, the virus around it. This living organism a billion times smaller than a frog and just as distant evolution-scale-wise, forms a collective brain that can end up controlling the huge jellyfish, communicating some unknown form of Intelligence with it. So why not a frog? Has the gut brain of the frog observed or knows the gifts of other nearby creatures, and then adapts them? Do they hear an ethereal summons to the lake for the twister to pull them up and dump them far away on streets? Or, more by accident, do they group enough together and are totally unprepared for the dimension change that transports them to the clouds until they rain down?

Imagine that day for the frogs. They're happily merging thoughts below the threshold, sharing stories, and BAM! A collective "WTF?" reverberates in the invertebrates! Collectively they pass into another dimension, where the constitution of Ether lets them fly, just like fish can drift in the sea waters. Then when returning to this dimension find themselves 1,000 feet above the friggin' lake they call home and traveling at a speed that pushes them 25 miles away?

Fish that swim in schools do not have sonar. And they have very small brains. So do frogs. Both behave in ways that reflect awareness of different physics. What does this mean? I mean, that is worthwhile to humans? They learn how to transform a physics energy into a portal or at least a communication and Gaia GPS.

Dolphins protect humans from sharks all the time and even get sexually frisky with some women during menstruation. They are aware of our beings, our systems, how we evolve, and our communication. They even mock us.

What land animal most closely resembles their communication? The elephant? Elephants use a sonar, and their voices travel over miles. They are aware of death. They remember people from over 50 years ago. They mourn. They cry upon heartache. But the trait that is most human? They get drunk and play drunk games and laugh while doing it. There are countless reports of elephants eating fermented melons until they are drunk. Then they take turns picking up the other melons with their trunks, sneaking up behind another elephant, throwing or smashing it against the ass of the elephant, they even sneak around like kids, and then they trumpet a laugh! While genetics and evolution, along with senses, can explain how they know the melons will get them drunk it does not explain why they laugh.

When a female elephant goes into "Must," what compels her? They still do not know. There is no sickness, no change in their body, food, or environment? Could it be a breakthrough then breakout reaction to learning about the greater world around them? Like suddenly they learn what man knows? Sounds farfetched? Then why do they most often seek out a human, even ones they never met and know are foreign to their habitat, to kill? Anger of a greater awareness mankind slaughters them for no reason? Where does this knowledge come from? Has a dimension opened for them to enter, look into, or gain a collective brain of Truth from? Or did Nature simply communicate a certain truth that drove them into Must?

When you juxtapose it with the protective nature of dolphins, it means more to me. Frustration on an evolution-awareness level. The female in Must just can't take it anymore. Why won't the humans learn? Why can't they see the energy within Ether? They try for centuries to break through. We're the ones with the big brains and fancy hands! Maybe they can even see into the future, and in a rage upon seeing an elephant friend killed without reason, can't keep their cool anymore, they lash out.

The final elephant story takes from the observation that some herds of elephants in Africa will come from all different directions, and walk in line, to a far location, where one of the most admired (mostly) female dominant leaders either died or was born. They cross mental boundaries in an even more amazing feat. More than once, when a human that raised and cared for them and loved them dies, dozens of elephants that loved her suddenly stop, and turn, and walk to where they last saw her. And why does this almost always only happen with female raisers? Why?

I believe they see the soul pass through Ether into other dimensions, and one of the last things the soul does is visit the elephants. Aware of her, they are so shocked to learn or feel her energy no longer in the Ether, they rush back to where they last saw her, hoping it is not true. And … they cry. A few have been so heartbroken their own zest for living leaves them, and they soon die, sometimes in that location. I think they know how to follow her, and would like to share that dimension with her, or come to her as she beckons from it. They … "let go" of their belief system, the constraints and limitations of gravity… trust their collective memory on how to cross dimensions.

What does that have to do with creativity?

Elephants paint! To music! Their skill evolves. Like we used our hands that were so dexterous to expand the ability of our brains, they use their dexterous trunks to expand elephant brains until it takes time to do a thing that has nothing to do with evolution or survival. They paint to share in a medium other than their sonar or biological methods. In hopes of communicating with us.

It is up to us to evolve faster to see what they paint clearly.

The picture exists or revolves within the realms of vibration. (Tricky to conventionalize because Matter does not always hold firm in other dimension that obey Laws of Vibration instead of just our primary Laws of Physics.) Herds of cows can be seen stopping and dancing to the music of a saxophone or clarinet. That vibration enters their minds and tells them to stop eating, to gather, and dance. Pigs paint (worse than elephants.)

So much of animal abilities and actions revolve around something we ignore. Vibration. Yet, ask anyone that *truly* has felt the pressure of Ether or "Dimension Merging," and they will tell you the fingers or forms "touching" them contain a vibration. An intense one. But the moment the mind uses words to focus on it, it diminishes. Not because we "broke the spell." But because we relied on a medium totally incapably stupid. Words.

One experience in particular – and I don't want you to think this is straight out of a scary movie – evolved my interaction with Ether and dimensions. One particularly intense merging Ether energy wrapped around me loosely, and a single word was thrust into my mind: "Expand." It came in so loud and clear it did what it was intended to do – blast words out of my Mind, leaving only Visceral interpretation of the moment. The sensation was of my atoms no longer firmly bound to each other. They moved like a flock of starlings or dunlins – the fastest of swarms of birds. They move and form objects that can be seen from a distance so clearly as hourglasses or anvils. The beautiful word for what they do is "murmuration." They "murmur" to visitors sharing a dimension or view upon their dimension – or maybe they are just trying to paint art as a group brain. Words try to capture it. The murmuration rolls "like a drunken fingerprint across the sky" as Richard Wilbur wrote, and included in his longer description, … "the quickness of a pulsating jellyfish." Amazing, huh? I.Q. Applying one idea to many, then returning it to his "one revelation."

The ancient Romans (2,000 years B.C. ago -- wow) said that these murmurations were messages from the gods. Packing a truth that could only be whispered. Which of us listens deeply enough? Or is that question just a human trying to make an advantage out of the art of their sky dance, unable to perceive Ether? And why in the world would one of their repeated images be an hourglass, invented in an era wherein our minds were perhaps so ignorant as to limit our limiting language? Are they telling us to look back to an era when we interpreted the world through Visceral means, through our Heart and Gut Brains? And, oddly, across the world, so did the Chinese, both cultures paying far more attention to Nature than most Westerners. But the coolest part of these ancient messages was that the wise men told others that the message had to be felt and not reduced to words.

Our modern day attempt was for scientists in the 20th Century to call it "natural telepathy" or "group soul." Isn't that an odd word or even odder concept? That large enough groups of animals collectively form a "soul?"

The chapter "What Language Will We Speak in the Future?" intersects with this entire chapter. One factor to consider is that the roles of communication might be reversed once Nature or Cosmic Elegance realizes that animals are learning how to communicate with us in an accelerating rate.

Koko the Gorilla died recently. He learned over 200 signs to communicate. His final sign language talk with his owner and trainer was his last attempt to tell us all he could to save us from ourselves. Basically he said, "Koko cries. Koko loves Man. Koko loves Nature. But Man is stupid. He does not learn from Nature. Nature will reward Man who does. Nature remembers. Learn to love Nature. Learn to see Nature. Learn from Nature."

The Hundredth Monkey Principle proved that when enough animals learn something evolutionary, that all monkeys or communicating animals learn it. The common link in the delivery of that step will be recognized, it is a step seeking an elegance and will create its own "Field of Information-to-Intelligence" to transfer Nature's reflection clearer.

Nature has an extremely long memory. That is why the long-term culture of a society – its national culture – is so extremely important to nurture. This is especially true when Tribalism challenges the value of national culture. Nature does not recognize Tribalism, as it had no identifiable process or direction toward elegance. It is not a Field creating any elegance, which Nature recognizes. Society knows this intuitively. A good example of how Chinese rising tech-savvy citizens merged cultural elegance with their new "Liberty with Responsibility" is proven in their "yu yi" consumerism. They seek a "metaphor" for luck, health, Nature, balance, peace, purity, and the innocence of children. In other words, they seek what all lasting Cultures delivered: "meaning." "Meaning" becomes the most significant element in the future of technology, explained in great detail in "Pillars of Encryption."

Elegance attracts higher elegance" and you can see it in the Arts of a national culture. When you add in the way a national culture appreciates wild animals and pets you use the tacit and intuitive communication of animals. A way to see this is in film and media: the Chinese prefer seeing superpowers coming from Nature instead of machines. "Kung-Fu Panda" is the best example.

I believe that storytellers of ancient eras that depended upon the generosity of the village to be fed after hearing an original story that explained the purpose of life and a person's direction in it realized that that gravy train would soon end. They became ... philosophers! (Silent laugh.) Few of them ever rose the level of Plato or Aristotle. And what did they do as their audiences over generations became smarter? They imbued those listening with the idea that the gods had not just amazing abilities in hunting and war but could communicate through long distance through Ether. They told the listeners that if they were "good enough" (generous enough) and followed the messages long and hard enough, they could rise up and share the gifts of the gods.

Usually, the gifts were valor and hunting. I remember a quote that said love was invented so storytellers would live long enough to become psychologists and philosophers. Love at first was spoken of as a gift first "of" the gods, and then "from" the gods. This flipped upside-down, until mankind, never ceasing his need to put himself above lowly animals, came up with the "selfish herd" term. They claimed the birds move like this so the middle ones can be more protected from predator birds, so it was a constant attempt to outsmart the bird next to them and expose them as a stray to a bigger bird spying them from a distance.

On the computer I recently watched two matriarch elephants rush to rescue a weeks-old calf that had fallen into the water. Not only did the mother know the calf had not been taught to swim yet, so did another mother. Meanwhile, on the other side of the fence, other adult elephants paced furiously fretting over the fate of the calf, and when the emergency was over, they rewarded themselves with a fruit-eating session.

Elephants may be one of the few animals that "knows" where souls breathe their last breaths. Their language lives in a sonic level we mostly do not hear, and comes from the frontal lobe of the skull, much like dolphins and whales. They communicate as individuals yet receive "hive" messages, which explains in part why all three of these animals commit suicide in herds. Their messages travel farther distances than human voices, over a hundred miles from some whales, and over 30 for elephants. What conduit transfers this thought? Ether? Does Ether look hazed or colored where souls have passed away? Do elephants recognize the "history" of individuals in the herd, much like grandparents know the lives of their grandchildren? If scientists have not figured out where in their brains this thought is stored, could it be in their Heart Brains?

One word enters my mind: Art. Birds dance. Crows use colored berries to paint on a rock, and then domesticated birds use paint on their beaks to draw on a canvas. Birds mimic human words and respond to them. Their songs are vibrations, we hear as noise, but they use not just to announce their location or scare off predators, but ... to merge or pop open portals in dimensions for their sheer amusement.

This chapter is a great example of how plasticity of the mind offers us chances to accept wisdom from animals and Nature on a level so high that it results in a mind that naturally accelerates to include new concepts.
Somehow, the elephants knew humans painted and once they tried it the act of painting was joyful because it expressed more meaning than other sounds or actions.
The art was releasing something inside the elephant.
And the elephant painter knew it did.

FAILURE IS NOT DURABLE

You get what you incentivize.
Chamandis

Focus upon the process of creating something from nothing. Don't look at the "end result" instead look at the moments before an original creative action blossoms into a greater work of art. Notice a rather stationary state relative to the explosions of artistic will and energy following inception ... the creator void of this specific impetus or vision ... movement the minimal required for a basic human function. The action might be sitting, standing, or a simple movement for a dancer. A blank sheet of paper for a writer. An idea for a computer coder. The pounds of clay for a sculptor.

Imagining an entire chess game -- or video game to an expert gamer. To create something lasting immediately upon initiation, to receive admired from it, emotion from it ... to make it durable – this does not require that the creator knew something others do not know. And the end-product does not simply conjure itself from the raw talent in your fingers and vision in your eye. It also requires believing in something few other people believe – that there is a message on the blank page, a form in the clay, an App in the code, and a dance in space.

"Belief" then becomes a fundamental element in the creative process. Belief on this level is entirely internal and cannot be provided by consensus or protected due to your own internal weaknesses. It is belief on the level the Oracle in "The Matrix" explains it.

Yet it is so commonplace it does not require "The One." It is what drives thousands of us to finish a single painting. Fewer are driven so cellularly by belief that they gather a collection of paintings to gain attention, riches, and the fame from that amassment. To drive public sentiment over generations requires the creators' works to be durable over centuries or even decades. How many humans experience this for months ... for minutes ... for seconds?

Now let's make this personal. Let's make this about you. Because soon you WILL be creating something durable, praiseworthy, inspiring, and sublime unto your inner being, your soul (Your Qualia Avatar.) What you will end up creating in any form is the expressed recognition of the gift your creator bestowed upon you.

For belief of that level you probably live on the fringe, because to see something others do not and the focal point you believe within -- that action separates you from others almost by definition. You might mingle daily with non-artist, might be quite social, yet must live on the outer limits of your circle of life, because "original" cannot be seen from the center. Why? because the complacency and distractions of common usages fill daily thoughts when stuck in the middle.

The mind on the outer limits seeing a vision others cannot imagine drives your actions. The routines of life helps that manifest. The moment you invent something in your mind, be it a new perfume scent or a new variety of flower, you moved to the fringe. You may not be aware of it, but for that moment you did. The creator of Post-Its is an example. Did that person realize he or she created something durable and so useful it is mechanically an expression of art, of unique vision?

The wind blows strongest against you when you are not surrounded by others. If you look at it like a revolving circle, like a record player, the speed at the outer limits is enough to kill you, as the inner circle is so protected and relatively set that those planted at that point feel none of your physics.

That is why original thoughts are so rare, because to live outside of society, to question the functionality of a device or even a government, seldom happens naturally amid the core of people happy to not be imaginative.

Yet, the very thing that drives most "Artists" defeats them: a belief they are exceptional. Too often they are not. An example helps.

I can speak on behalf of screenwriters for that is my longest career, and also in cooking, for I was a chef. Scratch the surface in the film world and 75% of all screenplays you read are "unreadable." Horrible. The next 15 percent are unoriginal or unmarketable. The final 10% on the fringe, outside rudimentary imagination, ring the bell.

The true talent required to write a durable, commercial, and magnetic screenplay might be surrounded by failures by the same author, yet the success sometimes requires the mountain of failures to support that one screenplay to be durable and last. It gets produced -- and produced well. It becomes immortal.

Which lends to the question of fame. Fame. How on Earth could societies come to the point that they do not just allow fame, but idolize famous people? This isn't new: Zsa Zsa Gabor was famous simply for her "famousness." From that she got acting roles.

I must ask that myself -- for I am driven. Am I driven for fame? Only in that it allows more of my work to be shared. I have never sought the spotlight or needed idolization yet know that fame comes from success for actors and even Directors. The invisible writer, the un-famed writer, is where so much of the magic begins. "Three things needed for a hit," Hitchcock said, "The script, the script, the script." But almost no one could identify an Industry-famous writer in a bar even if there were ten of them. So perhaps that is why, when I started as an actor, I moved within months into screenwriting. I had already written a cookbook and several fine short stories so I knew I enjoyed it.

Today Fame - with a capital F - comes with the trappings of moral destruction. You are what you consume publicly. Fame comes from a society that makes a fetish out of competition over nothingness. A celebrity is a person known for his well-known-ness. He is a human pseudo-event. His weaknesses excused, his dramas hyperbolized, and worse, emulated and envied. The celebrity is not always an artist within a field, such as an actor or Director, dancer or singer.

A celebrity proves his power and status among peers by consumption – what he or she consumes: food, clothing, cars, jewels. But they must hide this vulgar fact, because it prevents wealth through art from delivering "true power."

Is what celebrities do that bad? Is that a sin? It might be shallow. If not it turns into a whirlpool that celebrities drown in. No one throws them life preservers if they are not on top when they need it in the future themselves. And if there is one thing famous people in Hollywood know it is that they are targets for a takedown. That is why consumerism shifted to virtue signaling to those desperate to be lifted above other people.

While Fame can be described as the empty noise of madmen jealous, the status is insecure not just among peers, but in the passage of eras. The suicide rate for famous people is almost triple that of even the most damaging careers.

Perhaps the difference that prevents a balance of characters is simply that famous people surround themselves with an entourage that feeds their feeling of godliness. Of infallibility, and they know inside they are deeply selfish and flawed.

Do you ever wonder why famous actors love their pets so much? I often think it is because a lap dog or cat gives them unconditional love on a gut level, when the person knows inside they should be unlovable, a fact proven by their serial failures in intimate relationships. Again, failure is not durable.

This shows the difference between respecting someone as opposed to holding them in high esteem. The first presents what are called "Resume Virtues" – that can be respected. How much you make, how many fans you have, what you created. As opposed to holding that person in high esteem, or what are called "Eulogy Virtues." Plenty of Virtue Signaling, but who among them sacrifices money and influence to the point that it lowers their habits of consumption and self-absorbed actions?

The worst examples are the daytime talk shows filled with ultra-rich celebrities demanding that their political views be worshipped when they can't even get along as 3-5 people sitting in the same high-paid seats around a table. They understand one thing: win the audience emotionally, push buttons, and take the position of a victims because then they are not responsible for their failures or inabilities to uphold their own virtues.

If they cannot create an actual debate to solve a problem, then they are part of the problem. Not a single city or block of a poor inner city full of minorities has ever been improved, nor their education improved, by these selfish hypocrites giving one year salary to a single cause. And they can all surely afford one year off salary.

Hollywood is almost bereft of those with Eulogy Virtues. It is one of the few Industries that values self-destruction as an Artist. An early death immortalizes Fame in an otherwise forgettable actor.

On the fringe, durable Art manifests easier, deep where the exclusive universe of creative energy and use of that energy harnesses Fields and taps into universal Will. For those with Fame, how much original creation can be done when surrounded by the crowd? Dismiss film, television, and Social Media, and you see how rare these geniuses of dance, music composing, and fine art truly are. Or scientific minds, unraveling the mysteries of the universe … from a perspective that required solitude, on the fringe.

Fame, like riches, is like sea water: the more you drink the thirstier you become.

The fact that failure is not durable is proven in the "Fame Test" in our current age. But before entertainment was taken over by electronic media, that is when the most famous of all peoples really did live on the fringe. Writers and observers of Nature and life like Victor Hugo and Michel Foucault are examples. Go beyond that and read Proust if you really want to bake your noodle (as the Oracle says.) Would you have been the same feeling the same level of belief in yourself?

Excellence by its very definition requires a measure of exceptionalism, and that only exists when the individual is separated due to his or her talent from everyone else. Standing alone, Fame happens. The quest to be distinguished among the "many too many" has been rewarded.

When people talk about why there are fewer famous female artists, especially as painters, I tell them it is because females do not feel driven to produce so many paintings and show them in such a way as to become immortal through fame. This is not a slam against women. Au contraire, this praises them. This could be because women create life within them and know intimately they are immortal already. I believe that women actually channel more pure art constantly due to their reliance on Intuition Brain, leaving logic behind.

Now that we have shown that "Failure Is Not Durable," can we say "Victimhood Is Not Durable?" Nothing progresses for a victim until that is not how the person is defined by himself. Nothing "lasts" in victimhood other than a memory, and that dies with the last indirect person knowing of it. Victim events turn inert very quickly, unless it is channeled into a creative force.

Martin Luther King Jr. channeled his vision to create equal rights, for example. For centuries every culture in the world taught generations to deal with trauma, failure, and abuse as a learning tool. Only in the last 20 years has victimhood become "profitable" in politics, power moves, and ultimately money.

There is a connection here. If the victim is content with receiving the benefits that a victim gets and no more, then that is fine. However, history shows that public victims often self-destruct when the victim spotlight is not on them. And so many political causes require victims remain victims – so they can use these victims for votes, money, intimidation, and more freedom for them while restricting the rise of the victim.

In my opinion, Rose McGowen realized this from the start. While she did expose rapists and their victims rose up to protect women more during production she knew that the real goal was to go after the agency and casting director systems that funneled women to high-paying perverts that demand sex. When the MeToo Movement decided to stop at focusing only on the person it would fail to really complete her mission. She knew failure to do so meant the MeToo Movement was not durable.

What this means is that victimhood demands a relentless reinforcement of victimhood, or the creation and demands of another victim status, identity, or splinter group. It has to be a splinter group because politics has already gone through every single group that can be marginalized. Case in point is sexual identity, and how groups with the arena are now attacking each other as smaller and smaller groups are created. Cannibalizing victimhood. This is the ultimate replacement of "Social Media Tribalization" with Balkanization. It's happening right now. If the art of filmmaking has increasing severe degrees or decrees of politics injected into it, Hollywood will accelerate its fall from grace to bankruptcy that result in the destruction of the art of filmmaking.

In a society the promotion of Tribalism is not in and of itself negative. Tribalism in its most recent definition example is taking pride in an individual's heritage. From this springs many "cultural norms" or favorites. This influences the family unit. Tribalism often times gets reinforced by religions.

The Saturday family dinner in Judaism, church bingo or fairs, the celebration of Kwanzaa, Baptist Revivals are all examples of religious Tribalism that delivers positive results. Music transcends cultural and racial lines but often is Tribalistic. Bluegrass, Jazz, Hip Hop, Country – they link together people that share favorites and also often local family heritage. Union workers might enjoy a Tribalism and pride. Geeks and gamers form a tribalism of a specially talented extent. And college students can form allegiances to their alma mater. Academia and Elites surely know they form a tribal pack, and most often do only what makes the pack stronger.

These examples help identify each other. Working together toward a common goal around common social expectations helps people accomplish big goals.

But it does not serve the politicians, their paid army of protesters, and "experts." None of the protests or riots resulted in a fixed problem – not one. They incentivized nothing but anger and fragmenting society in smaller and smaller groups. The new group splintering off usually makes the larger parent group angry. The original call to action was replaced by money-hungry NGO's.

Study the emigration into New York by various European nations and you'll see their greatest prejudice and violence came from the larger previous group of immigrants. The Irish hated the Italians. Italians hated down the ladder to the Polish. Why? The newest batch of immigrants had new specialized skills, so they took away work, and also split government support.

When universities, high schools, and grade schools champion victimhood and failure, this becomes the reality children believe aside from what their parents tell them. So the higher goal for education is to destroy the family cultural unity – get in between the child and parent, because the parents know failure and victimhood is not durable and they love their children and want them to understand joy and pride come from within.

There is almost no level of danger, sexuality, violence, or stealing of rights that the teacher unions that vote in politicians won't go. Children become useful idiots, as Lenin famously stated about Collectivists. When they begin saying over and over these two phrases is when you realize we are in a new stage of loss of freedom: "We're in this together, it's for the good of everyone," and "Equity is more important than equal treatment under the law."

Splitting up the family was a key in how Hitler, Mao, Stalin, Pol Pot, and Castro rose to power. They first went for the writers and artists, to emotionally convince people. Then they killed them off, as it would kill off the art of filmmaking like it did in Russia and China.

Neighborhoods in major cities willingly "self-sort" and create areas called Koreatown, Little Saigon, Chicago's Polish District, and Little Italy to name a few. This helped them adapt upon emigration, and also helped people appreciate their culture and ways. It was a positive step. And most often, up until the 1980's, foreign tribal zones self-policed people within the community to avoid bad reputations. As long as they felt they were responsible for their actions and reputations these groups enforced the law with their own tribal additions. Simply being shunned and driven out of a community because of betrayal to morals and customs was enough punishment to keep most people devoted to building a healthy community. They welcomed curious outsiders and often showed them their best. The young were often reminded of the sacrifices of ancestors.

But there is a difference between healthy Tribalism and Balkanization. The Balkan nations make up Southeastern Europe and include Albania, Bosnia, Bulgaria, Serbia, Romania, Macedonia, Slovenia, Croatia, Kosovo, and others associated with the term Balkanization. A rise in population led to the "Slavs" spreading out into the neighboring nations. After World War I, most of the Balkan nations suddenly were no longer part of the great Ottoman Empire. It resulted in ethnic and political fragmentation.

This led to nations breaking into pieces. The former Yugoslavia and later the U.S.S.R. broke apart nations. The creation of a separate nation for people that had different religions, histories, or political goals made the smaller nations more peaceful. Yet, it also turned the nations against each other. They didn't trust their neighbors, isolated against them, and even attacked each other. The end-result was a failed state. Unable to manufacture, compete, or hold a stable currency, they almost all crashed into financial ruin.

It turned out that Tribalism in this case destroyed the futures for many generations. This also led to smaller and smaller nations. Rather than accept differences in culture and identification as they did between France, England, Italy, and Germany, the Baltic nations essentially "cannibalized" each other in a quest to "make things even" or just out of anger for revenge.

In the United States, two types of Balkanizations are occurring. Neither one is healthy for the sub-culture or region. The first type is easy to explain. It has to do with politics and the centralization of power. California now has Proposition 8, pushed by billionaire Tim Draper, who made his money in Silicon Valley. The argument is that California has such distinctly different types of people, it should be divided into at least two and probably three parts: Northern California, Central California, and Southern California. The deduction is that smaller states can better reflect the wishes of the citizens more accurately.

In reality the idea of splitting up California is for one reason: to increase how many Senators and Congressmen can be elected, to control who gets elected President. Powerful owners of technology giants think California was meant to centralize power around Silicon Valley and this helps them control laws, especially tax and social program laws.

What would happen is these states would turn against each other? Why ask why, and why think they would turn against each other? For the simplest reason, just look at water and power. San Francisco gobbles up water and power. This is all provided by Northern California, states North of California, and the Colorado River. Southern California should be able to claim rights to that water and power. Southern California also has the only Nuclear Plant providing electricity. Northern California provides an incredible percentage of San Francisco's water.

Shouldn't they be allowed to sell it at the highest prices to the best bidder? And what about fishing laws? Immigration is not as favored in Northern California as Southern California. Farmers are under-represented regionally if this happens. And the whole power grid would be fragmented. This would turn the states against each other. But the politicians also want this for Silicon Valley, which made them rich. There will be a tradeoff. Silicon Valley will get more control over laws nationally. This creates the sort of Oligarchs that appeared in Russia in Oil, and in the Balkans in criminal organizations and smuggling.

Eventually, the homeless problem in San Francisco will only be solved by moving these people out of the city. The same with Los Angeles. Who or what other state will they demand take the homeless they invited in before the split? The county policies created this problem. Who takes responsibility?

Tribalism of San Francisco power brokers has created an Elite Class that wants it all and condemns all who are not in their Tribe. They use power to fix things that creates more problems.

People are suffering due to the intervention of rich and powerful people into healthy society. History has proven that in every case of Balkanization, culture splits lead to war, unhealthy economies, and huge problems. The "Us Against Them" sentiment spreads to all factors. Will California withhold its food from the rest of the United States to punish them into capitulating to their political views?

The second case of "Tribalism gone bad" is in the United States coast-to-coast. The war against the longer American Culture has raged for about twenty years. This idea of groups and states turning against the "United States" really means they are turning against the proven success of "Out Of Many, One." That is our national motto. But politicians and people seeking power through NGO's and agencies stoked the flames of religious, racial, ethic, foreign, and even family hatred.

These NGO's did this to get votes and money. They lobbied for power to legally foment hatred against smaller groups. To do this they had to turn against what made the United States great: the Constitution, and principles in it upheld since our independence. Instead, these groups pushing Balkanization of society try to seed the history of the nation in its one main flaw at its peak: slavery and the Civil War. All the Balkan arguments spring from this idea. It negates history and the truth.

The reason these same people demand the end to Critical Thinking, study of European history as it relates to settling in America, and understanding of civics is because this also leads to study of the formation of a nation after a costly war. Reading the issues Washington, the Founding Fathers, and all that led up to Lincoln must be understood to end racism … but that is not the goal of people pushing Balkanization and leading the most violent cities in the United States.

Balkanization leads to perpetual victimhood, ergo perpetual Failure. And this leads to smaller and smaller groups to declare victimization. The goal becomes to be the "most victimized." But this requires constant feeding, because it is not durable in and of itself.

If it were, in quick time, the smaller group would do what all species and successful groups did since civilization began: they would stop Balkanized actions and work together with people that are different and have different skills. The argument that a system is so stacked against an individual simply will not hold up upon inspection.

People seeking a better life cannot reach higher goals when they are starved of the words and critical thinking to think for themselves. When surrounded by people forever demanding victimhood benefits, this seems to be the only way to survive, which eventually becomes the standard.

Being a victim justifies raining violence down on people not in your Tribe. Elites want this because it ensures they stay in power.

Look at music to see how much Balkanization would have destroyed its evolution. In the type of Balkanized world the Elites want, Rock and Roll, Hip Hop, and Rap would never have risen to the top. "Cultural Misappropriation" would have demanded that white people not abuse the rights of these Tribes and never been allowed to enjoy the music. Without the white audience and world audiences of many colors the rappers would not have earned hundreds of millions of dollars, which means that really the cause would hold Black performers down.

However, the Elites are actually very money and power oriented. Would Death Row Records have ever been possible today, when people would rise up and shout that using this music in a movie is a sin against the tribe that created it? This is important, because it was Death Row Records of the first that allowed black rappers to become worth hundreds of millions of dollars.

For now, Balkanization and victimhood point in only one direction: at the group on top. To justify this, they made up the phrase "Institutional Racism" and also "Toxic Male." This immediately allows each Tribe to declare others guilty of crimes they never committed. By reducing someone to a term rather than giving them the individual right to prove themselves is a way to ensure long-term Balkanization and a centralization of power and wealth around the Elites.

Why is this important to technology? First, technology mostly centers in Silicon Valley. If they want Balkanization, they can help it along through technology and never get any blame for it. It is up to the individual to keep his or her mind capable of being human and also capable of critical thinking – which means not allowing yourself to be put into a "Victimhood Tribe. "

At this point, you must ask, "Seriously, did any child come up with the goal to be the most victimized?" No. Of course not. A child wants love, wants to learn, wants freedom, and wants to explore the world. But when money is given to NGO's and therapists to increase victims they do this even if a child is happy and healthy. Guilt and a need for attention becomes addicting (I know because I inserted myself into this guilt system for my other book, "Lies My Subconscious Told Me.")

In the inner city, none of the politicians mention one of the key reasons for crime, violence, and drugs. It comes down to our cellular level. Every cell and organ in your body, as well as the energy in your mind, is created with the master blueprint to strive for individual high potential. If a person is living in a slum and sees no outlets for this DNA coded reality, they will find a way to satisfy this cellular drive. They can do it physically through violence or sports. They can do this by creating a drug empire or gang, which is the Balkanization within a single city of a group or race.

That is what gang warfare is – Balkanized justification of war. When - out of anger stoked by powerful Elites - inspired gangs or people are driven to burn down their own shops, businesses, and property, this is a Balkanized group being manipulated to self-destruct. And to remain poor and desperate. In Africa, warlords prevented global aide because starving people cannot challenge their Balkanized identity group.

Victimhood is not durable. So, Elites must keep presenting new outrage against "anyone that is not us."

While victims can surely be crippled emotionally and physically (which I am very sympathetic to obviously with my work with orphaned children as proof) what I find is that children that survived the most brutal lives imaginable (sex trade on Thailand streets as young as 3) they do not learn to "feel sorry for themselves."

Only outsiders try to turn the situation for the children into a "terminally-damaged" group. In fact, even I was poisoned with this victim idea when I first went to an orphanage that housed only children rescued from the sex trade. When I asked if they felt like victims and "forever-damaged-goods" the guides said they do not. They see no point in that, and instead are thankful they survived and are on their way to a life in which they can be free to create their own happiness in. Victimhood, they know in their guts, retards happiness and upward mobility.

And that is the very point. By removing, negating, killing, or replacing internal ability to find Joy, and the individual drive to maximize the potential in mind and body, a person can only find "meaning" in being a victim. This victim is herded for "consensus reasons" because power comes from how many people identify as this new victim class. And then how many people can be made guilty or made to feel guilty that are outside of that class.

While it is a hot topic to approach, the Gay Rights Movement is the perfect example. At first the Gay Rights Movement was primarily for Gay Men, even though Lesbian Women were always welcome and played a big role in achieving rights. Yet, being part of a Gay Movement offended some Lesbians, so they broke apart. At first the Lesbian group only Tribalized a faction of the Gay Rights Movements. This led to terrific improvements in society as they contributed so much to it.

Yet, from an economic standpoint, money had to be split and politicians bought or coerced for support. Then the Transexual Rights became the biggest cause to support. The optics of it all took opportunity, money, and political support from Lesbians, who rejected that Transexuals transitioning into women were truly women and should not receive any Lesbian rights. Then Transexuals Balkanized further, with those that did go all the way through hormone treatment and surgery looking down on those that did not.

These groups most often join forces and voices, but there are many issues that divide them. A key issue is that Women's Rights Groups and Lesbian Rights Groups achieved most of the equality and freedom for females in sports. Now, Transexuals are competing as women and getting all the scholarships and medals. Balkanization hits again.

When governments and groups give money to victims, and new groups spring up, a fight for the money and the power chasing it begins.

Once again, Victimhood evolves to prove it is not durable. It lasts as long as money and power can be milked from it. For proof watch how fast politicians and media personalities jump to the next new victim class.

The utmost powers in the nation or even in the world love this, because in splitting groups into smaller factions, they have no real power or voice to challenge them. To go even deeper, they gradually remove the words and process of logic that enable someone to form the opinion that they are not victims. This starts in education, and if any word manages to appear that does not appeal to the victim culture, that word is demonized. This can be seen in daily headlines now, so it is accelerating.

The elites, the true power mongers in what are thought of as free nations in Europe and the United States, believe the battle is won, and that individuals first, and human groups second, have lost any chance to challenge them. What was missing before is the power of technology to create instant and anonymous bully groups that can terrorize and destroy individuals or logic.

These Elite groups are never held accountable for the damage they do or the lies they conjure, and do not think they need to do so, because the narrative now is that, "the ends justify all means" even if it means destroying an innocent person. The most common is the individual capable of Free Will, that takes responsibility for his or her actions, that finds Truth in Nature, and within genetics or epigenetics.

We are all driven by Will to achieve our highest potentials. At the same time, human nature is to connect and be curious. The push by politicians and Elites to control the world demands that these two factors only happen in the direction that makes them more powerful. The worst of this whole dynamic is that the Elites never are held accountable for the damage they do to the very groups they claim to represent. This is seen where politicians stoke the flames of rioters, expecting that soon they will be held up as royalty within the movement … but instead are booed out of the crowd, and sometimes even attacked or protested again. When they can get you to refuse to believe what your eyes are telling you … you're too far gone, you drank the Kool-Aid, you took the blue pill.

Balkanization never works because it is not durable. Victimhood like failure is not durable.

By removing the glam of Victimhood
from your I.Q. action of applying one idea to many
you can focus your energy on success,
not Failure masquerading as Victimhood.

DIMENSIONS

"Applied Imagination" is the new Rock n' Roll.

Ask, and most people will say that the most fundamental element in our reality is Matter. The most-steady constant is Time. The most durable Field is Gravity (Big G Gravity because that is all they encounter every day.)

What if the most fundamental element in our reality, our universe, is the only element they all have in common? What is it is an element of consciousness that drives the extension of itself to the ultimate versions within our universe?

If so, consider…

Will

Imagine a position of it as the very center of everything… the very center before the Big Bang … the driving force of raw Information.

Information is on its quest to be Intelligence. That is why I state that the universal essence of anything in this universe is Will with a capitol W. It is the essence of everything from photons and viruses all the way to galaxies.

What determines our relationship to all this is our own individual Will and the cosmic universal Will. Even sand dunes have Will.

The next evolution toward what makes up our reality is when Will uses the "Information" of anything, like kinetic energy to drive engines, and even the creation of art and love, as tools to the highest Elegance.

To expand your idea of what ties our universe together in a way that obeys our Laws for this dimension, let's move step by step. The rewards will amaze you. We will discuss our dimension in a physical sense, then move on to how our Minds interpret higher dimensions because it all links up in the Blockchain Intersections coming to us soon.

Build upon what you were taught about in school. Matter is not the fundamental element, but it is what we see for the most part. Extrapolate this across all your studies and experiences, apply it to discoveries appearing in the paper, from CERN to Sci-Fi movies, to psychoanalysis, New Age theory and practices, cures, computers, hallucinogenic drug enlightenment, writing, sculpting, dance, wars, education.

Information is a sort of new measure of potential energy -- the new quasi- "Mental Planck Unit" of our journey to discover who we are. However, even Information comes after the essential we know as Will.

Apply this one idea to many. We mine information from DNA now. Information expands our computer searches. Dark Matter and Dark Energy in Space each has an energy. How is that turned into something that effects our life if all we have about it is Information? Simple. Information IS the effect and it manifests in agreement with the essence of its Will.

Look at our beginnings and endings of Love between a couple, a parent and child, people and animals. What transcends Time and Space? Will and Information - mostly in our gut or Intuition, captured in memories. When parents collect photographs of their children, they often are the key thing they carry out of a burning house? Why? Because of the Information reflected within them. We crave more Information of those we love, we use it to nurture our Love from seed to flower to withering stalk. We pass it on to our children through epigenetic inheritance and through experience. Our dogs and cats watch us so attentively in part because they are trying to capture simple Information on how to be a better friend, and new ways to show their love without words.

All we know of many unfinished Physics anomalies is Information, not conclusion. Einstein felt there must be one single unifying formula that will answer all questions about our existence but it can't be proven at the utmost levels, it can only be witnessed as a form of Highest Elegance. So many of the things in life important to us in our universe go unanswered … yet mean so much to us. The best example can be put in a question: what is Time?

All we know that determines our "being" is the accumulation of Information. Days, years, ages of Information compiled upon itself. Even our architecture is merely Information in a form orderly to gravity and the elements set on decaying it.

If God created us, and presented the complete answers to many questions to us … our question going both ways would be based upon Information: "Who created you?" If God was created by the combined mental search of mankind, then indeed God was created simply by Information, born from intellect and emotion. If God "is" all Information moved up to Intelligence, we are a subset of that.

Now, apply this outward, to all you know … reflect upon it … then reverse the process, and point all of that process inward. Information outside of us is no different than inside of us, yet we relate to our "insides" of both physical body and ethereal Mind in quite different ways. (The weighting of our experiences/memories is called Qualia, and there is a chapter just on that in our books.)

Could the gateway to other dimensions be simply the relaxing of the rules we apply to all the Information we hold inside our minds that demands we stay in this 4-Dimensional reality honoring Time as an arrow? "Deconstructing our Human Agreements" helps answer that. The chapter "Can We View Ourselves?" is another. It simply opens our minds to possibilities we then hit with our I.Q. pinball game moves.

Information gets stored in our entire body, not just the brain in our skulls. This is how we are able to throw a baseball or drive a car – muscle memory. Our lives are an interpretation of our sensory experiences. And intuitively we know the Laws of our universe and count on them (gravity, the speed of light, sound waves, etc.)

But now let's consider that we store Information in Six Senses: Intuition is the greatest sense but we don't rely on it as much as what we store in our brain in our skulls, or in our body as muscle memory for sports and driving. Is there a correlation between how little we understand our Universe and ourselves … and the fact that almost all our deductions come from just the Brain inside our skulls? Do all of our Brains need to be firing on all cylinders for the portal to other dimensions to open? Is the strange feeling I get when my I.Q. the result of applying enough Information diversely through my Intuition, Gut, and Smell? Is this how art creates something out of nothing? Is this how to knock on the door to another dimension? Is this my body almost slipping into another dimension because I relaxed my awareness and expanded it enough to let other Brains lead the way into the unknown?

Before deeper inspection, there is more than one way to look at dimensions or explain them. The manner in which dimensions are presented, or thought of, has to be spelled out. This is just one of many major chapters that will discuss this. Even more chapters are also in "Pillars of Encryption" because the future of Edge Technology requires a deeper understanding of dimensions. This is more technical. For now, as it relates to art, conceptualizing dimensions is a good introduction to expand your "Human Agreements."

The first manner of understanding dimensions has to do with Space or our idea of "perspective." Most people easily understand the 3-dimensional reality of Space, and the 2-dimensional reality of space as we look down on as a plane. We can look down on two lower physical dimensions. Most people know that the things in this world are in 3 dimensions, meaning it has length, height, and depth. For this meaning of dimensions, a higher dimension simply means adding a 90-degree on an axis, viewpoint, or extension of any dimension. If anything watching us is in the 4th Dimension, they would "see through matter" so that everything is sort of a Tesseract mated with a hologram. We see everything in "the next lowest dimension."

But the fact is that while we live in a 3-dimensional world, we see things in two dimensions. We do not see the opposite side of a block when we look at it. Only by shading and intuition do we know what we see in two dimensions has depth. If a sphere were coming at us at the exact right proportion it is shrinking, we would not realize it is approaching us. The more accurate way to describe this is that we understand the limits of our dimension due to ratios, or how one object relates to another object in its size, speed, distance, and other factors we need to know about it.

The other way of looking at dimensions does not have to do with just physical objects and the 90-degree added axis. The second way of looking at dimensions, which is what I am (mostly) expressing here, has to do with Fields, such as the Fields of Physics with its constants. We live what is sometimes called a 3-dimensional world, and by others another dimension of Time is added, so they say it is the 4-dimensional reality.

All it requires to go up a dimension is for the Fields and Laws that we must obey or accept in this dimension to cease to exist in the same way they do in this dimension. Many Fields we study probably do not exist the same or have our same Constants in higher dimensions. Light, and the speed of it as we measure it, would not necessarily be bound by that rule. Matter would not necessarily be bound by the same Energy Fields (gravity, electron, etc.) "The Janus Point" is an excellent book presenting a new concept of Time, but it is too complex for most people. In that book, the author talks about how some systems in the universe are not just in entropy, they are moving toward greater complexity. And that can be measured and also ratios can be used.

Think about this fact. Communication and vision would not necessarily be possible or reliable without what we think of as air or refraction of colors in prisms. Yet, neither air nor refraction is guaranteed in higher dimensions. Our very being could be all Mind, with no body and no Time as we know it.

This seems overwhelming, but just think of your dreams, and how chaotic they are, and how things appear and disappear. They don't obey gravity all the time, right? And you can sense far more than you can see. Going beyond that, where do dreams come from and what purpose do they serve? Are they a conduit or sharing Will and Information from higher dimensions? Now we move from proven science to concepts in our reality. If you leave behind physical aspects of our dimension we know, what is left is Will, Information, Intelligence, and this forms into an elegance we have trouble contemplating.

Is movement other dimensions or the edge of our universe only requiring three things?

Is one of them bravery? Bravery to let go of what we assume in this dimension? Recognition of Will as the essence of all dimensions? And removal from our concept of Time? Are those the three factors? In the movie "Lucy" she said what makes us what we are limitations: it's primitive.

Willingness to "Let go of structured explanations" as to what Time and Space is, and who we are, and what we are made of, requires unimaginable bravery. One level of bravery upon another … because surely our Agreements will be assaulted from many directions. Letting go of the words and ideas that limit our Minds, and cut off the Brains of our gut, muscles, heart, and intuition that open us up enough to get through that portal? And yet we do want some science, some physics, supporting our imagination. "The Janus Point" is the only book I've found that combines Physics with the concept of the creation of art, and a new concept of Time that fits my vision of our universe and higher dimensions.

[**Disclaimer:** I know that I am going to get ahead of myself in some paragraphs because I do not have the time to explain all details, but this leads to a way for you to conceptualize what powers a superior being might have in a higher dimension, and what they marvel over that humans can do in our dimension – maybe they are waiting for us to engineer a message.]

Time gives legitimacy to Matter. Or better said, Matter only exists within the confines (use of it by definition) of the "Human Agreement we know of as Time". The Information in the Matter still exists without Time, but we have trouble accessing it with the limitation of words that organize thought in our Brain Brain. Intuition Brain and Heart Brain both deal with Information that is Timeless, yet we haven't activated that translation within our Mind – that part that thrives outside the Brain Brain.

Case and comment on this point was given by Nikola Tesla. "The day scientists stop perceiving and calculating on a world based on material matter, they will leap forward more in a month than in countless years." This requires inspecting non-physical phenomena. And the ways to assay it (value it and deem it useful) will be through allowing the Nature living in the fields of our technology (Apps and Blockchain, experienced through Virtual and Augmented Reality Experiences and ultimately this leads to A.I. and Quantum Computing) to teach us Truth from Parallax Views.

The "Stream of Knowledge" in Tesla's sense flows more toward psychological breakthroughs. That means it is in what I often call "Conceptional Genius" more than Mathematical, Musical, or other formal "Science based Genius" that breakthroughs are experienced. In other words, an idea of an emerging truth about our reality begins as a thought, but not just one that evolved from our brain, it might be due to other stimulus – or even a dream, which Einstein used to his advantage.

Since most intuition and sensory experiences that leads to a breakthrough cannot be measured in a Physics Constant or unit, the experiences follow various paths to proof like Ontological Truths. The need to imagine higher or other dimensions will lead scientists to formulas they rely upon. The methods might include "discovery by accident" or might use Derrida, or ratios, or the Unitary Method to decode the Truth of new Fields, and even spontaneous works of art. Travel to a higher Dimension might simply be trying out a way to navigate thoughts, which as seen in dreams do not have to obey our Laws or honor the Arrow of Time.

You can think of it like teleporting your mind into the center of a lattice of Information that appears to expand, and as I expands it presents mirrors or views of life as you know it alongside life as you do not know it from other Dimensions. Most likely language or messaging will not be linear like a sentence, it will be symbolic, and the symbols will relate to each other. This is a bit more like how the language of Japanese works: you do not know the meaning of the word until it is juxtaposed with the other words. And the appearance of messages or even entire dimensions will appear chaotic, without obeying our Laws of Physics.

How do I know this? Intuition and this concept: because "Events have walls" and because the mere act of observing this dimension will change what we observe, and because Maxwell's Demon will help self-organizing which obeys Newton's Second Law of Thermodynamics … (did you get all that? If not read Pillars of Encryption Volume 2) this will result in incredible increases of potential Energy in the human Mind and being. How we use this Energy is up to us, but most likely, from that experience we will find material things and the power that comes with them less, or at least see it as diminishing. We'll be reacting to what we experience between dimensions. It might not be particles our senses receive to encode. In fact, think of absorbing this as waves, which means we'll see ourselves as waves when we recognize our reflection or footprint in this other dimension.

Matter will take on multiple energy forms on our whim. We become the engineers of Fields, but we cannot observe it because its Energy becomes Matter again – um, in other words, as we try to turn what we experienced into words and Mathematical formulas, it and we become "material" again. In a world of physicists, that is not so convenient. Just like how "Conscious Knowledge of Lattice" will arrange the Lattice pattern that engineers the Blockchain and in a few years A.I. uniting the Event Walls built on Fibonacci Ratio patterns that self-organize some great leaps forward can't be seen until the patters on the Wall are drawn to higher Elegance. Yet, I digressed. Back on point …

Embrace Information as the fundamental usable element of use for our Will in our existence and universe. And now, imagine that these two concepts are the only concepts that still function as we understand them in higher dimensions. In a higher dimension Will and Information will be organized into Intelligence based on difference Fields.

Unitary Method would ask, "What is the end game?" Give me the end game and I'll show you where your unknowns are. Our end game is this: our world exists in a galaxy that is very stable when its ratio is observable against the background of our expanding universe. Within a stable galaxy, our mission is to mentally and physically engineer "higher complexity." Within these smaller systems of higher complexity (even art) are quasi-Janus Points if you were to assume (unjustly or not) that the human mind is a reflection of our universe as a whole. Both the mind and universe are expanding and limitless. So, create a system of higher complexity within it, knowing that repeatedly we get a new perspective on Time. Then, take guesses, or just observe it outside our "human agreements."

Since we're speaking of dimensions ...

There has been discussion recently on how we, in other dimensions, would have enhanced senses. In particular, colors would not just be more vibrant, but there would be two or three other colors added to our sight. Smells would be stronger. And we would see the flows of energy at times in Nature. That fits perfectly with lucid dreams that give you messages that you must study to interpret. It is like Nature and dreams give us clues about our challenges, but we seldom study them.

I think a few key people broke through dimension limits. En masse, children with autism. Tesla. Ansel Adams. Sir Edmund Hillary. That dog that the Russians sent into space in 1965. Who knows who else?

I don't like talking about experiences I've had in letting go of our belief system related to Physics and Self. Primarily because 1) somehow I know it's not to be talked about to everyone until the Cosmic Consciousness or Unconsciousness rises or Vibrations change, because one thing I've felt for sure is that vibration plays a role, and 2) words we count on to explain everything are most definitely the key to "knowing" it and "using" key to open door.

Riding waves of Information keeps ringing in my ears and soul. And this leads me back to a single word I heard upon waking, when in a lucid dream, the word "EXPAND" etched into my mind. Over the months, the significance of this word grew.

Each dimension up expands its properties of Matter. Like water expanding as it turns into a gas, so life expands as you move up to higher dimensions. Or like an oil slick floating down a river changes its properties. But that is an illusion based on the Physics of our dimension. The reality is that the Information in the slick becomes more pliable, and all Matter working with it therefore must become more pliable.

Now, you can "move down" a dimension from 3 to 2, and a simple way to describe it is this: draw a circle on the ground in chalk. Step down into or onto that circle. You just merged with 2 dimensions. A circle could step down onto a point, or 1 dimension. The idea of a point, or actually three points creating a triangle that allows us to measure ratios, is fundamental to the study of Physics when it comes to determining a universe that is expanding, yet has pockets of stability.

What I am speaking of is more complex than our way of interpreting everything because it is not based on Matter but on Information as the fundamental element, and in particular how the spontaneous creation of art in a stable complexity presents the process of centralizing energy through our creative thought process, using that potential energy for the kinetic energy that results in a masterpiece of art. Not all the potential energy will enter a person in a way that science has formulas to measure it – a lot of "vision" will centralize in the mind of a person almost as if that single mind is in binary spin with a distant mind or source of Truth and vision.

Did that source of vision in binary spin with receptors in your mind exist within our Dimension? Or within a "higher complexity" far away in our universe? What put these two sources (the artist and vision source) in binary to begin with? Or are they the same, and Time sent it from our future self to our current self? Fun stuff to consider.

Can a highly-complex form of Art if produced with both "Spirit and Scale" send a signal to the universe that sparks the binary sharing of intelligence that answers the greatest questions of our existence.

I'll explain it this way. (Disclaimer: this is not Physics, it is how imagine higher dimensions to follow clues.)

We are in a 3-dimensional Reality, with Time being the 4th dimension. Laws for Time and Matter are set. If you wanted to move up a dimension, then first it would help to conceptualize it. Forget about descriptions based on Matter which include interpretations with our eyes and physical measuring of each dimension. Here is my (admittedly goofy and unscientific) list of what could be the dimensions that hold higher levels of Information and "Emergent Properties."

[**Note:** This is my vision of dimensions and is not proven by any science – it just helps me relate patterns and make sense of what is imperceivable.]

Dimension 5: Vibration – Chaos, Our Laws of Physics don't apply, Medium is Ether

Dimension 6: Soul – Will, Medium is Emotion, and Thoughts manifest Reality

Dimension 7: Information – Pure Information in E-8

Dimension 8: Creation – Cosmos Creation by Engineering of Fields

Dimension 9: Polarity/Entanglement – Entangled Sources, Math based Engineering

Dimension 10: Singularity - When the higher self passes through the gates of Hermetic balance and Humans merge with all Will, Intelligence, Creation.

(The current purely-tech notion of Singularity usually signifies when A.I. surpasses the human mind or can invent itself, or when technology joins many tech aspects into machines dominating mankind.)

Notice that as we move up a dimension, the next dimension becomes the dominant one when it comes to viewing and understanding what is going on in the dimension lower than it. Vibration as it is understood in our 4th dimension is so unexplained we cannot "play with it" (manipulate it to the point it opens portals or does much to change Time or matter, or the speed of Light and other Constants.). But in the 5th dimension we should be able to manipulate it and rise new Truths from its properties, as it is no longer bound by the Physics of this dimension. Next up is the Soul. It exists in the Vibrational Fields yet is not the central element. In Dimension 7, all distinct Wills and Information include a full understanding of the Soul.

We then seek the secrets of Creation through Engineering and Mathematics, we are one with the Fields that would govern the unthinkable. Creating that universe or reality would then inspire us to be communicating everything from everywhere at all time, not like a God but as a binary feedback loop. This requires two Photons put into initial binary spin, that when separated, not only always sharing data instantaneously by their feedback, but are capable of holding all information up until this point within them.

This is elegantly explained in the book "The Three Body Problem." [And again, this is just my perspective filtering through beliefs, on the way to more elegant organization that presents patterns I recognize, and from them form or see a message or truth.]. Understanding this leads to an accelerating intelligence. I'm writing this book so an Accelerating Philosophy can keep us Virtuous and able to think morally about what A.I. will tempt us to accept.

How would an "ascending" mental dimension system work? In Dimension 4, our current dimension, we learn the limitations, systems, temptations, freedoms, and mathematics of 3-dimensional objects in or Space-Time continuum. But having a body also gives us a chance to learn how the Ethereal unites with the "physical" and creates emotions. This is a priceless gift that is so unique that other beings in other dimensions probably watch us in wonder as we learn this dance of invisible entanglement.

We're bound by this physical form, and a mind is only partially used, but perhaps what is not used now is meant for when we move up to Dimension 5. There is a lot of talk in recent years about recognizing the ancient value of Vibration. We exist in it in Dimension 4, but can manipulate it in Dimension 5 and notice all its intersections in our Physics and Natural Laws in this universe.

Dimension 5 takes us out of our body in the way we imagine our Mind or Soul attached to our body. Vibration becomes the shaping mechanics of this Dimension. The Medium it uses to shape and transfer Information is the Ether. Ether is all places at all times, and communication of Information becomes limitless and possibly instantaneous. While we can look down one level at our body, and since Time does not exist in a fixed manner due to the impermanence of Matter, the "Vibration Being" can interact with Dimension 4. This is much like how we could add meaning to a painting in a two-dimensional state – we are God to that painting. But there is much more to this Vibrational state, for it can witness the energy of "Artistic Creation."

Dimension 6, the "Dimension of the Soul" where Emotions are created out of the winds of our Cosmic Drives: Love, Curiosity, Consciousness, Connectivity, Sexual Energy. The Vibrations in the Dimension below created a sort of Music of Energy that gives actual power to Emotions. Emotions can exist in even more of a vacuum of Matter or Measure, exist outside of Time, are the play toy of Dreams. And herein lies our Intersection. Dimension 6 travels lower a Dimension without losing any Information and can continue down into our Dreams.

Dreams exist in a Reality that do not have to obey the Laws of Physics or Time-Space-Matter. Yet, Dreams are largely Emotion when you think about it. Déjà vu is also the sensation of this

Dimension, as is premonition and the mother-child bond. It is through Emotions and their "substance" which is really just unbridled Information that the Soul travels.

What could possibly be a move up in Freedom and Value from a Soul and Emotions? Pure Information that can be analyzed on a higher level if we see it as a form of spiritual art . But then that means we become the conductor of raw information, and it is up to us to manipulate it so it creates beautiful "mental music." Information passed through every element, energy, concept, creation, and passage of Time. And it is driven to destinations according to its Will. Information moving instantaneously, as there is no separation of any one thing from another. There is no scale at all in this Dimension. Everything is One and Everything is Separate, infinitely large as it infinitely small.

There are Organs within your Organs we are unaware of because we have not expanded the Organ into other Physical Dimensions within the 5th Dimension. This is where the Akashic Library exists. It can move down all Dimensions to our 5th Dimension and seed knowledge, as if it is dust it can sprinkle into our Soul, then our Vibration, then into our Body … to see if we "get it."

If a few of us are fluid enough in our Vibration, if our hold onto this body and mind in our Dimension as it expands enough, the Body and Mind can open so the dust germinates. Gestates in our Mind, unlocked by Vibration and Soul clues. Dimension 7 wants so much for us to open more, to vibrate at a higher level, release our Soul … and is very patient because Time does not exist there.

Now for Polarity and Entanglement. Polarity is, well … as complicated as it is simplistic. Again the book "The Three Body Problem" explained it better than I can here, but I'll try to word it so that it's easy to fathom. As Information becomes individual energy, it reduces in size so much that the wavelength or vibration the Information is held in doesn't have a Medium to naturally travel or be shared through.

It's not that it does not have a Medium, it has all Mediums. {Information exists even in Time, and some argue Time is nothing but Information sliced to be viewed.)

Yet, if two photons collide and or mingle in an original spin, say, within an atom, and are thrown apart by a collision, then the two photons are forever in a tandem dance of energy. This dance continues even if a galaxy is between the two photons, though that would not usually be the case. As photons travel at just under the speed of light, and the particle is so small it travels right through physical objects, it doesn't stop. But the gravitational pull between the two photons gets incredibly strong over distance. The two photons constantly share Information in this spin, instantaneously. So, if one photon bumps into something the other knows it and stores that information.

In between is a Field of Information that is longer than a single wavelength of Soul or Vibration. Therefore, the Information in that Field is readable. This is how Polarity holds unimaginable Information that the Photons pick up from every interference. And since photons are so small, EVERYTHING is an interference. So it is the register of all elements and energies of all Creation.

Singularity. When all Information is pulled into the center of photons, and no Information escapes. There is no Time, there is infinite Energy and all Creative forces and history of both the past and future exists. But there is no coming back from Singularity. The only Information shared is on the Event

Horizon, which goes back to my chapters that mention why visionaries must be on the fringe of society and mental cohesion.

Their Minds move to the Event Horizon -- so the question is, does it do this because the next level up of the Soul and next Level up of Information share the secret of the Singularity, and urge the human's Mind to exist in the near solitude of the edge of society, on the edge of madness, where Yin-Yang plays him like ping-pong?

I hope that the brief explanation shows how each level up can step into the lower level. That it knows who we are and we don't realize them. This is like the Fish in the Fishbowl! This is like the Jellyfish and virus! This shows us what is North of North. It is the Truth our Subconscious tells us.

An even better explanation of what we would know reversing this is in that same book. Looking down upon the Soul, for example, Information and Creation would not just see the reality of the Soul, and not even the elements comprising the Soul, and not even the history of the concept of Soul … Information would see the life forms within Soul be created and expire made out of all elements of Matter and Vibration, both in and out of concepts of Time. When opening up a lower dimension, the higher dimension "sees" everything inside the lower dimension, even things that what exists in that dimension cannot see. Think of an electron microscope, seeing the creatures within a speck of dust, or the twists of a strand of DNA. Information would turn that DNA into a massive two-dimensional structure with endless facets and pockets. There would be no mysteries in the DNA then.

Ride the waves of Information out of this limited reality.

Brain Damage

All my life, I've read about people that get hit in the head and then have some near-death experience, then end up with some special ability, or they see some glimpse of another dimension, or a supreme truth about life.

On February 24th, 2017, I was hanging upside down with weights in my hands to do sit-ups to get that damn six-pack abs back. The bar broke. I fell so hard (and with my arms in the opposite direction to lessen my crash) onto stone. It was one of the clearest examples that Time is not what we make of it. In that eternity of the actual impact, I was sure I had died. The force, followed by a twist and compression of the neck, was so great the image in my mind was of me crunched upside down like Steven Hawkings. I lay paralyzed for a while -- a moment of clarity – "This is how life ends." The sound and blackout were followed by a void of Time in which I was no longer there, for I was everywhere. Then I came out of it, probably less than 30 seconds later. It took a long time to get each limb to move. Bleeding from my head, I staggered out to my roommate, said I was all right, but then collapsed on my bed. Within hours I knew my brain was damaged. It wasn't the pain that was so bad, it was that my mind stopped working normally. Compounding pressures and blood flow restrictions made it so that I had trouble speaking, thinking, moving, sitting, and seeing. Ether partially absorbed me.

The next morning I was taken to the hospital. Indeed, a cMRI revealed the impact killed off some brain cells. Only about the size of a quarter. But it was enough. And the cranial fluid was collecting in my brain, little by little. If it got worse they would operate on my brain, opening my skull up. Then install tubes to keep my brain draining. It took several days for me to see and talk normally without extreme effort. I could not move without major mind problems. Extreme vertigo beyond anything I'd experienced before, that kept me from moving even an inch for hours, took over my immediate reality and life within it. I faked I was passable, but, I truly wasn't "all there."

The first five days I moved like I had a stroke. Car travel was a roller coaster. It took 3 weeks before I could drive, and that was tough. Another two weeks before I could drive at night. This is because my brain had to re-learn everything it does. And to make it harder, it had to do it without full body balance.

For three months, I thought I would need brain surgery because the symptoms would not stop. Then a final CMRI said that the brain was draining normally enough to avoid surgery.

But, rather than think of the negative things from it, how the brain heals itself was being experienced. For within this self-inspection, secrets of how the Brain actually "thinks" and "remembers" were revealed in wholly new ways.

The first three days after the fall, when the hospital released me, my first goal was to keep moving to prove to myself I was not paralyzed. My primary fear – indeed my only fear out of this after paralysis – was that my brain was so damaged that I could not conjure the extremely deep and complex thoughts that kept me company most of my life. Sure, I would be able to converse, do normal things, but losing this upper level of thinking was literally what I think my life is all about.

I knew that when people would ask me about myself, many times it came down to me saying, "The only thing exceptional about me is my imagination." I'm sure most of them thought of someone like Tim Burton … that I meant I was a good storyteller with zany imagery. People that have known me for decades have finally found themselves in a conversation with me (usually inspired by new acquaintances or work) about concepts of the universe, mind, and being. I've heard often, "Where the hell did that come from? I've known you for ten years and you never talked like this." What they should realize is that they do not bring this out of me. I'm not like a Trick Pony that performs constantly on whim or command. And to be honest, my deeper thoughts about our role as humans or the edge technologies bore half the people.

So when for the first days after the fall, when I tried to focus on what I had to write, my mind simply would not get into it. I could see in my mind's eye -- the "box" that held all my complex thoughts … but my mind would not let me open the box. My mind felt either lazy or could not get closer, as if a force-field kept me from reaching the box. I at times realized that I could not recall my most common numbers and names: my Social Security number, driver's license, sisters' names.

Then on the 5th night, about 2 in the morning, a strange feeling happened in my brain. It was like a soap bubble was popping. Only the size of a marble. A soap bubble in my brain. Popping.

I know they tell us we cannot feel anything inside the brain so possibly it was the protective mantle around the brain I was feeling as pressure reduced under it.

I could almost hear it and could surely define its shape and degree of firmness. As this bubble burst, connections in my brain were freed, and a surge of information exploded into my brain. It was intense enough that I muttered, "OW!" The release of all this energy in my brain literally threw my brain and head to the side a few inches. The energy defused in seconds. Then oddly enough, I felt my brain feeling … happiness. More than relief … happiness, for accomplishing something grand. The return of a pocket of my brain to my thought process and memories.

The experience was intense enough that I was sweating and my body had twisted as I lay on my back. It wasn't over. I experienced 3 more literally mind-blowing events in my brain in the next 15 minutes. One pocket after another opened up to join the rest of my brain.

It was like opening a great gift! I didn't know what was inside! A gift full of much more than I thought could be packed into a tiny circle of my brain. Lost memories. Computing skills. Within moments I realized these are things that my mind stores, things my mind can use, things my mind can recall …. But it doesn't. Why? Because of my normal routines in life. I could sense my brain and mind debating the "usefulness" of what was returning to me or was opening to me.

We all think in patterns. Regardless of what is confronting us, our mind goes first to the normal, most open channels, in search of probabilities, and from them, deduces where to park questions and how to answer them.

I became aware of a feeling in my brain of Information either zipping around or being stored in my brain. The key difference was that I was acutely aware of the newly opened Information Highway between my left and right hemispheres of the brain. This highway seemed to have been widened by, of all things, binaural beats (also referred to as Harmonics.) I had been introduced to binaural beats years ago, during an investigation into how far the mind can be pushed, past our ego's hold on it. Hypnosis and other mind-altering methods use sound to widen the neural highway between parts of the brain and even between hemispheres, which is rare. And now, my brain was using it to recover from near death. I could feel my thoughts selectively using the highway normally reserved for binaural beats. This allowed it to access memory and high-level mental exercises. Energy pulsed from right to left brain hemispheres.

Now, I will state for the record, that we cannot actually feel in our brains. Maybe I was just feeling blood pressure moving from one side to the other. Who knows? All I know is it was like ideas were waves splashing from one side to the other side. And in these ideas were grains of insight or high I.Q. connection of symbols and patterns. It was like messages that needed to be pursued. I was drawn to connecting the dots of ideas. Like connecting the dots of our perception of an E-8 model of reality.

For over two weeks, each day revealed new memories I forgot, new ways of looking at things. In a word, the "artistic" and abstract way of looking at the world and my past. Highly visual, and also involving other senses. Requiring a higher degree of bravery, because some memories are buried for a reason.

The blow to my head dramatically increased the effects and programming of the hypnotism at the same time. After researching this I found it was common—that people with intense learning experiences in their past recall them when they suffer a serious brain injury. I am very confident these events were not false. Suddenly I could feel what someone that is the opposite feels. I could desire things I never liked, because a part of my brain that was locked up like it and is not free to influence the rest of the mind. For example, before I paid little attention to music, but one night a neighbor blared music, and it was as if I could see it as colored waves. This short recovery experience taught me more about people that are not like me at all, more than any other recent experience.

And I began to write this book, without a plan for it, and not knowing what my fingers would type the next day. I wrote over 300 pages, not once going back to rewrite any of them, constantly pouring out Information, in less than two months. Along with recovering and writing screenplays for clients.

Even now, I am still receiving thoughts or connecting ideas due to the brain injury. It's hard to explain how or why I know something came to me that would have been blocked had my brain not been forced to re-wire. And then there is the entire concept that near-death experiences open the portal to Akashic wisdom, or other entities in other dimensions that do not observe Time as we do, therefore cannot be within our dimension, yet communicate with us if we are open enough to it.

NIKOLA TESLA

Most of the world knows Einstein. Why not Tesla? My mind doesn't work like Einstein's or gravitate to his realm of breakthrough discoveries. But Tesla's mind was consumed with energy transference, often without any apparent means. This, it is my opinion, came through other dimensions.

Tesla clearly found Ether "Elemental." If the depths of his study were made public and followed publicly, so many mysteries would already have been discovered. In fact, his research and proofs were so ahead of even Einstein's in a practical way, that after his first announced discovery, he was pretty much shut down and silenced. His discoveries were often cloaked in "disinformation.": Think of that. The man that had the greatest understanding of multi-dimensional Information was the victim of disinformation. I can't tell if there is more irony here or more tragedy.

As he said it in a paraphrased way before he died, "The world can only take one life changing discovery from a man. After that, he's made impotent and destined to be forgotten." Simply breaking the barriers that prevent us from interacting with other dimensions explains most of his amazing experiments. His most famous is The Philadelphia Experiment. But he also discovered a way to transmit "free" electricity without wires. That would destroy a huge source of government revenue and control. He used a hat, then a cat, to move objects to and from our dimension. He spoke of the air around us of being a substance, though did not to my knowledge use the word Ether.

So, consider this. Many conjecture that both Tesla and Einstein were either chosen by other more intelligent life forms… or sought them out once they broke out of our limits in dimensions. They received different clues, then had to discover where they led. In Tesla's case it was not "explosive." In Einstein's case it surely was, with the release of energy from splitting an atom. The atomic bomb. Of the two geniuses, Tesla's discoveries came far faster and were far more diverse. They dealt with the interaction of a human being with new energy and awareness of dimensions. Einstein's dealt with atomic-level particles as well as study of space and delivered his Quantum and other proofs. Of the two, Tesla was far more threatening to people wanting to control humans by limiting what they knew (not just how their matter could exist before and after an Atomic explosion.)

Everything from DARPA and HARRP, chem-trails, aliens, and more can be understood better by studying Tesla's work and ideas. Unfortunately, a totally unexplained fire consumed all of Tesla's notes before he was about to announce new discoveries to the world. Funny that timing, isn't it? I don't like to add conspiracy theory to this book, yet it must be mentioned here that there are benefits for those in power to prevent the connection to Tesla's truths to energy. More complex, are how they would connect to our understanding of our limitations. Men and women without limitations … imagine that. Is there life outside of this dimension for humans? Almost impossible to ponder even through some observations that are as junior an achievement as my thoughts on the Human Agreements, which are in other words the reality we live in as it is agreed upon by the masses of humanity. Adding the ideas of Tesla, and what happened to him, reveals why my ideas could have been discovered long ago, but hidden from people. And Tara? She told me that the government did not want her to succeed, and even put pressure on funders to stop helping her discover more.

Creativity comes from both within and without. From our brains and also our hearts. This bond with the outer world, through Ether, can pull back the fabric of matter-defined space. There is a feeling to creativity just like there is a feeling to your Intuition. Everyone has had his or her hair rise and stand up. Our minds had nothing to do with this event, nor did our Brain Brain. It was our Intuition.

The feeling of another dimension influencing you is sort of like very soft fingers touching your skin, usually in the legs at first. Not sure why. But the challenge is not understanding "this" ... it is vanquishing the fear and also the attachment to the simplified agreement of Time and Space all humans are expected to uphold.

Two other realms that Tesla focused upon are motion pictures (because they are the primary source for storing experience and memory in his dimension) and memory. One question this adds to our building tsunami of questions is this one: his experiments proved that cells that go into other dimensions return with their genetic memory intact. Why is that important? It means that it is the dimensional wall of the cell that becomes pliable in going into another dimension, but not our own memories.

Memories are pure Information organized into an Intelligence pattern and are not bound by the Fields of our dimension. Otherwise, they would not survive the escape from it and then return with us still knowing who we are. Yet, we do that in dreams all the time, right? So much of our daily lives demands we accept general truths, such as the fact the Earth and Moon are solid and have a core. How could another dimension relate to its "matter" and still not relate to its Fields?

We can try to reason about this. To have to stop using the very words we are left with to handle such an extreme revelation if this is not the case would put our entire belief system and hour-by-hour survival in chaos. BUT ... would that be temporary? Can we transcend that with a sudden burst of knowledge and truth?

If the ultimate understanding of religion is love, then, if I could just merge with you for a few seconds, share your current life history and genetically touch the Truth and multi-generation memory in the nucleus of your cells, most of the goals of religions would be realized in a single moment! I could never hate or hurt you, I would not judge you, I would not abuse you, I would not care about superficial goals on the level that I do.

The ultimate theme park ride would be a cellular merge through dimensions with not just you, but the energy sources and senses of beings around us we do not see. But sense it we could, and this fact is why I am fascinated with something called Qualia. (It will take me years to figure that one out.)

Could it simply by the stubbornness of our Minds that prevents us from more honestly understanding Time, Space, Ether, and the merging of dimensions?

What then is the greatest human condition that prevents us from discovery: fear or laziness? Watch the movie "Waking Life" for the long understanding of what is at stake.

The Merging of Eastern and Western Visions and Data

This long title hints to the extreme variants and horizons that can merge the most useful positive accomplishments of tech for an Accelerating Philosophy that matches our desired goal: an accelerating intelligence.

Balance - as you know from my earlier chapters - plays a key role in my I.Q. search for connections of random data. I spent so many years in Asia I experienced the physical actions attached to their cultural philosophies. Each nation is quite different in culture and I could see how it affects their confidence to be creative.

Each nation has its own financial problems and not having enough money is a worldwide situation so you might find it unusual to include monetary policy in any philosophy; however, money is a basic need and shapes our perspectives.

While most people in the United States do not travel to Asia and rely on the news to educate them the reality among the citizens is different when interacting in business. I came over as the expert on Hollywood but needed to examine how the people and government set up for the future. Let me make this clear: it is good to learn from foreign cultures even if we do not agree with its leaders or politics.

It intrigued me to see that China's President Xi put special emphasis on "Balance" in many of his "10-Year Vision, as it becomes a pivotal point in China's goals to 2023 and 2030." Indeed, Balance plays a major role in so many of China's historic arts: pottery, sculpture, calligraphy, architecture, city design, education in and of the Martial Arts, political theaters, trade, and invention.

For expedience, this chapter will be divided into: financial, artistic, and technological. Each category will merge back to Ether, Derrida, and creativity through the natural pathways of I.Q., supporting I.Q. with "The Five Brains of Tara" and Chinese philosophies and art (almost the same as Tara's Five Brains.) This ultimately plays a part in answering the question of whether or not we are a "Computer Simulation" or "Matrix." Why? Because Nothing comes of it unless we go out to the "many points of relevance" and come back to "The One" (China, the USA, you, or me – your choice.)

In their own language of philosophy this method is akin to riding the energy of Yuen Sun.

There is a degree of "art" in each of the mentioned categories. Xi wrote about how to match the aggressive pace of Western media and entertainment arts with Eastern art (which has such a much longer history) so that the nation and citizens can keep a Balance.

I had already recognized the massive trove of art and the philosophies of China and Japan that dwarfed the same categories of American influence. But to me, almost from the start, my mind applied this beyond the obvious. It led to ask a question or two.

How far could Western arts, philosophies, and designs for the future get without balancing the most promising of it with Chinese art and philosophies? How much would it blind us to where the "touch" of the art gift takes us? How many doors would remain locked without that Balance? What parts of our brains could be activated or enlightened if we balanced the two origins fairly, that otherwise would remain dormant? And most important – what degree of Intuition (or Gut Brain) would be dormant or blinded by only using one country's source, compared to the broad path blazed when the two cultures and histories merge?

The goal of the Chinese Monetary Policy since 2014 resulted in several major social goals. One was moving people out of the Tier One and Tier Two cities. This required both the development of Tier Three Cities, and an expansion of opportunities that go along with China's Communism with new characteristics. Instead of stricter laws, a shift of focus to Nature, culture, philosophies, and art drew people willingly to higher Virtues. This allowed many companies to expand. It required "Liberty with Responsibility" as it was stated in the national address.

I am not sure, but I think that China's Monetary Policy experts wisely ignored the (absolutely contradictory and self-serving) Western Economists when they were planning the uncoupling of the Yuan with the Dollar. They had to, because the United States was diluting its currency at an accelerating rate. The 2019 race to print money in the Euro, Japanese, Korean, Brazilian, and UK markets proves that the U.S. Policy led to all other nations being forced to print money. By printing money it prevented the world buyers of Treasury Bills and the United States citizens from seeing how fast the policies of The Fed and Presidents were devaluing the dollar.

While China returned to a soft peg on the dollar, suffering a few setbacks on the Exchanges was acceptable short-term pain for long term strength. At the same time, China created both the PetroYuan and the "New Global Bank" … while also helping nations around the world. This was in some parts condemned as "spreading Communism" but that is not being seen anywhere since it started. The fact is that China helped nations that the U.N., E.U, and U.S. either could not or would not help. A good example is in Belt and Road money being used to improve trains and ports.

President Xi did it so well that the World Bank (and all the European and American tycoons and Feds running it and the IMF) broke a century of tradition and accepted that the Yuan could become the Third Reserve Currency (behind the U.S. Dollar and British Pound.) The Chinese experts had little to fear: U.S. "Economy Experts" (and I put that in quotes to show how ridiculous that title is) had failed to predict even *one* of the monetary collapses that hit the United States or Europe. Instead, I guess that China's leaders and economy specialists simply used two things: Common Sense, and the firm knowledge of how the Chinese people are a level above the U.S. citizens when it comes to being able to survive a tough time from a financial collapse.

This is "Financial Art."

First, let's use Derrida systems to see the Truth in this opinion following both I.Q. systems (that we must build invisible bridges to islands of excellence) and Ether systems.

Derrida (if you remember) tells you to take the *opposite* of "conventional wisdom" and view the situation from that angle.

Opposite view: Economists that the world relies upon are wrong, worthless, compromised, and in the worst scenarios they are dangerous and evil.

As the funny saying goes …

"Economists were invented to make weather predictors look good."

This is mostly aimed at U.S. and European Economists as they led global policy. In a different book I point out that The Fed was created after The Great Depression with one promise: it would prevent all future financial crashes in the United States. They failed many times.

Using I.Q., I will explain the roles of Economists by starting with another financially and socially worthless field of experts called Philosophers, but mainly of Western interpretation. (I think that true "Deep Thinkers" are extremely valuable, but in the course or trying to get rich they are poisoned, and are then used by Society and men in power to either distract the masses or abuse them for their own benefit.)

In the purist sense, Philosophers can do no wrong. It is only when men greedy for power use their ideas and experiments that Philosophers become anti-evolutionary.

Confucius lived in an era when it was much harder for men in power to distort the ideas for political advantage. Not one Western Philosopher was recognized in an era like this; ergo, almost all Western Philosophy ignores Balance and Intuition.

Let's look at how the common people and political leaders vest their value and decisions in Economists first, and Philosophers second.

First, let's look at a fact that exists now in the most modern of days to show why Economists are truly full of nothing but hot air.

- *People watch economists on television channels that only survive if they stay on the air, and that means they must secure advertisers.*
- *Advertisers on those channels are mostly companies offering investment advice, insurance, or banks.*
- *Banks, investment companies, and insurance only get rich when the viewer is convinced it is smart to invest with them.*
- *If an expert painted a picture of corruption, lies, doom, and a clear failure to predict the future, the viewers tuning in would never invest with the advertiser.*
- *Therefore, all televised and printed Economists must either convince viewers that it's safe to invest, or at least to follow them and eventually spend money.*

They most often refuse to speak the facts. If they make a prediction, most are quick to mention another possible result, just to cover their reputation. Here are a few facts.

Using Derrida, let's assume that China has the smartest approach and the United States has the worst. A few facts confirm this. The U.S.A. is going through a major cultural war that will rip apart the nation, yet financial experts say all is well. The dollar has lost all real value and has only an imagined value.

Yet one of the valuable contributions China brings to the future of civilization as it relates to The Arts comes from the same arena in which it gets its most criticism: censorship. On the surface, people can argue that no art or idea should be censored. Upon closer inspection, this proves not to be true, especially if the very reason the art is pushed into celebrity status is to gain greater control over the free will and minds of the masses. If the goal is a Western-oriented goal based in Greek and German philosophies this might be fine. But we cannot swap out thousands of years of China's past and culture related to each Dynasty. This is hard to fathom, so I will explain it.

As any nation ages, it goes through stages of social awareness and financial stability. Toward the end of a long run of prosperity, the elites and people in the upper 2/3 of any society – that section that can foment dramatic changes – gets simultaneously lazy protecting its freedoms, and guilty over its prosperity. The United States is going through such a stage, and it is not by accident, it is by design. And President Xi sees this, as does his Core Party. Both are often blamed for censorship. Yet, Google now proves to be a secret censorship board far more globally pervasive than China's Core Party.

Few people point that out. No one listens to China economists, and that is just how the United States must demand it. By censoring China's financial analysis, the United States can set world financial and monetary policy. Just look at how Europe was forced to follow the Quantitative Easing Obama continued for far too long? People that really knew the Truth warned this would happen – a wave of Q.E. would roll over the financial systems like the Euro. All currencies had to de-base. Making sure they did this were Western Economists.

Let us look closer at what really is going on in the United States that President Xi wants Chinese citizens to avoid. It's a good idea to look outside of both China and the U.S. for a better perspective if we want the knowledge to meet in the Gamer or film world … so why don't we do that for a minute or two? Maybe we start with another country and learn from that.

A very high-level Russian leader gave a great interview about the current self-destruction of the United States through its society. This informant named this strategy. He points back to two key reasons the United States could fall apart, but it would take decades. One of them was through art, the other is the celebrity as a political driver of sentiment. This was crucial for Russia to understand because the entire world was paying to see movies in which America saves the world, conquers space, or simply produces the highest-quality films about the hero within us all, and the American dream.

Around the time of Trotsky there were Fabians in Western Europe that wanted to turn the United States into a socialist nation with a dictator. They failed because the vision Americans thought was achievable in one generation and so clear to see. New York City and our capitol Washington D.C.

helped us see the true value of happiness, family values on full parade, freedom of thought to express oneself, all in front of our eyes.

The rise of anti-Americanism really began in the late 60's, rose up stronger in the 70's, and now is in a new era never before seen. It did actually started as a very slow national uprising in the hippie movement in the early 70's. Then voices against capitalism expanded in the schools, and also in the art world. That was when Modern Abstract Art appeared. Going against tradition, culture, and the history of previous eras of art, suddenly out of New York came a new type of art. Art of nothing. Nothing of true value or that reflected the truth of humanity. A painting might be nothing but a solid black canvas, or a fork, or random lines of colors. Yet it was heralded as great art. By whom? By leaders in the art community … that had an agenda that was not about art, but about politics and controlling minds in the future if you believe the interview with this top official from Russian spying and government control who had defected and spilled this story in great detail.

These art experts and groups would frequently "discover" an unknown abstract artist. They would declare him a genius and invite people to share in this glory, especially inviting the Elites with money to make things happen and donate to politics. No one dared to say that the so-called art was crap because sales proved otherwise. Art collectors paid fortunes and values escalated faster than normal in the art world. These sales were set up to make a "nobody" that had no real talent a famous artist. What purpose does this solve? Well, it goes back to the essence of this book. Creating and judging art is one of the few true Free Will events of the mind and as it is in our DNA we seek two things: the balance of Nature and the cosmos, and a reflection of truth we can decode in a deeper sense from art.

Its roots go back thousands of years genetically and it is based on animal instincts. These same animal instincts and epigenetic codes of civilization determine what is right or wrong, what is moral or immoral, what is truth and does a lie matter? In other words, creating and judging art comes from the same decision-making part of the mind that determines the righteousness of a nation, society, club, elite group, or global civilization. If only the Elites could trick the fiercely independent American human being to doubt his own animal instinct and opinion – it would lead to minds built on quicksand. This opens the door to total control, and also acceptance of tyranny. It is no different than writing a book of pure lies and declaring it the greatest book ever written, and can be considered flawless and the ultimate rule of law. If enough people swear by it, if enough are enriched by it, empowered by it, and detractors hurt by it, then fewer people speak the truth and point out the flaws.

So what exactly happened, in truth? Russians decided to reduce the American independent mind by altering what it decides is a truth, and they did this by confusing it. They did this by switching art with crap and then supplying rich buyers to pay more and more for it. This led Elites to think they had to also buy, and this led them to declare how much they think this is high art, when in fact, that actual decision was never activated from the normal spot in the brain.

The recognition of what is art was diverted or cloaked, it was then injected with the virus of consensus, and this was supported by higher prices paid for the fake art. It was a system feeding itself lies, and anyone around was afraid to declare the opposite. Now these pieces truly are worth a fortune, but only because of this first step set in motion by people that wanted to cripple individual recognition of truth and what is durable and valuable.

They were selling a lie. If enough paid for it, a lie had more value than tradition.

It is like telling a child that they are a better Russian if they turn in their own parents or friends, and that a better Russia would bring them joy and fulfillment. When Russia did not improve, when these same children experienced no joy or fulfillment years and decades later, it was nearly impossible for their minds to fully relate the lie with the unhappiness because they could not recognize truth or natural sources of joy.

This makes it a duty of artists today to resist the rewards of social justice acceptance. President Xi Jinping understands this. He is doing something very brave, which is opening vast areas of Free Will, artistic talent, creative skills (coding), and individual expression. He wants the current art to reflect what has made China strong for thousands of years, yet also evolve to today. There is no doubt when looking at a great Ming Tapestry or vase, that this is art. And we do want and need new versions of art. Yet, it must be valid as a piece of art. It has to have "Moral Roots" which means in its essence, "Rooted in Truth of What Makes Art." Balance, texture, color – things I mention in other parts of my video book, list the requirements.

As mentioned many times, the only true duty of any artist is to tell the truth as he sees it and express it through a chosen medium. When the vision for truth is cracked, all that comes from it is an untrue reflection.

**"Out of the crooked timber of humanity,
no straight thing was ever made.
No straight thing."
John Ruskin**

By forcing masses of Elites and college students to accept nothing for morality, they control the narrative of morality. You'll see how this really hurts a future for young people in the next example. "Moral Authority."

Recently the United States is besieged by protests, mostly against the President, but many now openly against the entire idea of America. Heading up major organized protests are chosen representatives to speak at the event. One such event was Women's Day. In this event they had 3 key speakers. These people were chosen to be identified as the faces, Brand, and image leading the cause, which, as a cause, must be based on some sort of "morality" or higher road to a better society. These speakers were Madonna, Ashley Judd, and a woman named Donna Hyton.

Madonna, for decades, led her fans with her "higher truth." She did it in videos showing obedience to Satan, acceptance of drug and sex addiction abuses, the killing of Christ and his messages, the primary goal of life is getting money and spending it without regard to the poor, and much more negative message songs. That day of the rally she got up in front of her fans and new followers and declared she had sweet dreams of blowing up the White House. This was cheered. Think of this. She is calling for the murder of innocent people, by the hundreds, and the fans cheered this idea.

Would they cheer it if they had to go in and clean up the bodies and go to the funerals with the surviving children? No, but they are not told to think about this, or any real factors in her speech, calling for mass murder. They are already trained simply to cheer and obey. They have now said that their morality is open to mass murders. This sadly is an almost daily occurrence in the United States. Politicians, celebrities, and reporters cheer on the idea to kill anyone in the way of their agenda … all in the name of fairness and a higher moral good.

It's so funny that Hollywood loved "Game of Thrones" right up to the last 10 minutes. This is when the female star justified killing tens of thousands of people. Why? Because she alone knew and felt what was moral and right. Hollywood didn't mind her killing those that strayed from her as Mother of Dragons – fans rejected the ending because she was held accountable to the greater truth. She was killed for this. In the ending, the key speech was given for the next ruler. The Hand stated that the most powerful thing in the world is stories. Humanity will follow a story long after they follow a leader. But, this Madonna stunt means that Hollywood is wrong, that consensus is not freedom, it is an angry mob, now ready to blow up the White House. One goal of this book and its philosophy is to prevent Social Media from bullying people into having to agree with Internet Social Media goals leading to unrest, hate, and murder. And to do this it must embrace a combined Eastern and Western Philosophies.

Next speaker is Ashley Judd. She is a famous actress. I think very talented. But she did not use that to promote pride in women that day, she rose to do nothing but strip away dignity from women and privacy from the minds of her followers. She ranted about her panties and sheets stained by menstrual blood, and how she wants to rub it in Trump's face, figuratively, as an abuse against women. Obviously, her role was not to lift the image of women, it was to force them to strip away privacy, to admit their own menstrual secrets and history. What this did was it used celebrity to force all followers to see that there is no limit to what you sacrifice of yourself if it follows our plans to destroy all your freedoms and this nation with it. Enemies of Trump applauded it and watched it, slowly getting over the ill feelings on the topics, and accepted, "If I'm against Trump I am like Ashley and I will let my privacy be destroyed by people in power to be like her."

Finally we get to Donna. Donna appeared like an upstanding citizen. She is a major force and face in the day. She specializes in serving the rights to women in prison and women sentenced due to social problems they cannot overcome normally. She said that Trump's war on women leave these women helpless and abused, and she rallied people to be like her, to embrace the ugliest of women's actions, as legit, and okay if it serves a goal of overthrow. Well, who is Donna? At age 19, Donna and 4 other kids kidnapped a 70-year-old rich grandfather and demanded $400,000. Over 3 days they tortured the man, burned him alive, cut off his fingers, abused him sexually, mutilated him, and he finally died. Donna did a lot of this. She was to be paid only $9,000 and used it to get a new model career. She took this man's life and got caught and was sentenced to 17 years in prison. During her stay in prison, she was accused of murdering at least two other women. Now she is free. And the Anarchists hold her up as a "Moral Symbol," a woman to be followed and believed. She tells people if you do not feel like me or do as I do, you are immoral and a bad person.

What did this achieve? It set up 3 women with absolutely terrible Moral failure and ethical disease in them as leaders, role models, and people to be copied. Given no true Moral examples, this enslaves millions of women and people to Moral destruction. A nation does not survive this. President Xi knows this. That is why he insists on Morality and also National Culture to be honored not just back to the Mao revolution but to the Ming Dynasty. Only Culture looks back far enough to fight the current tricks used to enslave the minds of the young people.

One purpose of this book is to have a Westerner warn against and prevent the "export of Virtue Signaling, fringe Political Groups, and anti-Cultural propaganda" from harming the excellent course that most all China youth and MAC's, scientists, and Big Tech leaders are on. This message has to come from someone in the United States with the writing skills to properly present it in its entirety. Only then is the immoral social twisting seen for what it is – anti-humanity, anti-Culture, and leading to chaos.

I am glad that Xi is setting up Moral Codes in entertainment. He is letting hundreds of millions of Chinese youths explore the world of Free Will in a limited slice of the pie. They might not get as much Free Will as other more open nations but the slice he gave Chinese youth was enough. I saw it when I consulted them in filmmaking. Xi is an iron ruler yet helping them by not letting them be tricked by too much bad information and bad leaders in social ways. He also has punishment for stars that give bad examples, even in drunk driving. This is a good thing.

The Perils of Academia

**For centuries travelers visited isolated nations
and brought innovations, new foods, and set up trade.
Art was a way to prove without words that
your culture was an evolved culture
worthy of learning from.
An Accelerating Intelligence needs to include
the ways Art is used to restrict individual
spiritual, physical, and mental evolution.
Art will then become one of the North Stars
that guides our Accelerating Philosophy.**

Only modern-day teachers believe that creativity is orderly. That it can be reached through their own brilliance if only you give up instinct and intuition for a while and in their stead, take courses. Child psychologists tow that line. They take the position of – "Study previous artists and do what they did and "creativity" will be summoned."

Maybe they used the word "summoned" because it might stand a chance. Harry Potter delivers better methods. The "intellectual yet idiots" group removes another layer of real-life experience from an arena that the elite wish they could master, but they settle on control and owning, the byproduct business event surrounding art and art education.

There are multitudes of reasons to study "Art Theory," the lives of artists, to visit museums, and form a foundation for the fundamentals of art. Having this base, you can venture far from the core. And as I stated before, you can only see and be the cutting edge when you are at outside the core. Culture is your gravity keeping you "real" interpreting the undeniable truths of Nature and your genetic coding.

But discovering Art through history serves up far from actual creativity. Seeking creativity through a plan hatched by academics who abandoned most artistic impulse is like restricting yourself to looking at a 2-dimensional map when you want to be out there exploring.

It's not just that you only study instead of making art happen – you're presuming that the other 4 Brains, your entire body balance, your physical touch releasing creativity comes after study. The Laws of our Fields in this Universe pull on you and force their rules upon you as you release creativity, in all its forms.

Tell a budding Stand-Up Comedian that the most important thing he or she can do is study the craft of creative comedy from the "Masters of Old" and he'll kill you after the crowd kills him. Living or creating in "reality" instead of a classroom involves all Five Brains and links back to primitive genetic encoding.

There is a deeper reason for putting creativity first and on its own plane - animal – the animal in us.

Art is one of the few things we do that separates us from all "lower" animals. Yet the things we create still go back to animal instincts throughout the creative process. The happiness we experience releasing our creativity in any art we do satisfies our quest for "happiness." Yet happiness is not in itself a sensation – it comes from sensation. The thrill we feel releasing our creativity, and all the applause we get for it, any wealth we get from it, rises up from our most primitive animal core.

Based in that core is an animal sense of freedom, and an ability to use what comes naturally to us to avoid starvation or being killed. The sensation from applause is from knowing that at least these people don't want to eat or kill you. If art were not encoded in our DNA as an expression of our ability to imagine and make a better life, then with the first applause we would feel satisfied and not feel driven to create more art. Yet the opposite happens. You're driven to do even more art by its sensations. A comparison might be the physical human orgasm. You might know it's a consequence of sexual stimulation, and seek it, but after you have an orgasm, the chase evolves quickly to a myriad of emotions and movements. Yet they are all tied to reproduction, one of our most basic DNA impulses. And like art, we want as much of this as we can get.

Exiting the Dark Ages during which one third of the people died of a plague no one understood came the great artists of the Renaissance. Still with a majority of people either starving to death or dying from injury, these bold artists bet that their talent would feed them and keep them from sickness by rising out of the gutters full of disease. These men and women were social *misfits!* How dare they assume their talent would make a superior being assign protection, food, and sex!

Look at our classrooms. Remember the students that excelled, that wanted to be in a band, that wanted to design, explore, paint, write poetry. We in a sense demonize the image of Artists as much as we glorify them. Their parents would maybe smile, but in the next breath say, "Well, artists live broke, they're losers, they become addicts, they never buy houses, and do drugs. Don't you want to also take Accounting just in case in your 30's you change your mind, because creativity is for the young."

And yet there is this myth or legendary lore about the people that created things in our recent history. The men that created cars, planes, computers – Jobs, Ford, Wright Brothers, Diesel – we're set up to imagine they are godlike, that they were born creative leaders, and they pull this magic from this sea of creativity. But it's not in all of us. It's for the misfits, the bold, the dead.

If the truth was simply told, that most Artists live a struggling life until they give it up, it might be more accurate than telling kids that if they follow a pattern dreamed up by teachers that the seed of creativity will sprout and blossom.

To start, you need a mind that does not notice rules as limits, or actually blows through the rules and constraints of their arenas. Rebels, yes, but some artists pursue this very calmly and sweetly. They know they observe the world differently and do not feel a need to explain this to those that do not.

But there is another reason why for centuries and especially after the 1400's that governments running schools wanted rules and also steps toward creativity to be seen of as how to nurture "happiness." Our most valuable innovation throughout history was in warfare. War was the predominant status for a nation – not peace. The idea of peace was simply a calm before the next wars. Wars lasted 100 years or more. Entire generations were sent to die, and did so because much of warring was to protect food for survival. And so, the powers of art and creativity were nurtured.

Leonardo Di Vinci designed incredible war machines. Machiavelli wrote of the strategy of ruling and conquering in "The Prince" and this led to the infamous "Machiavellian" label as a way to allow, "The end justifies the means." The first thing the Nobels invented were weapons of war. They invented dynamite, water mines, land mines, the rear loading rifle, better rifle design, a wagon wheel for war convoys that would not get stuck in icy mud during Winter and ultimately perfected the greatest tool of warfare: refined oil transportable around the world. It was the guilt of Alfred Nobel, dying in Italy, that drove him, through Emanuel Nobel, to create the Nobel Peace Prizes and other prizes that would exclude warfare designs.

Ever notice that the most significant award possible, the "Ignoble War Prize" does not exist? In all of anthropology, it surely would be the most universal and undeniable in how it changes the world. A list each year would shed light on the use of imagination as well as tactics and tools in warfare, and tracing it back to the inventors would expose a timeline of human progress. For example, a huge percentage of medical discoveries could only happen in war, in arenas so full of dying men that new cures or procedures could be experimented. The expanding Roman Empire brought water, organized crop sharing, and common rule of law to nomads and isolated fiefdoms that never had access to medicine, reading, and math. The spread of Attila and Genghis unified languages and trade allowing the Silk Road to reach Europe. And the wars fought for The Church spread Latin that communicated philosophies and science.

Until then, war between smaller groups or cities caused starvation and wiped out the dominant males, dooming the innocent people from surviving. With the power of the Holy Roman Army backing up a city it took over, no neighbors dared raid the animals and fields. The British Army through the Navy did the same by blockading trade until coastal cities pledged allegiance.

And in all these cases, creativity, through its inventions, gave the dominating power the advantage. This all changed when aviation outpaced the speed of conquering and could invade inland, behind enemy lines. The two greatest obstacles to an army advancement were food and horses. Napoleon solved the first problem of food. (Think how inter-related creativity and warfare are in this example.) He offered a fortune to any man that could invent a way to keep food fresh for long enough to feed his advancing army, especially in Winter. A chef from Italy discovered modern canning, which cooked out the microbes that spoiled food. The second obstacle had to do with keeping fresh horses so the army could ride and later charge enemies. I do not remember the historic battle that defeated a huge army, but the King demanded an answer. Other nations had already started using trucks and motorboats to supply and move ahead. The General stated that all he needed was more horses to win.

He was executed by the King that foresaw the future of warfare was mechanical, not biological.

However, we all --- whether you realize this or not – exist in that moment in our advancement as a civilization in which our creative force will shift away from the three things that played as tyrants over mankind: starvation, disease, and warfare. Almost all of what we created for the last 2000 years were devoted to these three tyrants – or to the things that illustrated power that might avoid all three. Bigger houses, castles, museums, collections, cars – they all signal others that this person rose above animal fear of starvation, disease, and warfare. BUT, with few exceptions, these people did not invent what made them rich and powerful. Those that did are artists and creators.

The two primary facts regarding who calls for war tell the story of how our animal drives and inborn gifts play a huge role in both the advancement of civilization and warfare. 1) Not one ruler of a nation with a dominant army ever created a single thing that proved essential or long-lasting to humanity. 2) These same rulers do not send all their sons to the front line to win. What does that tell us? That rulers are not the possessors of dominant genes or the most superior genetic leadership traits. Here again, corruption proves to be one of the core characteristics in our civilization. The corrupting of the artists and inventors led to empires.

Today, most all our major artists and creators become pawns or slaves of the military and/or rulers. I'll tell a simple story. I owned a restaurant near Malibu. Every day, a young man about 23 would drive a different exotic car with at least one if not 3 hot girls in it, and ordered the most expensive wines and foods without a thought.

We all assumed he was a drug dealer. Then one day a girl totaled his $200,000 custom made car out front. He didn't care. Finally, I asked him if he inherited a fortune. He laughed and said, "Funny, most people ask if I'm a drug dealer. No. I'm a genius in Math and Logic. For fun I invented a video game in less than two days for just me and my friends that turned out to be superior to a secret one the Pentagon was trying to invent. I was only 15. My parents made a deal that for the next 10 years I cannot invent anything the Pentagon don't own and approve. I get paid $2 million a game and create one a month. Truthfully, I create it in a few days. I create alternative energy sources, new ways to wage invisible economic war, ways to break any security code, games that show which government officials die in a plague or invasion, and ways to fill enemies with nightmares, ways to kill using swarm tactics and bugs. They never stop asking for more."

His humor comes from being astounded at the government employees' inability to imagine anything other than politicking – the method of making an issue emotional to get votes and money. He said the reason he spends all his time partying and having sex is because he already invented several games about his relationship with the Pentagon, and in everyone, they murder him to cover up.

Imagine what that mind would have created if the government had not kidnapped his mind.

Almost none of our political powerhouses created a damn thing. They conned the people into giving them power to enjoy what they allow to be created. They evolved twisting the truth to control people. And perfected harnessing the dreams and fears of people to keep them in line.

Google and Facebook prove to be arms of the government every day. The worst thing about this is that these two companies control the flow of Information – the basic element of the Universe.

Selectively, they allow only the Information limiting us to our governments. They played key roles in the "fake student uprising" takeover of Egypt, replacing a "Strong Arm Ruler" with a Socialist Brotherhood that at its core requires globalization and a reliance on Facebook, Google, and Global Elites for Information and truth, as well as popularity and success. Islam and submission of will and freedom are inseparable. But they bring the numbers of people both companies need. By controlling Information, the governments that now control Google and Facebook people. It is no wonder China outlawed both, not to limit youth but to protect them and allow the truth.

In a truly ironic contrast, President Xi in China realizes the future belongs to the leader that unleashes Information and promotes individual creative spirit – as long as elements of national culture, honor, and balance underlie the use and exploitation of it. So, the nation known as the largest restrictor of individual thought saw the light and set up a 25-Point Plan. And I alone recognized that it is based on a few "Universal Truths."

Here are ten of these points.

1. Information must be recognized in the purest form possible.
2. No nation exists long without strong Culture.
3. Recognize the new era of investment is not a slave to fear of starvation, disease, and war.
4. Allow individuals to "own" what they create, nurturing primitive instincts.
5. Recognize that "beauty" plays an integral role in a nation's survival.
6. Spread restraint on creation just enough to keep the self-destructive forces inherent in rebels and Artists from planting un-truth instead of long-lasting truths as they exist in Nature and our DNA.
7. Pride in Genetic and Epigenetic inheritance mutation that inspire people to appreciate Cultural Honor on a multi-generational level.
8. Unlocking the creative minds of youths.
9. Protecting the creators from speeding past the slower social trends that are truthful at the core.
10. Promoting Art for how it inspires and frees the spirit to create even more in even other forms of Art and even business.

So here you have a leader that recognized that the limits the United States and world place on creation can be managed or sidestepped for a short period of time. And it can also be funded. And the unleashed creativity in the genes of Chinese will outpace invention in the world for two reasons: the number of young Chinese, and the explosive power of their release of imagination once they reach a critical mass of "New Generational Thinkers and Doers" that embrace creativity as much as Culture.

Ultimately, creativity is "play" regardless of how arduous the task. To those that do it. I capitalize the word to identify what sub-category of play I'm talking about.

Play can happen at any age of course, but it is in childhood that play time embraces creativity. It does so in Art as its most obvious medium, but also in dance, math, language skills, design, communication, gaming, improvisation, imagination of our Universe, and so much more. But it is a certain type of play that simultaneously emerges from our most primitive genetics and uses the best of environment and tools to single out exceptional artists and leaders that would protect them. In other words, the Innovators and creators.

This type of play is called by people studying it for careers Free Play. Play without specific steps or rules. The more rules, the less a child creates their own social rules that enhance survival. The less rules, the more the strong protect the inventive.

Most games for boys revolve around a winner or loser. It is an elimination game usually, or a hierarchy. As teams win, they expand their territory much like an army for a nation. They accumulate money or tools or property. Then they harness the creative minds and commodities that stave off competitors. A team that goes all the way to the Little League Finals only to lose, surrenders its entire winnings to the winner, and it starts next year.

The leadership "Alpha" qualities are predominant in the games for boys. Who can run the fastest is, years later, who can drive the fastest or earn the most.

Girls admire both traits in men. Though both are dangerous. Organizing a hierarchy to protect the leader, like a team protecting the quarterback, appeal to the women as a protection factor. A quarterback uses the bodies of others to take over new territory. This is a structured game. While the quarterback calls the shots, they do so following rules set up by others to make the winner reduced in freedom just enough to give the loser a chance.

Nature does not work that way.

Games for girls usually revolve around everyone getting along. On communication and feeling. Absent in most boys' games. You never saw Cowboys and Indians – the ultimate land grab game decrying the Indian as doomed to extinction because they did not follow the Christian goals of worshipping money over spiritual gifts – with rules about nurturing the creative spirit of Indians.

Societies required the union of the male winner-loser games with the female get-along games. Together they had the chance at beating starvation (through popular winning gaining fans), disease (through superior mating) and warfare (by enlarging the front line and military strategy.)

But let's go back to Free Play – play without rules. This involves creativity as the primary force, with Information in its most true sense (genetic memory/talents,) items to turn into tools (pianos, crayons), communication (without using the Brain Brain because children are limited in words but strong on feeling), and primitive joy in creating something that is as alive as they are because they attach body system responses (higher awareness, creating something that defies structure but means something emotionally.)

The picture kids draw of a mother and father holding hands with kids in front of their houses is so universal it shows conquering of the 3 great limits or fears of humanity: a safe home, healthy parents, and nourishment – but in this case it is emotional and balance nourishment.

Playing with imaginary friends proves even stronger ties to the "5 Brains of Tara." The children talk without limiting words, feel other friends they create, believe in the power of immaterial creation, love or fear what they create with all their heart, and predict development through Intuition.

This means two things to me: Creativity is not taught (though fundamentals of any Art can be studied and it advances an innate creative talent) and that Ether connects young minds on a level that is "un-learned" by social rules.

A parent usually demands the child stop talking to imaginary friends, and when older, the child is ridiculed for it, so they are shamed to stop using this gift to see and interact in the Ether. My question has always been, 'What is it that these children know upon birth that they are forced to unlearn to fit into the template of extorting energy and conformity that society created?' Children trade the escape offered when breaking the rules of our social reality for approval from adults and to master navigation in the limited reality adults created thousands of years ago. The more computers evolve, the less Free Play with imaginary powers and portals children are allowed – for a reason. This scares the men in the Ivory Towers, the bastions of Old Money, of Hierarchy Rule.

Free Play is so essential to children becoming comfortable in society that numerous tests of fates drove entire cities or nations into retrograde without it. Romania under Ceausescu forced millions of children into orphanages without hugging or Free Play, and all the children either died from neglecting this, came out marginally retarded, or otherwise seriously unable to relate to society.

Now let's focus on a more specific arena of Free Play. The arena of art and creativity. The key global arena that made most Free Play and Art a sin punishable by death is the Arab Middle East under Islam. I don't state this to insult Islam I state it to prove how much the punishment for making art or Free Play results in holding back the nation or culture. The strict control over Art, imagination, and free thought as well as genetic and epigenetic inheritance freedom in mating (by severely punishing female creativity and value) resulted in the end to Muslim-Nation invention. Almost nothing has been created by this part of the world in over 1500 years. Which is bizarre considering it led the world in Art and invention for 1,000 years before now. The exceptions can be seen in Dubai and the UAE where they integrate shopping, entertainment, and tech in their Virtual Reality experiences. These two locations also excelled in architecture.

When Islam became a religion that demanded more submission of the individual and all his or her creative forces, it doomed all countries to a reversal of civilization's inventiveness. The practice of forced marriages, combined with how young the marriages were allowed, and the isolation of cities, led to massive inbreeding. The effects of this can be personified in a study done by the U.S. Military. They were trying to train Egyptian Pilots to fly supersonic and highly evolved fighter jets. But less than 10% could pass the visual flight tests in simulated combat. Then a doctor applied a recent discovery. It turns out that inbreeding within the Arabs led to the retardation of a part of the eye and our perception.

This affects both peripheral vision and association of what is seen to Time when used on a split-second survival mode. With the pilot unable to see enemies, and then frightened when they appear too close, the pilot was forced biologically to evade or fail to keep calm enough to win battles.

The same I argue applies to art and creation. Those Islamic children drawn to art and Free Play that did not follow the strict limits of the religion were punished. Later they would not marry as well. They would not be favored to get food when food is scarce. They might not be protected by the family. They feared their own instincts. They shut down all imaginative thought outside of how to now follow religious laws. Imagination totally died out in the Arab world. Their great buildings must be designed by foreigners. Their ability to discover cures in their massive hospitals funded with all the money they need is zero.

While the rest of mankind took the power of art and creation to ever-expanding tools and inspiration, Islam enslaved it to the past. So only the past reigns. While they send their kids to foreign schools to be able to relate to the modern world, the last thing they want is intelligence to be freed, for imagination to allow study and innovative ideas.

Which leads us back to why we fear the rising influence of academia, and their demand that creativity can be taught if only we follow their design.

They create relatively nothing. They follow patterns that disallow Free Play. They measure in words with a Brain when both are unable to put a value on or even see certain sources of creativity. They assume creativity is Brain related, when in fact it uses the other four Brains of Tara in most of the actual creative process.

To make it worse, politicians – those people who rise to the top by using data to avoid being beat in elections and then twist truth to hide goals – overwhelmingly rely on "Academics" (experts united in higher education systems) from only a few select schools to determine speeches, policies, elections, and more. Academics are the modern court jesters, or interpreters of Greek Gods that led to the demise of the great Societies that could have emerged. Was this planned? Yes. Plato was allowed because if you look at his work, it encouraged organized production for the rulers and aligned utopia with trading imagination with logic, freedom, and all beliefs in a source of joy other than accumulation of power and the steps within the government limits to get it.

Can we "conquer or usurp" Academia and how it twists the truth? Yes. Don't forget the eventual consequences.

"You can cut the flowers but you cannot hold back the Spring" (Pablo Neruda.)

Our creativity will lead to its final goal, which was as I said set up by the ancient story tellers: the quest for the gifts of the gods. Divinity. Divinity demands truth. It demands release from over-reliance of the limits the Brain Brain tells us.

I chuckle when I read science experts tell us that in 75 years we will be on the threshold of becoming gods immortal. Never sick. Able to communicate without words. It takes that long due to reliance on words and the Brain Brain. If people were free to access this it would compete with power-hungry Elites. They will allow this but only to the highest social groups.

Unlocking both the untold unlimited power held in a nucleus of an atom in our body will happen simultaneously with the deciphering the code of how DNA communicates millions of years of learning to our newly formed brains. This will not happen due to some discovery or invention of the Brain Brain. It will happen in an instant. A few times, then thousands of times. And the people experiencing it will be drawn to each other by some unimaginable in words way.

All that needs to happen are enough moments of interpreting our existence in an Ether. Communicating without words. Through sensations. Probably Vibration, Light Frequency, and an ability to "see" or "heart" the symphony unfolding around us.

The moment that happens, our introduction to the great unknowns – those beings sharing us or watching us – occurs. And Art – Music, Movement, Form – reveal the connection to and sudden freedom to our Fields of Gravity, life force, Time, and fear.

Why won't they come sooner to us?

I'll give a simply analogy we all can relate to. The weather. As much as it affects us, and as much money is spent studying it or preparing for it, weather is oblivious to our predictions. And we are in this position as "prediction machines." Our Brain Brain demands that as a requirement of its ruling over other brains. Yet, weather and its future should be connected through Ether and outside of Time, which is manufactured by us. The balance that weather caters to cannot be seen yet by us because like fish in water we are oblivious to it.

Weather is not a prediction machine. It creates a moment-by-moment interplay through the "Technology of God" exchange of Information, bound not by just one dimension with just Earthly rules. The interplay of Information in forms of mass and density of what we think about as air play too much of a role in weather for us to handle now, but we shall soon.

And who will be the first Interpreters? Scientists? Bwahahaha. In 75 years maybe. Academics, catering to their own imagined glory, thinking Creativity can be taught? No. Try again. Who will be the first imperators?

The Free Players. Those able to create out of the abstract. The ones allowing their Intuition and Heart Brain to follow Truths that defy categorization and dogma. Artists! The John Galts. I will challenge you as to why I know this is true.

If a scientist opens a nucleus, or suddenly decodes the genome/DNA (and forgotten -about RNA) do you think the discussion will be in words or even mathematics?

Think of a 3-dimensional Origami on the simplest level, and a 4-dimensional stella octangula shape-shifting as it rotates, and finally about the 5-dimentional appearance of Information unbound by Time and Space. What do they have in common? Abstract visuals deciphered by intuition that we feel through senses that we think of as skin but really are the receiving centers in every cell in our body.

When every cell in our body is drawn to communicating with every star in our universe, through Ether and unbound by words, a one in a trillion chance at a communication our beings can interpret turns out to be a thousand messages at once. But add on "Spirit and Scale" and the universe will notice and give us more messages that hold secrets to our existence. This is now Academia shaping the future it is the universe. The same source that inspired imagination for thousands of years.

The language will be as easy to decipher as us being drawn to the light in the distance. Tracing back the origin will break apart our reliance on Time and Space, as well as our dependence on words.

It will not be a spoken knowledge. It will be an "Expressed Knowledge," expressed through art, in all of the primary forms. What will this look like? An individual dancing amidst every-changing light patters in a reality that constantly shifts like the scenic CGI environment in the movie Inception.

Creativity is disorganized, anarchistic, Free Play.

Notice how nowadays schools do all they can to prevent Free Play. They find all sorts of rules against games like "tag" or imaginary survival games that had boys and girls sneaking around to surprise each other, or put silly notes on their backs to guess, with no winner or loser or rules. The game "What Am I?" was deemed racist or sexist.

Free Play fosters confidence in a single child being able to design his or her own little rules for a few minutes. They observe the adult world and make it tiny, make the rewards smiles and the good feeling of fast movement. The game forces them to develop what is called "a critical eye" which means they observe from a position of Nature, what is Truth, what is "fair"... what is balanced, what is special, how strong they are, how to survive, how to think like an individual, how to make the best of themselves, how to comparing with others in hopes of finding ways to compensate for what they do not do best.

This is the type of thinking that creates natural leaders, natural inventors, natural artists, and naturally compassionate people. So, it must be turned upside down, and the Progressives have to find ways to make it evil, because these games give them less control over how minds develop. If Free Play consistently presents the same winners, then the other kids follow the natural path set by the winners and copy what they do. This is no good for the "Thought-Control Progressives" that only want kids to think exactly what they say to think.

Vietnam, I have seen, expanded all forms of Free Play, all the way up to the beauty pageants and "Paris By Night". This put into the hands of each individual ways to enjoy Free Play as the keep growing up. This is a beautiful thing, and I am glad my daughter Miley Ngan grew up enjoying more of this, especially as a young female, because she shared in the building of confidence and identifying in herself what she is best at, and one thing is, she excels in forming her own individual thoughts on life, love, family, and her future. In other words, it made her a natural leader due to self confidence in observing the world and finding paths to joy.

MODERN GODS AND ARTISTS

For almost all of its existence mankind was motivated by three factors: avoiding starvation, not dying from disease, and surviving wars. Peace was not the norm. In fact, up until the 18th Century, most generations saw war as the norm, with brief periods of peace in between.

The Dark Ages began because disease spread so rampantly once the trade routes broadened that the key to survival was isolation. And almost the entire day of most people (excluding royal or ruling circles) were spent gathering food to prepare for the next famine.

The 19th and 20th Centuries saw the conquering of most all lands. With that came colonies or greater empires. And then global civility earned through expanding agreements between ruling nations to work together and recognize the sovereignty of countries.

Only since World War II has most of the world been free from wars. Yes, there is rampant warring in Africa and the Middle East. And East Asia always has tensions due to long histories of warring, not due to current war attacks.

How does this effect "Art" and "Artists"?

If you include technological vision into the realm of Artists, they allowed the rise of civilization above starvation, war, and disease. Yet, the science of these arenas seems more Math-based, or systematic trial-and-error at the least. So let us put that area in the undecipherable category for now. Focus instead on what Artists do.

They "create."

Technology allows what a "creator" delivers to reach a global audience. This makes the creator a god. I'll explain.

Think of most of the abilities given to gods.

Unlimited knowledge. That's Google. Google is the sword of equality of a god.

Seeing everything everywhere. That's satellite delivery and surveillance. NASA chiefs are gods.

Traveling like the sun around the world like Mercury? Passengers on a jet are gods.

Hera delivered more food than man could eat if he was "good" in her eyes. American farmers are gods, feeding the world with plenty. Indeed, starvation in Africa is not due to a lack of abundance, but due to warlords preventing the delivery of food.

But of all people, perhaps the easiest to label a god is Steve Jobs (although his death puts a crimp in his god ability.) Steve Jobs put faces, voices, and access to the world in the palm of your hands, on a level never imagined even 15 years ago.

This is a primary reason why religion plays a weaker role in guiding society. The key benefit of a religion is it could organize masses of people in distant nations behind a key ruler, following a dogma of control (though much of the lessens in The Bible are good even today.)

More than that, few of today's greatest challenges, and even less of tomorrow's challenges, can be interpreted, ruled upon, or answered by The Bible or the religious leaders of Christianity.

We will soon be augmented with electronics and we will soon have devices implanted in our bodies to make us even more godly.

What is lost?

Primary answers to what is right or wrong become obscured, not by the Nature in our hearts and souls, or the limits of our minds, but by the Elites that will inevitably covet the creativity capable of creating the tools that evolve their DNA into gods.

Will it be a sin to marry a robot and cheat on her, or covet another man's robot? If you terminate a robot did you break a commandment, especially if the robot has human tissue in it? Especially if the robot is coded to imagine it has a soul and connection to its creator? Should some humans with less potential for greatness or genetic differences be designated as a donor class that can live the life of the rich simply by giving up some of their naturally-born wisdom in their cells, or the resistance to a disease?

More important to me and actually to humanity ultimately is the question: is the godly gift to create something totally original something that can be taken by an Elite or ruling party? Is the "Spark of Genius" in the cells of each of us sacred? Does it reflect each of us as gods or define none of us as gods because a god gave it to us? Who shall be worshipped as a creator? Should it be the Elite that hijacks or hacks the Spark of Genius or moment of artistic creation, or the artist or visionary hooked up to the machine?

Notably, the inventors of the greatest things have most often had that invention usurped by Elites.

The freedom to be their own gods makes artists targets in one sense and rulers in another.

I cannot wholeheartedly state like some defiant beatnik that "The Man" or Big Brother has no right to what I imagine or what is in my mind any more than I can say that someone with a genius vision must be left alone. Because often if left on his own this visionary cannot deliver what his vision encompasses. What if this genius holds the secret to curing cancer, yet hates humanity and wants to see them die? Is this any different than breaking the will of a soldier holding a coded message in his head that will save lives on the battlefield?

And it will be a war zone, a zone of attrition, of supplies to the front line, of bodies left in the dirt. Yet at the same time it will be a time of celebration, of ease, of fewer limits.

Ultimately, this leads to immortality or getting as close to it as we can. Who then deserves to live? Who will answer what is right or wrong?

Artists through their forms of art.

Much of our inner voice comes from our relationship to Nature and the ancient understandings still coded in our genes. If we spend less time in Nature or aware of it, if we are not free to randomly mate with other genes to evolve our inner ancient wisdom, what will tell us what is right or wrong?

Will that come from within us, or from Ether or other dimensions?

ART AND THE "FIVE INTELLIGENCES"

In the last few chapters you realize the role Art plays in keeping us human at the same time it gives plasticity of the mind to form an Accelerating Philosophy. All this combines to give us a way to manage Artificial Intelligence. Art is the pinball bumper we must always be using to score points in real life.

How to measure the perceived Intelligence of a person?

One "test" surely could not encompass all measures or (more important) all forms or mediums for Intelligence. The use of computers to measure Intelligence adds several new layers to the challenge because it removes the Intuition of the Judge of Intelligence.

Recently, "Social Scientists" (uh-oh, that's a word seething with trepidation) broke down the Intelligence categories.

- LOGICAL INTELLIGENCE (Mathematics, clean deductive Science)
- LINGUISTIC INTELLIGENCE (Poetry, writing, speaking, tone, voice, semantics)
- EXISTENTIAL INTELLIGENCE (Spiritual, unknowns outside humanism)
- SPATIAL INTELLIGENCE (Visual, Object recognition, Form in Space, Colors)
- BODILY INTELLIGENCE (Muscle Memory, genetic coding, Natural Selection)

Notice how closely these follow the Five Brains of Tara concepts. Logical-Brain Brain. Linguistic – Gut Brain. Existential – Intuitive Brain. Spatial – Heart Brain. Bodily – Muscle Brain. Okay, so, it's a stretch for Spatial to be with Heart, but it breaks down again in what I call our COMMUNICATION INTELLIGENCES.

- MUSICIAL AND DANCE INTELLIGENCES (Visceral Expression)
- SOCIAL INTELLIGENCE (Social Intercourse, now adding Electronic)
- NATURAL INTELLIGENCE (Interplay and Observation of Nature)
- INTRAPERSONAL INTELLIGENCE (Communicating within your systems)
- EXISTENTIALINTEL (Communicating Truths of our Existence)

Art transcends all these categories.

Is Art the language then of Intelligence?

Instead, visualize the "Act of Art" – that moment of artistic expression – as the convergence of:

GENETIC WISDOMS AND TRUTH
SUBCONSCIOUS WISDOMS AND TRUTH
ETHER WISDOMS AND TRUTH
QUANTUM THEORY
UNIVERSAL SCIENCES DELIVERING TRUTHS
THROUGH MATHMEMATICS
WISDOMS OBSERVED AND
KNOWN IN THE FIELDS OF NATURE

The Five Intelligences and Five Communication Intelligences interpret all this wisdom.

Art is where Dark Energy meets Inspiration.

The Intersection is not a wide-open threshold, it is a "Lock-And-Key Threshold." As strange as this sounds, a great analogy is what a dominant and a submissive in sadistic-masochistic describe as their "Lock-And-Key." Their relationship and sharing of Information arena exists on the fringe of human psychological agreements, giving them a vantage point outward of society. As with everything in Nature, they must keep in Balance.

The "Dominant" has supreme power and intelligence and controls all events within their field of understanding. The submissive accepts this, yet is also expected to push limits, suffering in the process, to evolve to perfection. This never happens, as the dominant keeps moving the goals. The submissive has human limits to pain and mental expansion, so they have safe words form when it's too much to handle and the Truth won't be reached in that moment. Within their Universe, the Truths/Beliefs/Goals, and Methods of the dominant must align in many categories precisely with the same for the submissive, or their truth sessions end. This match up is like a key matching the pins in the lock to unlock their worlds of Truth.

The artist in all glory and humility embraces the truths of his being and talent, which he must match like a Lock-And-Key to the Wisdoms inspiring him and channeling through him. If the artist is not ready, the key won't unlock all the mysteries or pathways. But, if the artist is free to follow the vibrations or Wisdoms, then more levels of the lock open up. In a way, artists are on the journey, the progression, of unlocking the mysteries of life, while playing both roles. No scientific field has this freedom. Sciences are tied to many factors: money, religion, governments, corporations, universities, established opinions, prejudice, competition, ego, and much more.

Most of the creative work I've done in my life followed algorithm paths. In other words, I followed steps to get to conclusions. As a chef I followed a recipe, even if only in my head. In photography I obeyed the Laws of Physics regarding light and laws of Nature regarding composition. Only then was I following an algorithm in the developing room that ensured the best final results. As a writer, while I often know the formula and therefore the algorithm of popular storytelling, this does not always follow just algorithm influence.

In each art medium, there come moments of channeling pure Intelligence taking the Artist on journeys on unknown paths. When the Brain Brain does not govern all movements.

Receiving complete visions hits many artists and geniuses like a blast of light. It's shocking and most often they back off, wondering what the hell was that. It's sort of like being bonked on the head, but we brush it off as quirky.

Fear. fear of the unknown. The artist, like the Submissive, wants the pleasure of pushing beyond limits and tapping into a greater power, yet fear plays with them both. The shock of light transmissions or other visions usually causes potential geniuses to recoil. They're not ready. I believe that a sort of cosmic, multi-dimensional game of roulette (or is it Russian Roulette) are being played with the Wisdoms and Truths. Tapping into them surely blows your mind. And ... it's addicting because it triggers the most intense Truths and sensations locked in your genes, personal history, and subconscious.

This often requires switching, a term also used in S+M culture. (Not that I would know first-hand.) That is when the dominant switches places with the submissive. Usually if a balanced relationship looks like a lasting one, they switch to gain more insight into each other's powers.

In reality, it is the submissive that is always ultimately in control, because if not, more of them would die under the limitless power of the dominant. The submissive can always "take more" even unto his or her death.

Consider this in relationship to the artist. He can always "take more" from these wells of Truths he draws inspiration or visions from. Look at Bukowski, Van Gogh, and Solzhenitsyn. Yet the wells of fine writing themselves hold back until he's truly ready.

Look at what I wrote and see how the intersection of Truths emerges from me without pre-consideration – or in other words, in a limitless reality. My mind did not follow traditional ideas, it grabbed something out of the human subconscious, of a lifestyle I never lived, and applied it for what Truth it had in it. I didn't need to search for analysis to back me up, because the experience of releasing the art must be its own Truth,

I find so much new opinion and theory regarding Ether that backs this up that it is reassuring. The expanding fields regarding any "Akasian Knowledge" or "Library of Human Experiences" supports my ideas. I get specific with my feeling that the spontaneous release of energy by the artist – which is Truth and Wisdom – acts like a lightning rod for the visual Truths offered in the Akasian Library. Delivered visually, this bypasses the limitations of words and bypasses the over-used Brain Brain.

In all regards, the artist must be willing to go where no one else will go. If you went there, what would you perceive?

A cool example most people can understand refers to Einstein's discovery that Light is both a particle and a wave. A particle so small it is unimaginable. A Planck unit in its own universe of uses. He perceived this Truth not with experiments. Instead, he would do a meditation that made his mind a better antenna for visual Truths stored in the Ether, Akasian, Nature, and genes. His key fit the lock, and the gateway to visuals opened. The more times it opened the easier it is to open. Through experiences, Einstein transcended the limits imposed on him by the countless other rules and people around him. From his work, lasers and countless other inventions followed.

Could mankind be accelerating toward the mass visualizations that will catapult humans into the true meaning of life and the truth about our cosmos?

I agree that certain minds are tapped to lead in a "Re-Visualization of the Universes, Dimensions, and Time." And it is because I feel that vibrational frequencies play an integral role, I disagree that all Locks require a key type of meditation that slows down the body and mind. I believe that the Alpha highway can be turned into the Autobahn.

If you've ever driven on the autobahn, you're safe when speed relative to others' cars remains manageable. The flow of Wisdom and Truth follows that same law. As thoughts accelerate until they're racing faster than most people feel comfortable, the few artists and visualists on the autobahn fast lane will feel comfortable enough in the company of "their own types."

One reason such an explosion of Truths is happening is because the Internet now puts distant people on the same autobahn. It is up to them to race to the finish. Their lightning rod minds supercharge their engines, but the driving must be done by them once they get fueled. Fear holds them back, as they push the Rules and Laws of the Field in this "Reality Dimension" such as gravity related to vectors.

Speed of calculations therefore is relative to the environment and other racers. Same with Art, though the Art most implied here is art of Advanced Mathematics and Astronomy. How much of what they discovered through intuition points to how much art might play a role in their breakthroughs.

Bypassing the race completely are the Artists inspired by visions downloaded from everything. The racer must ask, "What is demanded? What are the choices and commodities?" Free of the constructs of society, the Artist becomes his own god of morality, ethics, vision, and inspiration. The storm happening in their minds delivers downloads, visions, and confidence to push further.

All energy moves to a higher level even if laws of thermodynamics state the system loses energy. What we are doing with computers now is accelerating the energy held in the nexus of our Wisdoms. Artist holding keys to locks outnumber scientists because Art doesn't have to tip the tumblers of limitation.

Einstein slowed his brain down through meditations, as did Steve Jobs. Tesla tried a different approach, adding Fields of energy to his meditation arena.

By "accident" I was drawn into the world of hypnosis. Little did I know that my hypnotist needed to put me through a Derrida type of switching, of opposite view of the world I live in, to break through to secrets. What neither of us could foresee was that I would evolve the lessons and keys from hypnotism, make it a lightning rod for explosive creativity and ideas. At first, hypnotism deadened my mind as I was put through courses, and challenges on spiritual and esoteric levels. It made me question who I am. This freed me and also tapped into a level of Alpha waves previous unavailable. The tests against fear I passed, though I missed a few big ones (the big black rat is an example.)

One more thing about Einstein. When he died he donated his brain to science. They discovered an abundance of Glea cells in the brain, which might act as an antennae for cosmic truths. He would not always look directly at his key target. Often, he would assemble or arrange countless side projects for deductions first, all of them clearing his mind of distraction or "noise." Then his vision would materialize.

Accelerating Intelligence

PART TWO

Using the formula to develop and live by an "Accelerating Philosophy"

Introduction to "Accelerating Intelligence - Part Two"

After reading Part One of "Accelerating Intelligence" a garden of inspiration showed us a path to accelerate our personal philosophy which instills morals and Virtues into what we desire as an inner guiding light as pivot at speed. Why? Because speed is just a Scalar. We want a Vector of Accelerating Intelligence and that requires a second force that aims us to our highest form of elegance in relationships, competition, imagination, and creation.

On a basic level humans already have an accelerating intelligence and thorough tests prove we understand what our senses perceive even if our brain does not have the information. The examples were humorous such as playing Peek-A-Boo proves babies understand Einstein's Theory of Relativity. The book "The Three Body Problem" illustrates what an incredible advantage this is for humans. We want to control the throttle and life will respond because the gravity of merit-based leadership qualities attracts every form of praise and trust.

The goal of this second half of this book is to create a merging of Eastern and Western philosophies to offer an "Accelerating Philosophy" of a righteous, enviable, a thrilling life. This in turn accelerates intelligence even faster to race ahead in ways artificial intelligence cannot compete. The simplest example in life is how the artist that created this cover was driven by the pure passion to create something out of nothing. The temptations for a quick fix using AI will always be there. But if you understand Set Theory and how AI is purely derivative you will aim for a greater destiny. AI has an end game and it is mediocrity because that is the byproduct of losing imagination and human emotional-driven morality and spirit-driven meritocracy.

How to excite into activity more neural reactions that help you take advantage of the accelerating intelligence is the first step. The phrases within chapters are to help you see you already engage in activities that accelerate intelligence. With specific goals the path is smoother.

The Age of Imagination must start with a book on making it safe for humanity before the technological breakthrough is released. These phrases will come into play. Ultimately, you will discover that if we are aware that technology brought us to this point – "The Final Filter" – we can aim a truly universal message with enough Spirit and Scale to signal the Will of the cosmos to reveal secrets of our existence. These alone will accelerate us to have what I call a Parallax View of humanity and our planet. This begins the hopeful steps to observing ourselves.

Let the mental quantum leap continue!

Key Phrases
Apply I.Q. on a much higher level.
Your brain is a pinball machine.
Plasticity of the mind can become a daily practice.
Create a Dynamic – be the casino.
Deconstructing the Human Agreements opens doors to understanding our reality and why we exist as a race and an individual.
Using the Derrida and Unitary Method to realize what's really going to happen.
An Accelerating Philosophy keeps us human.
Art breaking through unknowns that are beyond what AI can force upon us.

The evolution toward a goal to give every individual a way to preserve the joy of feeling human begins with an Accelerating Philosophy that can keep up with accelerating intelligence. That will be reached at the culmination of the daisy chain of these books. The technology books will blow your mind when the intersections are revealed to you.

A global change on this level requires two things: Spirit and Scale.

It is not enough to build a popular "system" or App, or even an entire Social Media platform. When systems are designed from the start to all aim to the same goal based on the same First Principles it creates something called a "Dynamic." The Dynamic then cannot be conquered by competition because all it requires is within the systems. Ultimately this delivers products like Games, films, series, and the future entertainment from Killer Apps. Investors realize this is a Next-Gen Dynamic and fund superior products that inspire "Crossovers" from other Multiverses. For example, Fortnite only wanted the very top characters and music featured in their Games. They needed to go to feature films (no television was powerful enough in a story sense) and musicians. They paid Arianna Grande $20 million to appear within the Game and the gravity from the Gamers pulled her as much as the money. Both the game and Grande added fans and fans translated to even more revenue from new streams.

The big take-away is that the right Gravity and use of Time use Energy on the ultimate level. The ultimate goal of Influencers and Gamers is to be in a feature film or television series as Livvy Dunn and Paige Spiranac prove. They top the Influencer charts but aspire to a higher level of elegance and this is achieved through Observation Skills over Adaptation Speed. They felt… they then saw… adapted… moved first. Now dominate. All others are measured against their success. They conquered the Influencer world through a meritocracy which all people respect and many envy. The only higher elegance is on screen pushing a new creative passion. Sydney Sweeney did this in reverse. A famous actress realized she could rise even higher through Influencer-like appearances in launching her own lingerie line. It was several systems working as she originally designed them that gave her the acceleration on a level it was unstoppable (no matter how much Cancel Culture tried to crush her talent, passion, and vision.)

A Dynamic is higher example of a vertical business model. This books opens your mind to creating your own Dynamic, and within that vertical is an Accelerating Philosophy that is your social (media) compass. The Dynamic makes sure the compass points true – makes sure you can stay aimed in the direction of greatness in a world where AI and politics try to slow you to a crawl. Get off your knees and sprint! AI is not built to stop sprinters, it is built to feed mediocre joggers in life.

Killer Apps are a sort of Physics and financial practice run for Blockchain: and Blockchain will show us the problems of Cloud Storage and energy use along with what weaknesses and strengths we have in a future where A.I. alters the truth and limits our lives.

Start by breaking down anything and everything to its Physics, then build up from there. This method comes naturally to me but not to most people and surely not to Hollywood executives. As the process expands in use people will learn this perspective just like they learned to use Apps on phones. The payoff will be grand: your avatar will reflect your heart and soul.

ESCAPING THE LIMITS OF LINEAR-TIME LANGUAGE

The human use of words in a linear pattern limits sensory interpretations and reality. The film "Arrival" presented this concept well.

"Accelerating Intelligence" leverages plasticity of the Mind, which is a start. Part One expands I.Q. and begins the merging of Eastern and Western Philosophies to create what might be thought of as an "Accelerating Philosophy." Part Two uses this to sprout "Accelerating Intelligence" with a conscience. Yet, the world needs products to make an investment real, which is why the third book delivers the Art that has Spirit and Scale and brings monetization. You escape linear language within your thoughts this way. The higher Mathematics and Physics appear in "Pillars of Encryption".

Circular I.Q. Think of it like a magneto using the "Flywheel Effect" to gain energy and the acceleration is enough to escape the gravity of linear language. And this will all happen as freely as playing a video game or posting on profiles.

Our concept of Time accepts the Arrow of Time which also limits our ability to "explain" our universe and belittles human potential on an individual and collective level.

To break beyond these limitations a game-changing platform opening a Multiverse using the Qualia-based "Imagii" Avatar will be ready in late 2026. In this Multiverse breaking the linear language barrier comes as naturally as playing a video game. Sensory perspectives increase with the Qualia Apps and the Unitary Method will guide each person in discarding flawed frequencies holding their minds, hearts, and souls back. The discovery will be both intellectual and visceral – a thrill. This activates your Gut Brain.

Readers and players will live out the meaning shown in "Arrival." As more people find someone who had the same dream or is a Doppelganger this builds Spirit and Scale on a higher level resulting in multi-sensory knowledge navigating patterns and ultimately evolving Will to the "highest state of elegance of the mind."

As wild and maybe even as impossible as this sounds the higher elegance will be like an antenna to the universe and secrets will appear. Art is the strongest medium.

WHAT GUIDES OUR EXISTENCE?

That question breaks down into three categories: External World, Internal Being, and Universal-Dimensional.

Internally, our genetics, subconscious, feelings, sense of freedom, and expense of energy and knowledge all give feedback we interpret as good sensations or bad sensations. If the sensation from applause feels good and also (by luck) removes primitive fears (disease, starvation, war, slavery) then it's incredibly powerful. Your conscious and subconscious desires lead you.

External World deals with education, politics, religion, systems, geography, social health, Nature, direction, invention, and more.

The Universal Dimension means all the energy, ideas, Laws and Fields, mystical influence, and the great unknowns related to gods.

Tackling Internal Being, control comes from who or what gives our life meaning. If this is love for children, faith in a religion, health, prosperity, and more options we adjust our Internal decision-making over to whomever "games" our desire to have many of our individual decisions made for us. That after all gives us meaning, easiest seen with religions. Religions and totalitarian dictatorships work hard to make sure all your answers can be found simply by turning over the meaning of your life to them.

But how do we get the opposite of this? Through knowledge and life experience. Art experience mainlines this into our being. Even just seeing great and moving Art delivers this in some form. But to trust your own knowledge requires confidence that your mind works at high enough level to give you gifts of the gods. The Internet and Computer age is giving most of this to us. This increases as Art and stories from the masses intersect, creating new Truths.

Knowledge used to be equated as
"Empirical Data x Mathematics."

Wise men said that enough research compiled together then distilled through Mathematics delivered the Truth. But as you gain confidence through experiences that your "5 Brains of Tara" could interpret, you realize that the new forms of breakthroughs mean:

Knowledge equals Experiences x Sensitivity

By that we mean sensitivity to stimulus registered by the senses and our other Brains. Observe and feel sensations is the Observation Skill. Adapting so that your actions and vision aims true to what your Qualia tells you is the key to winning the race of both self-respect and social achievement. These are "Eulogy Goals" not just "Resume Goals."

Any new experience could alter your entire knowledge base: delivering your first child, hand to hand combat to the death in war, skydiving, cheating death, surviving a deadly disease. Or even just divorcing out of your bad marriage. These are known as Qualia experiences that carry the most weight in shaping your choices and resolve.

So, what is an experience actually? It is a phenomenon made up of three ingredients: sensations, emotions, and thoughts. That's it. Our purpose in life is to "experience" then turn this into "Wisdom" with a capital W.

Does that mean experience is outside the Brain Brain or Mind? Is the Brain Brain just a receptor? Is the Brain Brain tapped into the Wisdom within our genetics to leap forward in evolution? Or do you tap into the cosmic or Nature grid and hope that these shoot arrows of knowledge through experience into your Mind or "Being."

If you look at half the people that changed the world many have something in common. They make tricky things simple for others to absorb as experiences that improve their lives. One tricky thing to master is to make strangers you meet on the Internet feel they have a true personal relationship with a celebrity. The goal in this book is to make this leap enjoyable and inevitable.

OnlyFans figured out the "game" is to let the content creator free to think up and do new experiences that engage fans. This freed up innovation so that the company doesn't have to figure out what works best. The OnlyFans model has all the power but shares the energy with each fan if they get on well enough. The fan feels it is a sincere emotion to care or even love the model. He is giving the energy of his attention to her, and she is transforming it into not just money but a lifestyle on its way to higher elegance.

All of this really comes down to what many deep thinkers now call "The Age of Humanism" in which humans choose between their natural gifts and the artificial knowledge offered by Big Tech. Attain god-like gifts and talents.

Three things fuel this: the computer delivering more Information faster, the cosmos downloading more supernatural "Light Information" into more people faster, and our physical experiences expanding enough to saturate us beyond normal limits with Sensory Information.

Art unites all three of these sources for most new artists now. It delivers a universal vision or truth. Interpreted by people that now want to think they are their own god, able to apply this artistic truth to evolution.

For now, the great challenge is to look at humans and ourselves individually to give our life meaning. Meaning lies inside, not in some church or business or product. Yet, we ourselves need to honor the ideals of Nature, genetic history, and Culture to be stable enough to expand.

Don't look at it as taking meaning and faith away from our one almighty God, instead look at it as putting more faith in humanity and our own ability to interpret, our own ability to find a lock and key fit with cosmic truth experiences.

If we do not cross over this divide,
then it will be harder to remain flexible
enough to absorb the accelerating
pace of Information.
Modern life is a "deal."
You give up meaning in search of
power and regret it later.

Farming Imagination: The Rise of the "Imagii"

Most people can remember that scene in "The Matrix" when Keano Reeves first sees himself plugged into the Matrix like a human battery. The shock upon realizing he has no free will or reality to use it in sets him on the path to be "The One." The One ruler of the reality behind the Matrix fake sensory inputs.

Imagine if your ultimate dream was to be a high-ranking human body plugged into the Matrix like Keano Reeves was in the movie. Imagine you'd be a rock star, higher than a hot DJ., or a celebrity, all for what?

You must ask yourself, "What is the end game of the powerful people giving me technology that controls me biologically, mentally, and spiritually?" (Because that is what they are doing.)

They goal comes sooner than you imagine. The ultimate experience for those rich and powerful people?

Farming Your Imagination!

De-coding Dark Matter and its integration into the multiple Matrixes of the multiple dimensions merge religion and spirituality so they can support the accelerating pace of innovation and communication.

The Brain Brain relying on words and mathematical equations can only search to a minor level into Dark Matter (and are at an even greater loss when it comes to defining Dark Energy.)

The other 4 Brains engage through art and imagination. The new era of genius merges the 5 Intelligences with accelerated communication systems (visual, physical, genetic.) This won't be your normal I.Q. test, nor will it be the normal use and abuse of your genius.

Genius and Imagination soon become contact sports. I am 100% confident this happens by 2030. Read "Atlas Shrugged" to see what must happen immediately before this contact sport era begins. It will happen when people that cannot create have all the status symbols and the idea of paying money to artists can be rationalized. Remember, if the Elites can make people believe absurdities they can then commit atrocities.

How to imagine this… this arena where Imagination is a contact sport.

"Burning Man" meets
"The Alice in Wonderland Rave" in Las Vegas
meets "The Hackathon."

We need something, some innovation combined with a parallax view of our reality that so futuristic it seems like science fiction.

Why is this absolutely inevitable?

Because as Humanism (the worship of the Individual as a god) evolves to Tech-Humanism (Dataism) the meaning of life will be challenged. For 5,000 years, civilization found the meaning of life in gods, Pharaohs, Saints, Popes, Mother Earth, religions, stories, myths, and rituals. All religions have two things in common: They determine current rules and rituals upon ancient texts, 2) These ancient texts mention nothing about technology and how to find meaning as it evolves humans past religious dogma.

Apply I.Q. (this one idea to many) and you'll see that the fundamentals of religion must either damn innovation that makes us like gods or become irrelevant. Yet, religions' greatest attribute is its ability to organize people, their thoughts, and what they find meaning in. In a way it gives most of the people living in the world a rest from needing to answer why they are here and what "here" actually is.

A new "pseudo-religion" already emerged, based on the limitless powers of the computer, human brain, and innovative technology. "Experts" now say attention is added to the three keys to survival (food, water, shelter.) Attention for letting someone farm your imagination will deliver these three.

[Disclaimer: This paragraph was added in 2022.] And the push by Bill Gates and tech to begin transhumanism and the implanting of chips or technology into all human bodies proves my point. Already I believe that there are many science experiments unofficially testing this out on unwilling victims. Experimentation is a fundamental aspect of the advancement of science: after all, a woman in a coma from COVID woke up when she was given a massive dose of Viagra!

There are also insidious experiments with animals and people in which the person doesn't know all the life-altering elements in a shot he or she receives to prevent some local disease. Haiti and Africa have long been testing grounds for vaccines that proved so deadly they covered the experiments up. The reason this fact is in this chapter is because some ways to farm imagination and control thoughts will be introduced to society as a way to remain healthy or to "be a moral person." This will lead to control over someone's health or wealth. But these might be needed to anchor future true chips that "farm imagination" for the hierarchy to use and enjoy at the expense of who gives the imagination.

Within years, the most powerful companies and people on Earth will begin "mining imaginations." Yet, they must sneak up on what they are doing, what they plan for our future, because some of it will require the use of humans, souls, and brains in what now would be seen as repulsive and almost criminal ways. For example, today this just looks like fun with AR, VR, and role play. These adventures into Edge Technology Entertainment remain fun until electronics augment our senses and bodies. It also jumps forward when Apps talk to Apps on a whole new level and they are based upon an Organism-Environment creation.

How could this sound so impossible to you? You watched "The Matrix" right? The scenes where Neo learns Kung Fu in seconds? Where do you think in the future they get that skill from? Not a machine. From a human that mastered Kung Fu. This new era will begin simply with volunteers, then move to volunteers that get paid, and finally move to people "chosen" much like "The Hunger Games" to "share their talent, because it is only fair." Fair. That word comes up so often, already coming from Intellectuals, from those that talk of social justice. Now why do I single them out? Because Intellectuals never are expected to pay the price for the people they are paid to serve, or the public they dupe.

Following the example of religion, the powerful companies will "make deals" with humans gifted with superior imagination. Trade "meaning of life" for power, fame, and money. To do this – to bypass the levels of rejection by humanity – the move has to be fast and overwhelming in its rewards. Overwhelming yet fleeting and shallow, just like we adapted to putting so much emotion into "Likes" on Instagram, and gaining followers on Twitter.

Why? Because upon the very launch, the Intellectuals, heads of Big Tech, and Social Media news and Apps have to make everyone watching wish "they were that mind."

The closer mankind gets to unlocking the keys to our existence (dimensions, Time, Space, Ether) the more they must rely on pure imagination and interpretation through more than the Brain Brain. Artists and Influencers have respected titles now. I'll add a new one.

"Imagii"
The first Artists and Influencers
that embrace Qualia tech and expand to include
Imagii avatars will harness the Maxwell's Demon
energy of the entire mass acceptance.

Is it an accident that my new super-humans have a name as simple as Emoji? No. No one feels they have to worry about the soul or meaning of an Emoji. Because, I mean, why? They were just created for us as shortcuts to imagination and expression. Through a sort of quasi-art design invented by computer code.

Oh … wait. Isn't that what these superstar super-humans are? Exactly.

The same Elites that send men to fight the wars that make them rich and powerful and keep humans in fear and anger mode will devise ways to make the enslavement of human imagination "a great deal." Think of this social engineering like the "precogs" in the Tom Cruise film "Minority Report." With all rights of the Declaration of Independence, innocence and guilt were given to a few trapped visionaries called precogs. The problem was precogs could be wrong, which is the point of the movie and the opposite conclusion politicians and Silicon Valley and bank gods want us to realize.

How will they achieve this? By promising the New Era of humanism. If they just let these "Matrix Imagination Sources" be harnessed, the Elites will deliver what spirituality promised them really, really fast! They will promise "The Deal First and the Journey unfolds within the vicarious experience fans enjoy. Trading individual humanistic values of a few new rock stars for the meaning of life for billions of fans.

Your memories will become a chip in mental poker. Your Qualia will be commoditized. Too wild? Let's simplify it with a reference to something you already experienced. If you ever had realistic sex in a dream then you transposed sensations from "someone or somewhere else" and woke up with the after-sex energy. One goal of all the books is to end up with a tool to find out where the dream came from.

[SIDE NOTE: The original chapter was written in 2017. An event recently happened important to add as an aside. This December in 2022 a video emerged to present a massive artificial womb factory to pump out newborn infants. Each building can produce 30,000 fetuses at a time, and the entire complex can create 400,000. They framed it as a blessing to barren women. While they speak about how the fetus can learn additional languages and knowledge while developing to nine months, they leave out how this programs a brain to believe anything and remove the mother from the equation and thereby remove the nuclear family. I doubt there are 400,000 barren women or women unable to carry a fetus to delivery in the world that they will serve. This birth pod artificial advancement is to serve the more evil power to replace humanity with programmed human work organisms. Ask yourself, "If they can choose what we get to learn how could we limit what they refuse to install such as Free Play and Free Will?"]

The "Four Orders of Gamification of Society" already began and has been experienced (therefore absorbed in Qualia) by two generations now.

The Four Orders of Gamification lead to "Farming Imagination. "
Order One: brief, fun, no rules, like playing a game on your smart phone for 10 minutes.
Order Two: transactional, has goals, and completion of a task, sometimes requiring money, always requiring attention on a level that if reversed, it would allow the company to map your Mind.
Order Three: social online "glamification" that requires knowledge or skills or looks that get you approval from people. This leads to money from them so this stage builds in rewards for showing more of yourself over time, and you get used to giving all you dare.
Order Four: "Gamification" on a global scale, tempting people to go more into Data-ism.

Just like DJ's a decade ago and now Gamers and Influencers are being offered fame and fortune, people with minds tapped for experience will organically go through this Order.

As more senses are added to Media experiences, this makes the first ones using this more visceral and the rise in virtuous sides of their soul an epigenetic desire.

The body will create endorphins that get a player high, and big tech will lobby for permission to make it even more intoxicating, addicting, and paying off. The minds they want to cull now are the ones that they want to farm soon.

This is not a big leap from celebrity today. Celebrities prove their power through consumption of what the Elite want them to consume so that millions of fans will follow. That is "Their Deal." Their journey, shared through their art (Actor, Singer, DJ, Coder, Inventor) promises a thing of Meaning to people they never meet. Because this does not have to happen overnight, the Elite have a year or so to addict young people to this Game, with promises of the celebrity journey meaning more than anything their parents ever promised. They control young people through emotions or things they cannot define yet, like a need to be accepted, which goes back to our primal urges.

Consumerism has morphed to include Virtue Signaling.

What we see now is that the Elites have replaced consumerism with causes, like Climate Change. What you Like on Facebook and what protests you support is your new accessory celebrities flaunt. They can afford any number of causes with unaffordable consequences. That makes them worth idolizing.

The appeal of anarchy and outgrowing their parents is the most highly-addictive social theory drug anyone could offer youths. Freedom and coolness. Impossible to resist. Delivered in a Game that does not demand anything from billions of youths … except abandonment of bravery and individuality. A transfer of energy, free will, and physical experience. If you've ever seen a movie where powerful people try to implant the youth, brain, vision, or innocence of youth into their souls, this is a way to do it.

The trick is – how to make it appear good rather than evil?

They need to continue to erode religious hold over mass humans because it no longer applies to this new Reality. Meaning must be found in their Human Source. Who did this before? Marx and Stalin, transplanting meaning from religion to the Troika.

Apple and Google do this today. Both have god-like powers, and both offer endless ways to find meaning through them.

But they lack the human component essential to power through progress.

The human component supplies three things an Apple logo cannot: energy, the human brain, and the mindscape of consciousness.

The Game probably will accelerate the I.Q. process so fast most people can't see what's presented to the Imagii. But the Imagii sees more and at superhuman speeds. Second, the Game will immerse players in changing Fields of Intelligence that are integral to The Ether: Music, symphonies of Light Pulses, still and moving visual images, immersion in Nature, Muscle Brain interaction with the fundamental shapes leading up to the 3-dimensional stella octangula.

There is an irony here. The Elites sucking up all this Imagination will need to use what gives us meaning as an individual with Free Will to tap into the visceral energy of the donor.

Plugging the human brain into sensors tracking the electrical impulses of Imagination -- mapping this as the geniuses open portals to knowledge, meaning, and creation -- is the process. In time, enough algorithms under study allows the Elite to concoct recipes for imagination. Something they can inject or apply from an external device, implanted in their own brains or nervous system.

Think of this like a Game that creates hundreds of young men and women with the observational powers of the star in the movie "Limitless" after he took the pill. Or of "Lucy" after absorbing CNO-4. Only it will be a massive Game, shown as VR-Interactive Entertainment. Artificial Intelligence will hit society with this so fast most minds will be overwhelmed and begin to embrace this new Matrix.

Yet, it is important to see now, before it happens, the peaks and valleys of A.I.'s rise to control our reality. At first A.I. will create off of what humans created. Flaws in the A.I. will be noticeable on a technological level. Faces or movements will be obviously engineered and not real people. Then, A.I. will improve to where viewers cannot tell the difference between real and fake.

The fastest way to this is for A.I. to ply and play the weakness that men have when it comes to sex. Men are more visually taken over than women when it comes to sex. Men in their hunger will consume what they see as fake and then love it even more when they cannot tell what is fake. But at this point A.I. is learning off of previous A.I.

The stories, pathos, Telos, and emotions will deviate from human-inspired ones. This will quickly result in a type of inbreeding that makes future generations plagued with flaws in what human relationship with each other and the natural world.

The world tuned in to "American Idol" mostly for the emotional door it opened to see overwhelming pressure from a God-given talent and Will-honed performance. The human being won. Will the world tune into creating "Imagii Avatars" based off of the "Qualia Kluster Apps" for the same reason? I would hope it is more of a balance or that is a counterbalance to how A.I. tips the scales against Humanism.

The audiences will patiently go through the journey because it's both physically addictive and spiritually a necessity.

Artists will be the highest-prized players. I predict combinations (or teams) picking players from one of 5 out of 11 categories.

- FINE ARTISTS AND SCULPTORS
- MUSICIANS AND SINGERS
- CODERS AND HACKERS
- MATHEMETICIANS
- DANCERS
- CHESS PLAYERS
- RUBICS CUBE PLAYERS
- GAMERS
- WRITERS OF SCIENCE FICTION

- PHYSICAL BODY CROSS-TRAINERS
- HEALERS, MOTHERS, ANIMAL TRAINERS
- SCIENTISTS IN PROGRESSIVE LABORATORIES
- TAI CHI MONKS

What is missing? Any religious representatives or spiritual guides. Large Think Tank geniuses. Teachers and Professors. Social and Political opinion makers.

Art free of society, religion, politics, university, or philosophers. Why? Because of their lock step reliance on words, drawn from century-old social structures that cannot possibly play a role or help in interpreting new eras of Information.

They also need to remove these influences in order for the Players to become demi-gods.

Just like top athletes become role models, icons, and celebrities, so will the Imagii Avatars representing each individual as an Imagii. Just like no one can play golf like Tiger Woods or basketball like LeBron James, no one will be able to play "Imagination" like our Imagii can play our newly-fulfilled Qualia Apps and the ultimate avatar of the future. This is at least 6 years away and up to 10 years away. I know because the final tech innovation will take me this long to present to you.

We want to accelerate the philosophies we live by to give us a way to steer this race across the seas of Dark Energy and human traits. If the universe reveals secrets to us as we prove worthy by engineering such an Imagii Avatar we want them to send us the secrets that nurture what makes us human.

Each team will be "coached" by a highly-creative innovator that keeps his eyes on the endzone: "Power Through Information, Wealth Through Application."

None of this sets anyone "free." What it does is something essential to the progress of humanity. It replaces seeking meaning from outside sources, with finding it internally.

As higher discovery leads to even higher discovery, as visions merge, tracking the power and truth portals will be bid on by Elites hooked vicariously to the Imagi of their choice.

The brilliance of the artistic creative moment will be harnessed, then used, free of outdated rules, allowing movement without morality.

Then the big quest will be taming human corruption of purity.

The only one to "Farm your Imagination" should me YOU.

[**NOTE ADDED IN OCTOBER OF 2022**] This chapter was written in 2017.
Around 2020 the Metaverse proposed by Facebook
failed to get approved. That's good because it would be
the precursor to farming imagination and having the tools
to harm people that do not allow it.
But their devices that strip privacy

are now sold openly, replacing what
we see naturally with data we consume
that Facebook/Meta designs for addiction and power.
If this one example so easily proves my vision for the future,
it should help you trust my other visions on both
the past and future…
(which is an odd thing for me to say because
Time is a construct for the human mind
to deal with our reality with such limited knowledge.)

WILL ART EVER "CREATE THE ARTIST?"

Physicists currently run systems proving
that Causality goes into the past
and can change the present.
What might be missing is
"Spirit and Scale" unified on a single
known outcome in the future.
Want to be an Artist?
Deconstruct the Human Agreements.

Kleptocrats hope their global domination will make this the new normal.

From the first drawings on the walls of caves, through the gilded Egyptian Age, art reflected mankind rising above animals, and gods rising above mankind. The devotion to Christianity led to The Renaissance, yet again worshipping an influence from outside ourselves because this gave life meaning. It is not until the computer age that Humanism – worshipping human ability and finding the meaning of life within ourselves – that Art indulges the race creating the art on a level that steers the very creation of Art.

Bio-Science, computer technology, and New-Age Philosophy based on the acceleration of imagination join forces that will quickly erode the concept that the human being is indivisible … that just one soul inhabits the human body, and through this immortal soul total knowledge can be found if we simply "slow down and reduce stimulation to zero." (Buddhist meditation.)

We do not need external forces dictating quantum leaps in our understanding of the universes and dimensions.

Liberalism won out over communism because humans like the idea of Free Will, owning their own success, beating the challenges of failures, and finding versions of the meaning of life in the value of every human soul (being equally worthy of life.) But now Liberalism in its true sense is sacrificed by a party that is Socialist which means eventually those rising in power will kill off the true Liberals. Why? The first enablers know too much, they cannot be allowed to live. Artists - when in the creative moment - are an ultimate example of Liberalism, which is why this matters to the art of filmmaking.

We create the cinematic art but do not have the power or right to determine how it impacts the audience we create this art form for. Until now. Many in Hollywood demand the power to do this, and at the same time they remove people for small infractions, they embrace true monsters. Mel Gibson gets crushed for what he said while drunk while Roman Polanski who bragged about having sex with literally scores of girls under 15 gets a standing ovation. Why? Because Polanski was turned into a god of victimhood which is an absurdity because he made victims out of countless children. He serves the purpose of sexualizing kids at a young age.

Remember when the film "Thirteen" came out, about kids that age discovering sex and how much it infuriated conservative people? Whatever happened to the uproar over "Aerobicize" on Showtime? Polanski was idolized and applauded for his potential to make life imitate art, or as in the paragraph before he was a first enabler. They applauded him because they knew how close they were to being an outcast from Hollywood for doing something wrong. Polanski enables their salvation from their own sins. But maybe it was more than that, maybe the actors at the award show knew the real criminals were the systems that fed immoral behavior.

Fast forward many years, and even the slightest comment ends a career … meanwhile, everyone knew Weinstein and others were date raping actresses. The enablers were the ones that should be on trial: agents, casting agents, managers, and the other actresses that said, "Oh, get it done, it's worth it, it's how to get to be famous." Is this the action we want a newcomer to go through to create an artist of a good level fame? No.

Polanski's movies did not abuse children or turn them into Mia Farrow in "Rosemary's Baby." (And Woody Allen was destroyed for falling in love and marrying his adopted daughter.) See? Technology will spread the power of wicked desires just as the Internet already did with porn.

The Internet needed porn to grow, and it did indeed inspire a flood of girls into porn. Was that the first example of an art form creating an artist? Or were they simply seeking sex, money, fame, and drugs? I think they began with some whisper of an idea that they are performing a skill that requires art, and they learned it by observing art (in that form.) But this idea of the art creating an artist I'm talking about is much more evolved and soulful than this perversion and Polanski example.

As the Technology Age nearly removes fears of starvation, disease, and war, it simultaneously reduces the need for (therefore the value of) lower income and no-knowledge populations. As long as people were needed to feed the manufacturing of goods, the delivery of services, and the human fodder for wars, the Elites took care of the lower masses that did all of this. And to be sure, all levels of societies around the world were lifted. Yet, slowly forfeited is the liberalism idea that every soul deserves equal credit, value, opportunity, and a utopian existence.

While the religions organized people over centuries before, giving followers a God to worship that could do incredible superhuman feats, and Royal bloodlines also supplied an idea of sacrificing a lower life for a higher leader or Creator, these examples are losing their relevance to the advancement of mankind. Too many questions and choices fall outside of the entire spectrum of Christian, Catholic, Muslim, Judaism, Hinduism, and other religions. Instead, our gods are computer algorithms, entities, Apps, software, games, realities, and direction in life.

Within computers we now find the meaning and direction of our lives. To deny this would be futile, for everyone around you dives in to augment their reality and capabilities through the computer.

Artists are no different. Ask for a custom artistic logo, and impatient and untalented graphic designers simply cut and paste from a template. They find meaning and social value through what a computer generates. In sound mixing studios, music editors sample previous true musical art creations, add them to synthesized band instruments, adjust voices to be on pitch – and create an idol or demi-god that delivers the new meaning to millions of fans.

Up until now, even on these computerized levels, the true artists brought their gift to the arena. Their vision poured out of them, their muscles following instructions from the 5 Brains of Tara. When asked where this sound or vision came from, they reply with various references to the soul, the core of an artist. Some famous as well as millions of undiscovered artists, innovators, and scientists swear they tapped into a source of inspiration and vision outside of themselves.

A source living in the cosmos, or deep in their soul, or in the powers of Nature. Perhaps this source comes from within the nucleus of every cell in their body. Maybe the Divine it from another dimension, or from aliens, or the human consciousness swirling in the Ether around our planet.

Regardless where it comes from, one thing you can be sure of … the powerful Elite will one day come for it. They covet it, and will use their resources, both financial and political, to allow them to harvest the actual "Spark of Creativity" and all its gifts.

Already, Apps talking to other Apps and complex algorithms zero in on the parts of the human body holding the secrets if creative thinking, artistic expression, and access to the portals to other dimensions and truth.

Hacking the Artist in a person will be the biggest new market in the future of the Tech Age and the creation of the super-human.

Quickly, the Elites will guide us through laws allowing this to happen, but more important, they must make it appealing to the masses that don't have the power or money to do this. Just like organ harvesting led to poor people selling organs to rich people, Artists will sell their soul to the elite.

Eventually the pool of true Artists in all Fields will get bought up like sports talents by agents and ball teams. The bidding wars will lead to the next step: Identification of Art in the genome or DNA of a child. Already they use this for early warning of inheritable diseases. Great, huh? But wake up. This step to help us is a deflector for where it leads us.

The Elite will still face the same obstacles that the Medicis did in Florence in 1400: a limit on Artistic Genius, and their ability to inspire gifted people into the levels of immortality occupied by Di Vinci, Michelangelo, and later Raphael, Descartes, Shakespeare, and all others.

A vision came to me that took a few hours to distill to its meaning. The vision was based upon seeing the Truth and Beauty in the explosion of the arts in the Tech Age.

Very soon, within years, technology tools will allow us to create blank slates and spaces that either channel creative genius vision into a child artist or coax it from within the cells hoarding it.

Let us look at a moment in the life of a young Artist. A series of tests identifies great musical or painting talent in a boy age 8-16. The breakdown of his genetics, along with the 3-dimensional monitoring of their brain waves and system communications, feeds into a computer algorithm. This already happens in some cases, on a small level. Usually child prodigies are discovered, then a team of experts work slowly to advance the skill. So, this isn't much news. Until you see where it is headed.

Imagine that boy standing in front of a computerized blank canvas. The canvas reads brain activity in real time, predicting the impulses and visions of millions of Artists that preceded the boy, and injects codes that force the spark of creativity between neurons from within the soul or from externally in the great unknown. Whether channeled in or drawn out from within, the process is the same. The object upon which the Artist delivers his creative spark hacks the free will and forces it into action.

While this sounds bad, it's what makes Olympic athletes. The best of the best. Trained. Pushed. Timed. Schooled from young ages when the cells fire their healthiest, their most varied innovations in muscle memory and human endurance. Things the coach does not have within his body he just brings it out of the athlete or pushes more of it into the athlete. Electronics taped to the body send all the data from the inner core, the soul of the person along with it. So, this is all endorsed, accepted, and more important, exploited. Nations revel in the Gold Medals. All benefit. Yet, the athlete is "used up" by 25. Put to pasture. Maybe he or she lasts until 40.

Now think of a book. Kindle already knows when you read your book, how long you read, and what other books you might like if you finish it. Soon Kindle will know how to read your face, see when you laugh, trace your body systems to see when you got nervous, it will know when you last exercised, and what the weather was that day. With this data, the Kindle can start to groom your reading systems. It will pick what you read, when you read it, what you retain from it, and know how it changed your life.

To make sure it is a positive change, it couples with other technology and rewards your body with feelings of happiness, sensations of pleasure, enlightening comparisons. Imagine reading a book you end up loving and then going to a lounge.

Along the way your Kindle told your car to alert you when you see objects related to the book. For example, your windshield might highlight an ancient vase in a window that is from the same era as the one in the story in the book. Or might identify someone from the same country as the hero. It might spell out the object in the home language, to make the book immersive.

At the bar, people that read the book will be identified, along with their current social status. You could then barter for what you want from a stranger through the book story. Think this is far off? I have just one word. Pokemon. People needing to buy Pokemon images were matched with people that would sell them.

Soon, this athlete will enter his workout arena. Control over Fields and the reality of the room, companied with Apps specific to the athlete and routine, will attach to the body. These wires not only record what the body is doing inside, it charts the brain activity, and the Intelligence of all 5 Brains, then applies it to the laws of physics and this reality or dimension (gravity, space, time, force, vector, etc.) A few things can happen. Technology could alter the physics by increasing gravity which is like adding weights to body building. This triggers reactions in the body that through electronics can be zapping visions and Information into cells, timed to vibrations or songs. Much like a gymnast does already. But in this case, the technology goes right to the sources of art: the muscles and the brain activity.

The room itself can tell the gymnast what part of the mat is ready for a custom move or combination. It will replay a hologram alongside or after the routine that the Artist then copies. The process continues until the brain fires the optimum sequence guaranteeing perfection. Everyone wins, right?

But what are the long rang effects of this on the human. Could be great, could be terrible. Look at steroids and what they did to football players. Or helmets making them think they are safe from brain damage when elites knew they were not.

Now let's take this from dance or gymnastics into painting. What if the canvas flashed a thousand possible images into the eye of the artist, so fast none of them took over the vision, it just sped up choice selection. As the boy painted, his brain is monitored. As it reaches the optimum level of artistic brilliance, the firing within his brain is captured on many machines. The chemicals produced are harnessed. The brain alters its way of processing, favoring innovation, free play creativity, and vision execution. Along with this, the genetically stored sensations associated with bliss and power to be immortal pour through his body.

What if an Elite of society, with all the power and money he wants, hacks into this? Sets up a program inviting young people into his laboratory? Paying them enough to make their lives better. And if the child becomes a rock star of creativity, the Elite sponsors him like the Medicis did. But while the Medicis thought that by being close to the true Art greats, they would be closer to God, get into Heaven, and live perfect immortal lives – the Elites now get God, Heaven, and perfection immediately from the Artist. The one thing that kept the Elite from all gifts of gods was the inability to create!

The canvas will hack the painter. The room will hack the dancer. The blank paper will hack the writer. The flow of Information will hack the coder. The mixing board will hack the vocal cords of the singer and fingers of musicians. The film camera will compensate for the new Director and Actors trying to transcribe their visions and hack into them the talents of the great film artists of all time.

The camera will remember that this new starlet looks like Veronica Lake and Faye Dunaway, will remember the look, angle, hair, and time of the edit, and time the starlet, unknown to her, to repeat fan-winning moments. Editing bays will edit within minutes the entire film into 7 or 8 versions and let the Director pick from them, compensating for the changes automatically.

Who then owns the byproduct? The Elite or the Artist?

Better question is, "Who owns the Fame?" Fame has power, the actual Artist is the source, but the benefit comes from exploiting Fame and what is created.

Art in its process of completion will "create" the Artist. Like constantly playing against a computer player, the chessman learns by playing. And up to a point will beat his opponents if he or she follows the algorithms, and lets them become second nature, which in fact is probably the most accurate use of that term ever.

To encourage this, society will form Olympic teams of artists. They already do in symphonies; they just don't look at it that way. This leads to separation of societies, just like it separates famous movie stars from fans via bodyguards and big fences. The fans want to think they are inferior to their idol. Just like ancient Egyptians felt better giving superhuman attributes to Pharaohs or Popes.

What is it that allowed civilization to innovate more under capitalism than socialism? Simply decentralizing the source. This augmented artistic talent, and the hacking of it, decentralizes the source of Art.

Is this a bad thing? No, it could even be a good thing. First, it will set up labs around the world in which Artists surround themselves just with Artists and innovators. This could lead to an awakening of new parts of our brains. Then it would spread Art to the people with the money and power to deliver great visions otherwise overlooked or thought useless when looking at it from the bottom up.

This breeds coveting. The Elite will still covet the true source, or at least the channeled person, the source. The future of Artists therefore depends on their power of "distributed" Artistic vision over centralized artistic vision.

Yet, this will allow Artistic geniuses in remote jungles to have their gifts and brains harnessed and added to the collective genius. Why is this important? Well, think about it. Most university studies leading psychology, invention, and Mathematics go to the nearby brain pool to study brain activity. NOT to a jungle in the Congo or field in Hunan, where a local healer-genius might locate a source of Truth and Intelligence in plants. By not studying that brain up until now, as it finds genius or innovates from signals from other brains and systems, we prevent ourselves from unlocking that wisdom and Truth.

Through hacking the Artist or creator of things mankind will begin journeys into the unknown, much like ancient explorers ventured out in ships onto oceans unimaginably limitless.

While this looks like a severe reality to allow Elites and governments to establish, really this is quite tame compared to the social engineering that the Industrial Age put us through. And the Industrial Age improved life for all of humanity, as is the Technological Age. When the Industrial Age required more factory workers, they took them from the farms and wilderness. The old senses and skills these people knew became useless. Why have a nose able to smell fear or identify animals from a distance when there is no use for it? We do the same with most skills babies are born with that we do not perceive.

This is why the Elite will hack into younger and younger bodies and minds… to gain entry into what I call the "Portals of the Innocents."

Just as medicine upgraded from helping save the sick to helping make the healthy immortal, so Art will evolve from presenting Art for pleasure to presenting the artistic process to god-like Elites more core-human.

One unexpected but inevitable new commodity will arise from this: security for artistic Minds, souls, and sources. Protecting a young Artist from experimentation might be the most valuable offering.

Imagine how much deeper
you would gaze into Art
If you knew it could generate
an artistic talent in you
or add the abstract process to your
decision making.

ONE UNIT OF "WILL" – THE PLANCK UNIT (OR QUANT) OF CREATIVITY

"Will"…
the essence of the universe.
How small could it be?
How do we imagine the size of
the essence of inspiration or imagination?
How do we quantify anything that small?

[**Disclaimer:** This chapter will make de-facto Physicist go bonkers. I'm making up a new Constant because it's a way to offer more perspectives because the properties of higher Dimensions so it is fathomable. I am even wrong in some cases connecting concepts that are whims from actual scientists and are not proofs. Considering we are in the generation that will probably interact with UFO intelligence it's good to read some about it.]

Planck is the distance covered by a photon traveling the speed of light in a vacuum considering gravity and the Planck Constant. To give you an idea how small a Planck Unit is, visualize the tiniest dot your eyes can see. Look to the left and pretend all you see in your perspective is not even the end of the entire universe. Think about how much bigger it is than the dot. Now look to your right and focus on something in the far distance. Imagine that just as distant away sits the actual Planck in size. In other words, the smallest dot you can see is halfway between a Planck and the end of the Universe you can see. It's SMALL. And it's considered the fundamentally smallest unit of measurement of distance or size. Within it lie the secrets of how we exist in our universe in the 3-Dimensional reality set in the 5th Dimension.

There are several Planck Units in various Fields – or as most prefer to say it, there are several Constants we rely upon in formulas to investigate Fields.

But why would I even mention Planck Unit in regard to humanity and technological progress especially in Virtual Reality, Artificial Intelligence, and Blockchain? Plasticity of the mind and de-constructing human agreements. The topic plays such a major role it is covered in more detail in my other book Pillars of Encryption.

"Small" Talk

Consider Planck's unique recognition. Planck is itself a Unit, yet when present in equations it represents the other (very few) "Constants" in our known universe. Art lies in the abstract, in the unmeasurable "spark" that results in the next Claire de Lune, The Marriage of Figasro, Swan Lake, David, Catch-22, One Flew Over the Cuckoo's Nest, and Dali's The Persistence of Memory. Would we even want to apply Constants to that spark… to that human moment of "creating something out of nothing" that culminates in a masterpiece that inspires all of humanity to seek the highest form of elegance?

While minds get on the path from Will to Elegance it is primed for imagination so let's get more into Planck. See why and how it was inserted into the cultural learning in advanced Technology, in the study of American and Chinese Culture, Stoicism juxtaposed against Tao, in accepting the creative responsibility that comes with Killer Apps, then Blockchain, and ultimately A.I.

In some form or another an unnamed Plank Unit of creativity lies in the I Ching as much as in The Declaration of Independence. The emphasis by the current China leadership on appreciation of Culture and taking on "Liberty with Responsibility." In America we seek a code of our existence. These evolutions require a cosmic and social awareness of balance starting from our original sources of Truth, which are Nature, genetic awareness, and family. Planck literally makes Nature what it is to us. Now, before you damn everything related to Chinese Party leadership, remember that when their young people were dying at the computer due to being addicted to video Games they passed laws limiting Game time.

The suicides from self-abusive levels of computer and smart phone use dropped. Education scores went up. Less deaths from playing a Game in China. The United States didn't follow. Political leadership is flawed by either demand or definition. If you see China as the enemy then be smart and learn about your enemy to defend your ideals. If you do not see China as the enemy simply appreciate the addition of their artistic and cultural accomplishments like the I Ching, which is actually the oldest example of mathematical coding.

Balance keeps our universe from thrashing itself apart. And in other chapters, we discuss one brilliant hypothesis: could the existence of mankind literally be acting as a balance to Entropy, or the increase in chaos in our universe in this dimension? In some ways, mankind records the universe's evolution in its ability to turn chaos into order. This can be seen in music, pottery, architecture, agriculture, nutrition, observance of the cosmos. And even more so, in the advent of technology organized in 1's and 0's.

Blockchain, Artificial Intelligence, Quantum Computing, and Big Data drive us in the reverse direction of chaos. And they use Planck Units at some time or another to base their formulas. As Blockchain self-organizes according to the Fibonacci Ratio in a Lattice seen in Geology, this organizing will quickly become evident. So let us explore what this evidence is based upon. A.I. will take over as Blockchain popularity wanes.

There is a limit to what people will do for Blockchain because it does not tap into a primal urge. Adopting Blockchain will become a hassle to anyone that isn't into it for money which serves a primal urge ultimately. A.I. will expand playing on different weaknesses we have like laziness and phone addiction so it will scale.

The Planck Unit was discovered in 1899 by Max Planck in Germany. From there, the Planck Unit played roles in scientific breakthroughs for the other "Planck Units." These Units are also called "natural units" because their origin comes from our observance of Nature on our planet and also from what we deduce from Astrophysics.

There are many studies that spread out from Planck Unit truths, such as "Planck Scale." That is when the magnitudes of space, time, gravity and other units goes beyond our quantifiable standards in this Dimension. This is one of the many reasons it is important to understand: forces we don't understand will rise up when we get so dependent upon A.I. we'll suffer consequences. You'll find out in the study of Blockchain that the Intersections with Artificial Intelligence, Apps, and IoT will blow beyond the limits of our computer science. It will increase Plank Scale, and in ways count on Plank Scale.

Self-organizing "Lattice" and Smart Matter will expand out like a universe from Cloud Storage. So, there is a parallel to Planck scale to computer science. The question we will ask too late is "What message do we want to send to the universe so it doesn't think we're unprepared to learn more secrets of our existence?" Ponder this a while… are superior beings watching to make sure we can qualify for upgrading through the eternal and universal test called The Final Filter?"

There is a lot of talk about needing to "scale" in Blockchain. We want Blockchain to be infinitely scalable. But we demand it without studying the primary rule of reality in our dimension that defines this: the Planck Scale and all the Planck discoveries we can count on (that will come into play with Quantum Computing and the latter stages of A.I.) Why does the Planck Scale apply here? Because it gives us a test example across several different other human Sciences that is very important to remember. What is this, or, what measurements in what other Sciences, you ask? Planck Scale refers to magnitudes of space, time, energy, and other units, that are beyond or below which the predictions of the "Standard Model" including Quantum Field theory, and General Relativity. The next-level reality will be no longer reconcilable, and Quantum effects of gravity are expected to dominate.

In other words, this chapter is warning that improper Blockchain and A.I. Cloud storage can result in a new (Planck Variable) of gravity and other Causality will arise. This gravity draws Information that compounds beyond the structure of Cloud Storage as we know it. All we need to do is recognize this fact, work backwards in the Unitary Method through the Sciences in "Pillars of Encryption," and then build the Cloud to naturally expand in complexity according to the Fibonacci Ratio. Right?

When I list the universal Planck Constants you will recognize them instantly: the speed of light in a vacuum, gravitational constant G, reduced Planck Constant "h", the Coulomb constant, and the Boltzmann Constant. This works with Quantum Mechanics, electromagnetism, and thermodynamics. [Okay so you may not know of these Constants off the top of your head – it may not be so clear to you. Let's look at it this simplified way.]

Quantum mechanics, electromagnetism, and thermodynamics all play major roles in our evolution of technology with A.I. and Quantum Computing. You do not have to understand higher math to know that processing data creates heat, that magnetism plays a role in your codes, and Quantum opens the door to faster computing.

What you need to know in an unexplored level is that all the Planck Constants will add up to "1." If you read the chapter on Unitary Method, you will see why this is so important.

The problem we're trying to address and then evolve into Elegance and Eudaimonia is that Blockchain and AI as of now have almost no Constants being related to its evolution and future, yet we plan as if we can predict its Vectors after the "Big Bang of Artificial Intelligence" and later Quantum Computing godliness. AI in the hands of Wall Street and even Genesis Code alone cannot be valued. And we do not know – or rathe there is no - "Vector" of society. We only have Scalars of society, and this will get more turbulent once AI is applied.

Our agency within our Human Agreements are not ready for this or not evolved enough for the task before us, which is to measure AI so thoroughly from so many angles that we keep it from harming us. We might name a discovered version of a Planck Units within it as AI evolves or as Quantum enters the arena because it makes us feel as if we comprehend the incomprehensible. Or, AI might lead us to a discovery that is indeed a Plank Unit of intelligence (relative to the intelligence of the universe.) If so, can we trust the leaders in A.I. to tell us about it? Probably not.

Remember Kipling's
"Six Blind Men and an Elephant"
and you will get an idea of
how humanity will make deductions.

The world population needs to leapfrog over customary Q+A given us by "Academic Experts" and go fully into the vision seen in this book and in A. as it is investigated in "Pillars of Encryption". One way to leapfrog is through Art, another is to have a philosophy for living that pivots and accelerates to match the misuse of AI

Let's get to this key fact. Almost all metrics and confirmations in Computer Science and Cloud Storage must eventually deal with Planck Scale. The compression of pure Information on an individual through or by AI will warp one or more of the Constants, such as gravity. The other Sciences prove the Laws that will govern this event. Blockchain, A.I., V.R., and Data-ism itself due to its Intersections will create an opportunity for us to determine a new Unit, one that will challenge our control of the Planck Constants. We should have seen it coming but nobody did or will.

AI will create a "gravity" as "Smart Matter" influences "Dumb Matter" and the Voids of the cosmos and then surely all of our Clouds will be touched with the Lattice of Self Organization – all happening in Planck Constants first, then interacting in what will become the Planck Units of A.I. as it is compressed. Then we have to prepare to use these Planck Units of A.I. as it explodes (hopefully not in our hands.)

Of course, society cannot be reduced to a Planck Unit or constant. It is evolving. That is the point of this book and "Pillars of Encryption." We move constantly toward more elegance while we counterbalance chaos/entropy. If we cannot determine a Planck Unit, is all hope lost? No. It is found in the Fibonacci Ratio explained in the other book.

Somehow the Fibonacci Ratio can be applied with both Spirit and Scale to some specific human gift from our Creator and be used to evolve humanity in a way that keeps balance with AI (It played a role in keeping a quant open longer and we did not expect that, right, so there's a chance...)

Now, that is not to say that we can suddenly just create a new Planck Unit ideal just to make us happy. Instead, I am trying to open your mind to the idea that a new "Constant" will come out of the Fibonacci Ratio expansion in the AI era. If the universe notices our quest, then the more people knowingly searching for this Constant encourages the universe to reveal a hint.

This is where we have two options and we must honor them both. Paul S. Wesson wrote that in only the realms of Mathematics can Planck Units can be used as a convenient trick that saves labor brainstorming Mathematics, but it shouldn't be so reckless in any other science. And not without the consideration of other dimensions. Why? Because in reality, playing around with Planck Unit interaction or using it as a base of complex equations can lead to a loss of Information. Information never is destroyed – it simply transforms, effects, or slips into another Dimension.

My argument is that AI WILL create a Mass of algorithmic templates large enough to create a gravity so that all the Planck Scales can equal 1 as they do in our physical world. To think that this will not happen to me is absurd. AI will be many times this measurable Mass. We will be given multiple opportunities to notice ways to reach this end goal.

Shrinking the distance between AI data in the current Cloud storage system will affect gravity, thermodynamics, and will bend the Speed of Light as pure Information draws in Energy from Dark Energy that is used to turn Void into Lattice. The only way for it to do this is if it "Bends Space" as A.I. and its V.R. build (or that is) "rebuild" our past to engineer experiences on a human level. While this was all set in stone before AI Intersections, the idea of Time will alter, thereby changing the Speed of Light. This will not create a Black Hole but instead a White Hole. That is too complex to explain, but it will happen. Not due to our planning, but because Self Organization due to Dark Energy will use Planck Units to come up with the number 1 to keep balance with the cosmos.

Before I move on into some personal visualizations and uses of the word Planck in ways that infuriate real scientists, it should be stated that in Information terms, the Intersection of Blockchain, Apps, IoT, V.R., and AI should be seen as "The Big Bang of Artificial Intelligence" in which the total measures at the Planck Epoch were/are all equal to 1: 1 Planck Length 1 Planck-Time old, expanding. From there, it will accelerate, expand in a design according to the Fibonacci Ratio if we encourage that, and by using the Unitary Method we will discover the moving parts we can control. More likely we as a culture will notice the Big Bang of AI. Hopefully, we recognize it as a Janus Point and choose which side of the explosion to be on (or what direction to focus our expansion.)

This is all very cool and exciting. It puts the individual at the center of the sciences, and through interaction and the abstract "creating something out of nothing" it allows each to be the architect of envisioned personal history and future. You need a tool that can Scale and by good fortune I will reveal that tool in the future when the evolution is right.)

Now, back to some less complex science. Personal relationships will show how to borrow the idea of a Planck Constant or Unit to visualize the future of AI and how it relies on The Age of Imagination designing from Information. How it all comes back to “creating something out of nothing” and the Spark of Genius of Art. Remember, the universal essence is Will and its journey is toward the highest state of Elegance.

Even the Planck Unit and photon are comprised of Information helping us to see reality in its higher Elegance.

Planck Constants replace unknowns in equations because they are set. But now I veer back to a way to explain our likely future reality. A.I. doesn’t follow strict Laws of Nature. And each company Brand is different, as if existing in a new extension of our reality with our Laws. So, actually, AI couldn’t have real Planck Units or Constants. They should be instead named “Planck Variables” even if this term negates the concept of Planck Units. Why? Because it might help us fathom a new Planck Scale perspective.

This chapter would fall short of its potential if it did not list the examples of Planck Variables determined in the future. They come from the diagrams covering the other sciences: Physics, Astrophysics, Biology, Geology, Electronics, and Humanities. The cool thing is, they must all add up to “1” … isn’t that awesome to know?!!!!! By knowing our goal is to add up to “1” designers can work with the calculations in each of the other sciences to make sure a template is safe to let loose to “Scale.”

AI AND BLOCKCHAIN PLANCK AND QUANT UNITS

Begin with the principle we want all we innovate
to nurture humanity within our universe
so that we "do no harm."
Sounds nice but do we even have a choice in
This on a Planck Scale?
The singular Planck Unit or quant of the universe
may be referred to as Will.
Next comes raw Information, wherein Will drives it
until this becomes Intelligence.
Will drives the Fields within this Void toward greater elegance.
Elegance can be realized in a dance - and also in an algorithm.

Imagine this. A company wants to create a new user experience for a fan base for a band to increase profits by adding to the connection fans have with the band. The easiest science to start with is Geology. Begin with knowing that Smart Matter will influence Dumb Matter and as the Data-ism expands until it is all compressed for ever-increasing storage it will naturally form a Lattice pattern and fill in Voids with its Lattice.

Geology is "Elements of Information" reorganized due to pressure from external forces over Time. By knowing the Elements such as a person's history then likely the pressure and Time will next be known, creating a quasi Planck Unit.

Let's move on to Electronics just for the fun of it.

The "Source" is the Brand, such as a record label.

The Current is the band.

It releases albums at a certain Voltage, gets past Resistors like expenses and logistics, until it delivers the music or video the Source decided would be most appealing to the audience. Once this is figured out, Electronics has a Planck Unit for this in theory. Can it be used by the other sciences though? It surely is more responsible than only Mathematics.

Now in Physics, the designer must take into consideration the Natural Laws of our universe and Mathematics to store the AI Data so it doesn't morph. Basically, this means a company must design the next evolution of Cloud Storage just to be safe and to assign Planck Units that are truly infinitely scalable, which means they will follow the Fibonacci Ratio and may or may no be shaped like a stella octangula.

Astrophysics delivers the raw creative Information. This will be the great unknown that will gradually become a "known." In simpler language, people will learn where Information comes from, dreams come from, and where the spark of creativity resulting in Art comes from, and this breaks down our Human Agreements.

It is in the Social Sciences of Culture, observance of Nature, our relationship to our DNA, and epigenetic memory that we get to choose variable Planck Units. This sounds like it is contradictory. It is not. We all are human yet all have unique fingerprints. Quantifying our human Social Sciences will be exciting, and a "magic ingredient" in the equation.

Ultimately, we want to come to a Planck for AI that is secure, scalable, evolving, and observant of Culture and Nature. Just as we all have our unique email addresses now, we will fill out our profiles in avatars, with the Cloud company and Virtual Reality designer knowing the total of all sums will be "1". This allows for Self-Organization and Information evolution to happen without us monitoring every single interaction – which is as impossible as monitoring every interaction in Nature.

Luckily, the universe will see to it that we are protected like babes in the woods at first.

Tara (the scientist) makes her breakthroughs in bio-genetic medicine by imagining the human body as One Unit. Its own Planck in size, infinitely small as much as it is infinitely large. The lesson here is that in the Intersection we design the Planck Units first, we don't let the explosion happen without a sublime design that observes Nature's design, which is after all the source of all Planck Units.

It is useful to remember that music gave us clues as to where to find a few missing elements in the Periodic Table of Elements through the Law of Octaves.

By viewing each cell in the body and each nucleus in the cell that contains a whole string of DNA, as its own Planck, its own limitless resource of Information, halfway between the almost-nothingness of Matter, and the expanding universe, you can conceptualize the freedom that our minds offer out bodies.

There are a lot of dark/invisible objects between visible planets and stars, right? They call it Dark Matter. Then we discovered there is Dark Energy. I call it Ether.

Dimensions are using "Fields" joining forces within what can be perceived and also not perceived: Fields of Gravity, Energy, Mass, Measurement, Mathematics, Waves, Quantum, and … Fields of IMAGINATION.

Where does Imagination come from? Here are possible answers, yet they all are answers.

1. The Past. It is passed down to us, genetically (within our genes)
2. Spiritually (our ancestors place what they realize as other truths and futures into our heads)
3. Desperate survival (danger forces innovation, the most primitive source)
4. The Cosmos or giant landscapes that inspire Awe.

5. The Egyptians were given secrets that triggered massive global imagination into our Being, is another example.
6. Other Dimensions (cells from other dimensions merge with our Unit, carrying artistic talent and vision)
7. Ether -- (All Wisdom, Vision, Art, and Escape exists in the Ether, that we barely translate with our Brain Brain.)
8. The Consciousness of Humanity (each human's brain is a processor in the supercomputer of our existence since the Hominid, and as the speed and versatility of each drive increases, we both solve new problems such as how to heal ourselves, and absorb the truth locked in masses of brains creating "Chunks" of inherit knowledge.)
9. Other Artists (the Mimic Society, in which others' arts trigger a chain of events in the neurons in the brain exciting the zones and connections in our brains that suddenly move beyond limiting words to emotional (Heart Brain, Gut Brain) and bring into this dimension that reality.

There are several more sources, but you get the idea that Imagination is more than a lone brain coming up with a lone concept and sharing it via artforms or not.

"The Hundredth Monkey Principle" illustrates how #8 comes to reality. The clearest way to explain this is to go back to its origins. A group of anthropologists were on a group of very small islands near Indonesia. On one island, they taught monkeys to wash their potatoes before eating, and it would taste better and be easier to digest. It took two generations and 100 monkeys, then suddenly every monkey on that island knew to wash potatoes. But even more amazing, all the monkeys on islands they did not visit knew to wash potatoes. This is the clearest example of both Ether and the concept that all our brains are drives in a supercomputer sort of reality/universe of Information. This was negated by many scientific hypotheses until recently when use of Quantum proved that what is observed now can change the past.

Number 6? Imagine Star Trek, when they "Beam Down" to a planet. You see their cells turn into particles that do not obey the Laws of Physics. Did they turn into Planck Units that do not observe Time? When they are reconstructed, they merge with cells from the dimension they used as a medium, to travel unbound by Earthly Physics. In those particles is of course Information. That Information changes the Unit. It must. Then end up in a dimension merging again new Information.

One cool test of this, that some people claim to have achieved, is experienced through another Sense that tells the Truth clearer than the Brain Brain. Taste/Smell. Transfer a steak to another dimension and back, then try to eat it. The characteristics of all the cells in our Dimension share to create a likeable taste and smell. Add new ones or take some away, and while it passes the matter test and visual test, it does not pass the body system of Taste/Smell, the origin of one of our little-used Brains.

The greatest obstacle for the freedom of not just going through other Dimensions but bringing something back from these realities in the form of Information (even Imagination) is how desperately we hold on to this Dimension. I'm not talking about religious or cult mumbo-jumbo. It's a simple logic equation, that has so few variables at the basic equation level that it is simple to solve. Mental Chains, Emotional Chains, Survival Chains. Within these three categories you can place pretty much every reason you can think of: religion, love, cell decay, word limitation, etc.. All but one, that is.

There are some things that mankind simply is not wise to know.

Whether a Creator created us and our universes, or we created the Creator and universes – this warning holds merit.

Art traverses the dangers and brings us back something small, an abstract Planck Scale of connection to a greater Truth, given from a source so inconceivable until we evolve (I think through processes that increase us) that the energy to be unlocked from it must be served in Planck Units. A single painting like the Mona Lisa. The incredibly executed dance or gymnastics movements. Poetry that lasts eons and becomes passed down through Epigenetic Inheritance, as they have found is the case in both Shakespeare and Mozart. Movies from "E.T." to "The Matrix" to "Forrest Gump" to "Arrival" each with such useful messages now set in a form to reach billions at once.

Perhaps you've experienced it before – that surreal feeling leaving a museum of Art that displayed works that touched you. A levity in your Brain, perhaps?

Or have you experienced this when you witness something that inspires "Awe". A natural wonder like Half Dome or the great wonders of the world? Mount Blanc? Angel Falls. A tornado. That is how it feels to me when I witness Awe. And with that, the energizing from the excitation of the other Brains. Excited they think you're on to that place where you activate Them (the other Brains) for daily existence.

When I watch films often times a kernel of Truth breaks in it. "The Matrix," "Arrival, "Back to the Future" and other Sci-Fi films are common examples. (Almost all we enjoy being invented now was first invented by Sci-fi writers and artists.) But Truths can also come to me through separate Cinematic Vision Arts … through "Editing Vision Arts" … through "Music Vision Arts."

Now I'll take a break and ask you why I think that Plank deserves its own chapter in this book or vlog? Because Plank applies to several aspects or chapters. In Tara's Five Brains, we are ONE UNIT. In religion and humanism, you will read how mankind sees his body as "Indivisible." One undivided, not able to divide, soul or conscious. The Planck is a scientific measure. The Human Being seen as its Planck, its smallest unit, shows us that scale becomes irrelevant in data, science, conceptualizing, and Art.

We live in a "Mimic Society. Mimicking rather than originality is such a dominating factor over our progress as humans we don't realize it. I'll give some examples that relate to film making, the human mind, and our evolving senses.

Children suffering from Downs Syndrome do not hold their faces in place like other healthy kids or adults. They usually cannot close their mouths, hold their chin taught, focus with the same movements, and more. They lack the ability that all healthy children have, which is the ability to know when their face reflects their parents'. Oddly enough, Stem Cell replacement finds the faulty cells in the brain and replaces them if the treatment is done before age 5 (the younger the better.)

Let's look at this deeper. Cells "from another dimension outside their body" are beamed (injected) into the universe of their body, and merge/replace what is wrong with information held within a healthy Stem Cell, the origins of all cells, or, the origins of that other universe or dimension. The Stem Cell itself, in its DNA, has its own brain for its own unit and universe. Within a year the Downs Syndrome child usually holds his or her face like healthy people around.

Now I'll apply it to my art of filmmaking and Directing actors.

As we grow up, we watch the faces of our parents and friends and mimic them. When loved ones or friends frown telling a story, we don't realize it but we frown. For ages anthropologists said we are simply sympathizing. That our brain tells our face to frown. But now, they found out that the opposite (Derrida) happens. Our frown tells our brain to feel that emotion or fight that emotion. The Muscle Brain helped the Brain Brain, due to the prehistoric/primitive lesson that brain learned in the Survival stage (by smiling or frowning with a dominant male or group, this displays shared ideals and emotions without words, which allowed them to form groups better to hunt, defend, and multiply, thereby making it a dominant trait.)

For the first time, the use of injections of Botox or filler to lips and foreheads, as well as nose jobs, face lifts, and even colored contact lenses, is commonplace enough to do reaction tests. A thousand viewers are told to watch an emotional scene (or even funny scene) with actors that have obvious facial work that changes how their face would normally move while expressing that emotion. The results were overwhelming. Actors with unnaturally moving (or not moving) faces failed to reach viewers as well as Actors with natural features. The Brain Brain knows this is not "human." The other Brains hold back attachment or make their own similar deduction. All this simply because the Mimic characteristic of human society was excluded.

What is Mimic? It is simply Information. Information displayed through an art that stretches the existence of the Artist (person or Actor.) Each day we perform the art of our interaction with people. Mimic gives clues and lessons.

My own personal experiences escaping the chains of words and our concepts of reality in this Universe hit me with examples of all of the concepts listed above. That is why I wrote them, with as few words as possible, to inspire more visceral experiments and experiences. Again, the hardest part of this is the limitation of words. And the survival fight or flight comes into play when deconstructing Time and Space. Fight might be the Derrida opposite of what I should feel, because my greatest moments felt more like a release of fear, with the simple word "expand" penetrating my body.

My "A Human as a One Unit" refers to the human body and all each individual has as One Unit. I don't see us as a whole bunch of parts ruled by the Brain or Heart, as most prose and examples limit us to. The over-reliance on the Brain Brain to interpret all the Information can only hold back the powers of the other Brains.

Look at a cat, the symbol of dual dimensions for Egyptians. The cat seems to me to use its Intuition, Heart, and Muscle brains more than its Brain Brain. Relying on these helps it always land on its feet – does it also open the gateway to living nine lives? Personally, I think that cats can see more than one dimension, and see vibrations in visible waves of energy, explaining why they can get spooked when controlled experiments prevent them from seeing or smelling a dangerous predator, yet they react with escape movements as much as fight movements. Escape to where? Maybe the cat is on to something. Which makes me wonder … why am I allergic to cats?

WHAT IS "NOW"?

"Lucy" and "Arrival" present the
riddle that urges us to watch
for clues to escape the limitations of "Now."
In both movies the hero's intelligence
broke out of its shackles and her accelerated
beyond accepted understandings.
This is Quantum Tunneling:
a cosmic patterning that
delivers a Mind to a clear
perceived outcome,
not as a hope but as an
already established reality.

Whether it is our Creator's design for us or our own ego we survive imagining ourselves in the center of the universe we navigate, living in a moment we call "Now".

Like the fish in water that is unaware it is using water to propel forward, we use the Now to propel into what we call the future.

We believe the past happened because of expansive agreements: between ourselves and Nature and universal forces. And between those that know us, and between our memory and things in books that confirm it all.

Before I delve into this any deeper, it is important to mention two extremely weighty aspects of "Now." The first thing to mention is that the concept of Time and Now breaks down in the Quantum Realms of our reality. Second is that in a purely Physics or Metaphysics sense, accumulation of new Information is what happens when Now becomes the Past. Our universe is in a disordered state, moving toward entropy for the most part, and this means there is less Information about the Future.

Every moment we meet the Future and it becomes Now we have more Information. This is both measurable and part of Ontological Truths. When we remember we are organizing Information in a way to create a memory from it. When we take gas and turn it into a solid we change randomness with Information organization.

We also need to deal with Maxwell's Demon. Déjà vu is simply a clue given to us outside of the Arrow of Time. All this results in a transfer of Energy or turning Energy into something else because Energy is never destroyed.

We're told to be "more present" and "enjoy the moment." I've found the best way to do that is remember the following concept: we do not experience Time or life, Time and life experiences us.

To fully even begin to grasp how awesome this is required my study of Dante's Inferno, the Hermetic wisdoms, the Emerald Tablets, the Tao, the I Ching, and Carl Jung's final works. Now is far more complicated while also far simpler than we imagine.

So many Eastern religions, gurus, and (supposed) enlightened ones instruct us to "Live in the Present … Be Present" (Live in the "Now.")

As mentioned before in dimensions, Time gives legitimacy to Matter. What does that really mean? Perhaps now is a good time for the quote that was altered for the film "The Prestige." An original quote stated that a man's reach should exceed his grasp.

The Tesla character of "The Prestige" stated to the magician lead Hugh Jackman, "Man's reach exceeds his grasp? It's a lie. Man's grasp exceeds his nerve." And… "The only limits on scientific progress are those imposed by society. The first time I changed the world, I was hailed as a visionary. The second time I was asked politely to retire."

His work was destroyed upon his first breakthrough, and he was killed before making his third breakthrough. With this in mind, we accept that society only wants us to realize a limited aspect of our reality. Are we ready to learn what "Now" truly is in a cosmic and spiritual sense?

I understand their principles, which many do so I won't get into that quagmire. Instead, I challenge the inadequacy of words to fathom "Now." Much less teach it.

The very words needed to share information between all the branches of sciences, primarily Physics, discount both physical experience in shifting "Now" … and the fact we interpret out reality through other "Brains" that have the ability to transcend Time. Love transcends Time.

The ferocity with which we hold on to Matter (our bodies and our understanding of the Earth and nearby planets that effect our lives in ways we can appreciate) prevents our other Brains from experiencing the solution(s.)

Approaching it Derrida-style offers solutions that were proven on a physical level with horrific effects in Nikola Tesla's Philadelphia Experiment, using the U.S.S. Eldridge. Hundreds of sailors on a Navy Ship were instructed to partake in an experiment and return with notes. Surrounding the ship were huge generators made according to designs by Tesla, their purpose to (as the Navy defends) make ships invisible to radar. What happened was that this huge ship disappeared completely. It reappeared hours later in a shipyard in Norfolk, Virginia. The entire area and all witnesses were debriefed, forced to not believe their memories or senses. The U.S. government made it Top Secret and refused every report or fact until the story got left behind. It has been shown in films and tv shows.

There was so much effort to debunk this and cover it up, Derrida tells me that it happened. What did it test?

The impermanence of Matter when Time gets manipulated or escapes our Arrow of Time concept of it. Literally, it refuted the principles of the atomic bomb and much of Einstein's work. That Tesla protested Einstein's Theories and tried to challenge him in debate with proof that would derail the atomic bomb as the world's most dangerous weapon and atomic mass as the greatest source of energy.

Many people think he was killed, easy to accept when it's proven the government rushed in and confiscated all his work, made it Top Secret, and then worked on it. Then sent him off to die without attention.

Tesla proved that our concept of Now is an illusion, an elaborate "agreement" between us to simplify our existence to a truly stupid level. But that is changing. Why?

The pace of discovery is uniting with the practice of exercising our other Brains, accelerating invention that dumbs down the tools of visceral experience to where it can be taught.

We agree that the past can be proven with memories and friends, and accept that we cannot see the future, and believe we exist in the Now at the very nexus of past and future. Yet the discoveries about both Space and Time, defining Now, tear all our theories of reality apart as much as they prove them solid.

This is especially true when new measurements of atomic particles proves that they come and go constantly. Go where? To what Time? If the particles simply move to another dimension, then it does not affect Time necessarily. But if it moves, then relative to our understanding of Time, the seconds it is gone could be years in other dimensions or our own.

We count on memories to prove Time exists with a past and present. Yet all the atoms in our bodies have died and been replaced when we remember something that happened twenty years ago. Except cells in our brains. The withering of the brain distorts memories. Yet memory itself is so unreliable.

We remember our childhood much better than the events filling the decades after our mid-teens. During this phase, the events that make up memory expand so fast due to, as children, our school and hobbies expand so much, that there is a bottleneck in the process of memory. Does that make this time less "real?" I argue that it is because brains in the 12-20 range apply so much of the body existence related to the Future that our brain also decides the Future is more important to store than the Now.

Yet what memories stick throughout all times of our lives?

Physical ones. Intense ones. Stored in other Brains. Falling in love. Your first child. Tragedies and accidents. Losing loved ones. High-Qualia moments.

They discovered that girls start puberty younger and younger, and that different races or areas of the world effect it in girls. While there might be many reasons, one reason stood out to me. They discovered that in households where the biological Father left, the girls started puberty much earlier. Their systems could detect the pheromones of the Father. When it was not around to sense, this meant that the girl was either not protected by him and should get down to mating, or that his dominating pheromone overwhelmed the mating data in the pheromones of other men around her. In either case, their bodies got ready to reproduce.

So I ask, what brain in the body deciphered this? Not the Brain Brain because if it did, we would be able to write and talk with definite conclusions. We first had to be able to measure pheromones before we could find this answer. And these Brains, like the jellyfish and virus brain, measure Time not in years or in memory stored in this lifetime, but according to clocks measuring Time on totally different standards. In genetic and celestial standards. The quartz measures time due to its reliable

Vibration as an element of the Earth in the galactic energy fields. It has no past or present and is unaware of the Now. Does it have a soul? We say no, but some scientists argue that all objects, including minerals, have something akin to a soul.

Why is this so important?

Computers, and their evolution relative to two things: our mind's ability to expand reality, and society's ability to change social rules that have dictated our actions for thousands of years.

If Art presents Truth, then Art plays a role in deciphering Time, especially Now, because art only "happens" in the Now, in the dance from vision and emotion to application of the art process.

And it becomes of paramount importance if we allow ourselves to base progress in solving this riddle on Information being the fundamental element of the Universe. Don't stop there.

Will is the essence of all things in our universe even before raw Information, the essence of all things whether Dark or Light Matter, Dark or Light Energy, a Field, an atom, a sun, or weather.

Information and surely Will transcend Time and does not cease to exist in past or Now. If viewed the right way, our actions Now can change both past and present.

But words are not the medium to solve the riddle. Far too much information is shared in the universe without words. With Fields and forces. Just look at the forces of Nature and the methods of sharing information must be in the trillions!

How could we be so egotistical and naïve to think we should use words to define and solve Time, Now, and Space?

Art evolved right when mankind became aware of its mortality – that it cannot escape death. Was Art a byproduct to bridge immortality?

While it seems like I'm all over the place describing Now, it all centers around the same principles.

1. That words are inadequate in explaining complex reality
2. There are several brains in us, we just over-rely on the Brain in our skulls
3. Information, not matter, is the fundamental element of the universe
4. Loosen your desperate grip on what we think of as reality and the physics of the Universe
5. Believe in Ether
6. That parts of us come and go constantly to other dimensions
7. That other things like animals and virus use different brains to relate and survive
8. That some mysteries can only be solved by experiencing the moment in a viscerally overwhelming level
9. That Art reflects the Truth that the other Brains know, and is a window to the Truth and our evolution over a false measurement we call Time
10. First came the Fields, the engineers of our Reality

The concept of Now is so extremely complex that it evades the brightest minds in Physics.

Or … does it?

Does it escape the words used by these men and women?

Does it escape the relationship they have with their ego, which demands they relate Now to themselves in this dimension?

Using the Derrida Method, accept that Now is incredibly simple. That we are instead chained by the old and limited methods approved by teachers we use to communicate "Now" concepts makes them almost impossible to fathom and this is what the status quo prefers for us.

Let go.

Remember the one word that penetrated my body and mind when I had my most extreme experience shifting and touching another dimension.

"Expand."

Everything after that had to be interpreted physically and understood by my other Brains. The learning by my Muscle/Body Brain was undeniable. I feel the aftereffects many times each day.

Intuitive Brain confirms that was a real experience. And who's to argue with that when words used to argue have no usage among the other Brains?

If our universe exists only in our minds, or better yet that our collective minds created the universe, and the primary function of the brain relies on Information, and Information is the fundamental element of the Universe, not bound by Time or Space/Matter but instead is what it is made of in both real and concept levels …

Then what is the purpose of Art?

It presents information for all our Brains to assimilate, helping perhaps the weakest of our Brains, the Brain Brain, learn beyond words.

Does this have an end game? By that I mean, can I deliver something that proves this or evolves toward breaking free with the proof experience we experiment?

Yes. So then … How do I describe it enough if words are inadequate?

Imagine … Imagine if I.Q. expansion (beyond what you live like at this Now) suddenly came to as a mental Cirque d' Soliel performance shared instantaneously with a billion people around the world by riding the crest of an ever-evolving technological tide pool. And the kicker is, you "live it" instead of learn it.

If you really want to blow your mind, read the massive book "The Matter With Things" by MacGilchrist (followed by Proust's "In Search of Lost Time."). This door to multiple dimensions cannot be unlocked with words. Art presents the medium. Tech floods us with the essential element, Information. Now happens past, present, and future, no longer bound by our shallow concept of Space/Matter (Dark or not Dark.)

Hint: Our brain is but a neuron in the brain of the galaxy, itself an organ in the body of the Universe, and suddenly this isolated neuron connects through the "Information Feedback Loop" in the Multiverse so our dualities share what they know without words.

And it will all be experienced through Art – Entertainment Media requiring a balance between Eastern wisdom based much without words, and the word-laden Western wisdom. This is why it is so crucial to me to balance Asian cultures and Hollywood.

It shall be as difficult as it is effortless, a true union of Eastern and Western, and ancient and modern philosophies … that can pivot and accelerate yet still be tied to what matters most to us – the greater gifts given to man. A soul. Indivisible and conscious. Creative in the realms of various arts. Curious. Needing to connect with others, to share the experiences. Memories holding our concept of Time in place.

What Don't We Know?

Vision without execution is just hallucination.

Even the most illiterate and limited person revolves his life around the profound idea that he or she knows life around them enough to make predictions and plans, even if the plan is to not change or alter course.

In the case of Event Horizon Film Studio the execution is in the films and series with all their Games and Apps that give it Spirit and Scale. But to create what I understand as a dominant environment means correlating the strengths in all these books into one "Dynamic." A Dynamic happens when all the "Systems" involved in an industry reform and function on a superior level. This results in either wiping out the inferior past business model or they run parallel to each other and feed off different energy. It's becomes the ultimate disruption example. More on that will be explained in other books on the Science and global politics.

My Dynamic allows the vision to be executed.

The Royal Astronomer Martin Rees loved to point out, "A chimpanzee can't understand quantum mechanics." His point was not that they (or commoners) are too stupid, it was that the chimp is unaware quantum exists. Then Rees applies that to our ideas and even laws of Physics and the cosmos.

We glorify ourselves with this idea we know so much now, we have so much figured out. Though, what in reality remains completely unknown, unnoticed, and unused? This goes back to my thoughts on Dimensions, Time, and Space. The greatest obstacle to discovery was never ignorance, it is the illusion of knowledge. To think that our self-centered existence in our fragile bodies with such limited senses is capable of deciphering or even noticing all that is Truth in our universe seems absurd.

Our Brain Brain extrapolates on the Fields and Laws using Mathematics, formulas, and observation. Our Intuition Brain tells us some more. Yet our entire existence comes from a brain giving orders for simple survival for the last 500,000 years – it was not meant to be used for all decoding. Like the fish, we are unaware of the water we swim in.

Art gives us an abstract way of pursuing Truth. If science and math have limits, abstract discovery and Art is a door through that wall.

When people read something I wrote and ask, "Wow, where do your ideas come from?" The obvious answer is, "I think of them." But do I? Put into question Free Will. Question whether Art is something exclusively created by our Brain Brain. Then question whether the Feedback Loop and the "Technology of God" "inserts" ideas and visions.

Too many of my more brilliant phases of creating "something out of nothing" went so far beyond what my Brain Brain could assemble that I believe our Being becomes a lightning rod of Truth and visions from other Dimensions, or from things in our Dimension we cannot observe. For those that believe in the E-8 model of Dimensions, points of Will and knowledge became points that touched my senses. My entire Apps Project came to me in a flow like water that resulted 3 weeks of furious writing energy I had never felt before. And when I reviewed it, I had no idea how I knew some factors. They just all ended up working out.

We all believe in Love, but it is not a mist that connects people. It is an invisible force that carries with it powers beyond our building it from scratch. I don't think anyone said two thousand years ago, "Hey, I'm going to keep advancing this idea of Love and see how far I can push it." Just as amazing is the vision and execution of all the arts. While logic can tell us that the dancer or pianist composer uses fundamentals (usually) from their studies, many greats in dance never studied for the specific dance that made them famous. The "Original Content" conjured within their mind and soul drove their bodies to move upon Will.

This gives me hope that as we accelerate our communication to the point that words slow us down we seek Truth and lessons from Art – the process as much as the end products.

CORRUPTION

Is it more than a social deviation?
Is it in our DNA?
If so it must be recognized in
the pinball game of our minds
as a pinball alley that cheaters
use to steal points.

Does corruption play a role in art?

Observing history and extending it to people around me, the conclusion that "corruption" must be a fundamental factor in the survival and success of the human race rose suddenly. I'm not talking about the corruption we focus on most in the news. The bribed politician, the corrupt drug enforcement team, the Ponzi scheme surfacing, and even the fake news of today reveal mostly just what is trending. Though these examples may result in the deaths of many people, still this is not as significant as the positive effects of corruption. This sounds impossible – to make the jump corruption being a benefit. Let me explain.

Go all the way back to the nomads, 500,000, after the Hominid. Our bodies have many unique qualities among mammals. Two of the most significant revolve around nomadic survival.

First thing is, we are one of the few true omnivores on Earth. The pig, shark, and bottom-feeding fish are the three species that come to mind – animals able to eat a wide variety of foods and process them for nourishment. Most animals are either herbivores or carnivores. Some carnivores like big cats can eat both meat and vegetables, but the cat systems are greatly in favor of mostly meat. Through adaptation they expand, but before that, their body chemistry/biology had to be set up to digest many different foods. Indeed, some foods like chocolate are so hard to digest that remote villages that never ate it can die because the stomach cannot break it down and it clogs the intestines. In the 1960's, a food-relief program for Bangladesh delivered thousands of candy bars. A great percentage of people died after eating them.

But Man needed to start as an omnivore and then evolve even stronger ability to eat new foods in order to survive. As man walked from Africa or China to Scandinavia, or took ships to islands, they had to test out what would sustain them.

The second unique attribute our bodies evolved centers on the adrenal glands. As nomads, we might score an abundance of food and stay. That is why so many cities started around water. We could always catch fish to eat. And trade for distant foods.

But the human body explored so far from its original habitats. Our ability to cloth ourselves with the skins of animals, then later make fabric, allowed us to explore new cold worlds. The invention of the scratch plow allowed one man to make more than one man or his family needed to eat, allowing the man that could not farm but was better at making sandals to barter a trade. This established cities.

But the nomads. Imagine. They would walk days, weeks even, between good food sources. Consequently, our bodies needed two ways to produce the energy for the long walks: food, and adrenalin.

The same source of energy that was in our DNA – the energy for fight or flight – produced in our Adrenal Glands, given signals from our hormone messages, allowed us to nomad. But instead of releasing all the adrenalin at once, the glands would begin dripping it into the blood stream about 2 hours before dawn. So that humans would wake up with the energy to forage and explore. The human body can live for days on this, though it gets weaker and more prone to sickness.

The subconscious right side of the brain was always acutely aware of the balance it must navigate… like it constantly tightropes. The balance between our need for food and our inbred desire to explore.

Imagine that you're in a clan of nomads. And upon reaching a field, the bright red or blue berries drew our attention, and with that, hopes that it would be nourishing and digestible. The smart nomad that went on to birth a long blood line would spot the berry but probably not be the first one to try it. Instead, he would "corrupt" the safety system of another nomad, convincing him or her to try the berry. Or mushroom. Or puffer fish. Oleanders, poinsettia, and other tempting foliage. Once the new food was tasted, and the taster did not die, then the more dominant nomads would also eat. By law of attrition, this had to be either accidental, or as I think, planned.

Ancient cave drawings, painting, and messages prove this, as they would illustrate to following nomads that the red berry was good to eat, the blue seeds were poison.

Indeed, royalty had "tasters" especially when accepting new food from foreign lands.

The concept of corrupting others to either survive or prosper led to legends. Men and women, both bad and good, corrupted others for power and longevity. Bathsheba. Cleopatra. They corrupted men with their beauty in exchange for power. Not just sex was in play, but the superior minds of both of these women realized that sexuality was a great trade for power.

When you advance that to the modern age, where corruption lies behind all major Industries and commodities … "Corruption Pays."

Or does it?

A great study performed in the 1990's compared the Catholic nations with Christian and Jewish nations. The defining difference was corruption. Catholic nations that have leaders and people that follow the doctrine of the Catholic Church never developed the global power or success that nations following basic Christianity reached, with only a few short lived exceptions. The reason why lies in how it treats sin, and causing others to sin for you. For Catholics, you can do the most horrible crime, yet repent, donate to the church, and be spared hell and gain entry into heaven. This allowed the corruption in Italy by the Mafia. And the corruption in South America from both the Mafias and drug dealers. This occurs in all the major Catholic nations today: Mexico, Brazil, Italy, Spain, and more.

In contrast, Christianity separate from Catholicism does not use Confession to absolve sin. You gain entry into heaven by living a Christian life of hard work, fairness, generosity, compassion, and honesty. Failing this in life dooms you to hell. Therefore, they tolerated less corruption. Without the Roman Catholic Church benefiting greatly by its coercion there was no unified driving force making corruption acceptable.

The examples of corruption propagated by the Church are endless, but I will focus on just two examples. Coffee and cocaine.

When coffee was first discovered and quickly spread through the developing world, the effects of caffeine were so extreme that the Church leaders were challenged. This drink made people feel better and more productive. This was not good if you are the Church and want total control over peoples' sources of joy and happiness. So they outlawed coffee, as did the Muslims, for the same reason. They issued a Decree damning those that use coffee and damning coffee. They said anything this exciting has to be the work of the Devil, and they spread rumors about how coffee causes giant cysts, disease, and ugliness, as well as making women barren. But, productivity fell dramatically, reducing the money – tithing and taxes – given to the Church and Muslim leaders in the Middle East. Within two years had to reverse their Decree! How obvious it is to want to corrupt people in mass for power?

As for cocaine? Hah. This one is funny. Cocaine originally was used as a miracle drug "cocaina" in South America. There are 14 alkaloids in the coca leaf, the weakest of them is the cocaine we enjoy. It healed, numbed, gave energy, and balanced other toxins. But C17-H21-NO4 (cocaine) had one property the other 13 did not have – it could be synthesized so it would last. Combining a coca leaf in your mouth with ash from the fire gave you added energy if you sucked on the leaf as you walked the jungles like nomads in search of food. When the traits of cocaine reached the Church, they started through missionaries distributing cocaine to African colonies and Europe. The benefit was obvious.

People would work longer hours and be happy. So what if it addicted people? (Though cocaine is not addicting in the same ways that opium is – opium was outlawed because it made people work less, not for its health reasons at first.) But with cocaine use came sins, mostly sex and violence. This was still seen as good if all the people using this are African gold and diamond miners, given a bag of cocaine, a jug of water, and some biscuits to last an entire day in the mines of intolerable heat and foul air. Just keep those miners blissed out! The life span of these minors was only 20-25 years old mostly because their hearts raced constantly.

Then based on half awareness that cocaine increased gold production the Pope issued a Decree that cocaine was a sin, and like coffee, a gift from the Devil. But wait! Gold and diamonds mined in slave labor for the Church dropped insanely. And the miners they addicted to cocaine rioted which slowed down delivery of treasures even more. In a clear case of international corruption they announced a reversal of this rule against cociane. This was way back in the 1400-1800's. They never explained the addiction and connection to early death for miners. The ends justified the means.

Then late in 1800, cocaine appeared again, in the United States. After slavery, African former slaves wouldn't work hard. But give them a bag of cocaine that only cost a dollar for a few days of work and they worked happily and still had energy for sex and socializing. But then came the Reconstruction of the South, and new generation of Congressmen.

One Congressman focused on a few isolated cases of a white woman being raped by a former slave. He won his election on the motto, "I will end the terror of the coke-crazed n***er." This led to it being dishonestly included in the Harrison Act that formally made any heroin derivative illegal. But money was lost, so the same corrupt lawmakers established the first covert drug dealing channels we fight today. Corruption pays. Imagine that!?

In Mexico, drug dealers go to mostly Catholic Churches – no problem getting into heaven despite murdering thousands directly and millions indirectly if enough cartel money is donated to the Church.

So where does this come into play for art?

The Renaissance Era marks the major move by Artists to depict the world as we perceive it. Before that time, artists were commanded they could not paint a drawing in 3-D, as we see it, it is only for God to create such perfection. But, the global power that these emerging artists gave anyone benefiting them (the Medicis and Rothschilds) was coveted. So the Church instead took over and corrupted the Artists to glorify their religious leaders and symbols.

Even with an Artist as great as Leonardo Di Vinci, he was banned and assassinations were set up for him, all because he took his freedom beyond the power of The Church. The artists' survival gave confidence and a lot of competition for centuries. The Renaissance was inspiring to Artists around the world.

The Church had to corrupt Artists to control the masses that admired them more than God. It was not until Savonarola that deeper corruption was exposed. In the heights of the rise of the Church it was Salvatore that convinced the rich that if they craved independent sources of pleasure, even Art, they would be damned to the fire and brimstone of hell.

In China, only their cultural concept Great Creator could make a perfect work of art, so all art made my humans had to have a flaw in it – ergo the famed flawed Chinese ceramics and fine art. The rulers had to corrupt the talent of the Artists.

The Age of Enlightenment freed the minds, but rulers corrupted as much as they needed to keep control. Any Artist that delivered any Truth in his work that was contrary to the ruler was killed. This led to a sense of defiant pride in Artists.

In the modern world of filmmaking, film was used as propaganda to corrupt the minds of all citizens, and especially children. "Potemkin," and many other works are examples. And just as often, works like "The Little Dictator" by Chaplin caused Hitler more anger than his military losses.

In the United States, society by and large didn't like to embrace "dark" subjects like violence, sex, or the vices. It was a natural repulsion amidst such a long era of prosperity and cultural pride.

But then came the 60's, and factions contrary to the government found social arts, especially music, as a motivator for anarchy and "more freedom." While people did get to make and enjoy wilder art, the freedom was a two-edged sword. The "freedom" presented caused disease, violence, formation of corrupt gangs, and broke apart families with the addiction to drugs and vices glorified in songs. This too through has to be seen as both good and bad. As Don Juan said, sometimes a drug is needed to break a person free of ritual habits to make way for mental freedom.

The music of the 90's until today glorified gangster life. And death, murder, riots, protesting, cop killing, rape, and, most of all … a sense of lack of responsibility for ones actions or the state of an entire group (teenagers, ravers, blacks, anarchists, Goths, Satanists, and all the rest.) Rap corrupted young minorities.

Now why would a government allow such a thing, if indeed they have the power to control Art? Because it served two purposes: control and distraction … from the real corruption crimes of the government leaders. Keep the people high, they forget to look past the rhetoric and lies. Corruption allowed the formation of a far more controlling New World Order. While I don't dwell on the NOW, it is there, in many forms.

It allows new Art, but cripples new artists and fans of art.

China is in a unique situation. The leadership knows the history of both banning some forms of art and of thinking, as well as the allowing of new Art and ideas. Most of the time, the people consume too much and lose their moral compass. President Xi must walk a balancing act between Artistic expression in movies, games, literature, and comics, with the fear of how it corrupts, to the point of chaos and lost generations of addicted youths. The corruption of the English in Shanghai led to the Opium Wars for good reason!

Is a larger force of corruption occurring now in the Art realm of film, media, television, and games? Yes, very much so, but it does not come from the higher up source of power, such as one person, despot, religion, or government (although politics plays a big role in this.)

Corruption of artistic freedom is a tyranny against Truth. I'll point to popular examples of this to prove it, or rather, to unveil it for what it is and what is truly going on here in Hollywood.

Michel Richards (of Seinfeld) was doing standup in a comedy club about 2005. It was the night for raunchy, in your face comedy. Other comedians used the N-word countless times, yet they were black. Richards got on stage, and was heckled by someone in the audience, and not just once, but a few times, and in bad heckler taste (if there is such a thing.) Richards had two choices: ignore it and fail or confront it and maybe fail but maybe win the crowd. So he began dishing back to the insulting heckler, who was black.

After enough trades, Richards made references to apes, the black man pushed it further, and suddenly Richards said the key word that you can't say when you're white – the N-word. The black man and others made a huge publicity event around this and ended the career of Richards. This is tyranny and the start of Cancel Culture. This is witch burning. Did Richards do something seriously wrong? Yes, of course he did. Was he pushed until he gave this fan what he hoped to get? Yes, according to witnesses. But we simply cannot defend him as a comedian that made a mistake.

Instead, the power of the mob demands is unforgiving. If not, we also must be racist. We corrupt our own judgements and sacrifice Richards. This paragraph does not justify racism - it highlights the ease at which we point a finger at one guilty person and ignore millions of people guilty of worse crimes like pedophilia and Polanski. The difference between these two men is Polanski bragged about actions of serious sexual abuse while Richards apologized but never was forgiven.

Suppression of Truth and Artistic Truth. Dozens of other comedians found they could not speak their comedy if they are white. Blacks corrupted the field of language, from which comedians for centuries were allowed to pull material from, and blacks still can pull from any time. The N-word in all the rap music, in daily language, was a corruption of the codes of decency, that the blacks argued was their right to throw around and also into the face of white people, anywhere, anytime.

Richards was the headliner that night, meaning, the black comedians only got their chance at the big time because of Richards. Deepening the betrayal. Almost no comedian or friend came to Richards' rescue. In fact, dozens of comedians gradually announced they are quitting standup. But a few did explain the fact that, in today's trend of demonizing anything said, if it does not tow the line of radicals, they are afraid of the Richards Effect. Seinfeld himself said he won't open tour in standup.

There is a reason why Hollywood emerged in the United States. It as the very fact of freedom of speech, and the freedom to own what you create. There was not just breakthrough art, there was a "grace" among performers and creators.

Grace is the act of allowing another person to grab for more than they deserve, and to do it so it does not merit you or belittle the up-and-comer. It is what made movies work. Grace between stars. That grace has been killed off by the P.C. culture first, then finished off by the micro-aggression trend now.

The Bill of Rights and Constitution were created to prevent the tyranny of the Big (government or royalty or powerful financial people) over the Small (the person, artist, or comedian.) Lenny Bruce fought the ultimate fights against the government when they tried to prevent him from telling the Truth.

What the press and activists are doing now is a disgrace. It is turning the idea of the Constitution upside down. Now, there is tyranny of the Small (small fractions of society) over the Big (the majority of people and their culture or beliefs) and the government (who twist it around to get elected and then will turn on the activists when they are no longer needed.)

The question that is prevented from being talked about by the activists seeking power and the politicians helping them is that the world they create would never allow the freedoms the small groups now have. Proof of this is in every moment in history when small groups demanded censorship. There are no exceptions for many reasons.

Politicians, race baiters, micro aggression groups are being used as fools, corrupted by Politicians they serve by promises of power. I don't think most activists are bad, they just cannot see the obvious, they are too close to it.

The corruption of the art of film, performance art, comedy, and music brings an end to the freedom of expression so elementary to creating a film or other media project. Yet like most corruption, the power it gives the tyrants is short-lived and rootless. It will backfire on them and already has.

The "Corruption of Hollywood" extends into the betrayal against actors in how the studios mainly focus on the tent-pole films that do not require such levels of acting, and in the bad deals they dealt

China in the first round of expansion. This will change. If we are lucky, so will the tyranny of the micro-aggression censors.

To make sure that the politics of creation in a tyrannical use gets the most power, the activists are exporting micro-aggressions and unrealistic demands to Asia, Australia, and Europe. Some argue the strange social evolution of Europe is the origin of the leverage against artists and free thought.

What they do not realize is the precedent they use has to be "allowed" by the people high up, who are using them. Why would they use micro-aggression activists? One, to distract from serious issues. Two, to make the hardest groups to totally control (white men and women with education with or without powerful connections) easily intimidated and questioning their Intuition for what is right, and what is freedom.

Ultimately, it comes down to control. The first stage is the fear when it rises up. The goal is to condition citizens to accept less freedom, be it through corruption, art, or fear of some disease. I saw first-hand how fear of bird flu allowed rules no one questioned.

The laws and social rules will eventually be turned against them. This is how tyranny evolves – seen as a benefit to the humble, the poor, the lower classes. But once the power of tyranny above them eclipses the tyranny of the corruptors over comedians and writers, then the true goal will be shown. Which is extreme tyranny over the very people that delivered the micro-aggression activism.

This starts by creating a splinter group, and that group has to splinter.

So you see the Corruption aspect of humanity hits the Arts very hard.

"Victim Mentality" - like Failure - is not durable. The idea that Artists should champion victim identity as a winning or even fair strategy goes against everything Artists built in social currency and relevance since before Christ.

Artists must be brave, they must identify with themselves first and recognize the culture they grew up in as a foundation. Trends come and go. What words and images artists are "socially-allowed" to use changes every few years, as it runs out of "Trend Power" or "Social Steam." But this recent tyranny waged by politically-motivated and highly funded splinter groups is censorship on the highest level, a level that Stalin would have envied.

It corrupts the idea of personal identity and cultural foundation for all Artists, as they champion the idea that for a citizen to be a victim is a sublime goal.

There is no animal or plant that thrived playing the victim. No society group that used censorship that was not a puppet for a global elite demanding total loss of self-confidence and identity.

Now in filmmaking, I try to only chose films that are not corrupting. But what is corrupting? Corrupting can also send great positive messages and glimpses of the future. "The Matrix" is a great example, as is "Arrival." Confronted with mystery, the imaginative part of the human mind melds with the hopeful part and gives a simple person the traits of a hero.

For me, in fact, I turned down many jobs that wanted me to be part of the opposite results of free art: chaos, anger, prejudice, hatred. It is easier to go with the corrupting flow and say, "Everybody's doing it." But that is not my way. Hollywood and the Music Industry get very vocal against anyone that speaks out saying that the freedom of Rap music led to the glorifying of gangsters that spread violence and made urban black hoods dying grounds and fills the prisons.

They refuse to tell the truth, that the greed of the music makers gave the financial and political means for great wealth and power, but the urban minorities paid the price, delivering music that might be more "real" to them, but results in the decay of the family unit.

This is how Corruption influences The Arts now.

BALANCE

Ahhh, so now we counterbalance Corruption?

While "Will" is the "essence" of our reality, Information proves to be the fundamental usable element of the universes and dimensions. As physics events occur and life evolves, the Information manifests from these moments comes into creation, gets extinguished, or maintains over Time.

Focus on the last example: Information that maintains over Time. Earthly animals, plants and minerals. Fields and Laws of Physics in specific Dimensions. Love and its affairs. Creation of Art. The processes of the Five Brains. Expansion of the Universe and the cosmos within it. The gifts of perception and transfer of knowledge. Emotional bonding with what motivates you.

They all share one thing in common. Balance. Nothing last very long out of balance. Like a whirling dervish, unstoppable until it slows down and loses balance, the gyroscope of who we think we are revolves fastest when in balance with Nature, people around us, and the Truth in who we are and what we create. A balanced mother or parents have far more chances of raising the child they created than parents that are out of balance. All we build, all our immortality projects and our businesses, seek balance to survive.

The gap between what Artists can envision and what they can create depends on many elements and that all must stay in balance. Herman Hesse went insane because he could not balance the new and troubling insights he saw when looking deeper into "man" – essentially his own person genius as an Artist – with the lies society and rulers forced upon us.

Truth is the ever-present fulcrum we move to keep balance. Stay balanced over Truth, and you can extend the length of this teeter-totter as far as your flow of Information, translated to your vision, as far as you wish.

The challenge comes down to using I.Q. to flip over this teeter-totter. Put Truth above your flow of Information and shorten the length. This allows precision to Balance with passion in much finer increments, much finer detail.

Your systems internally add their measures of Balance. Your Five Brains work in synch, your Intuition add emotional tests for Balance, as Truth feeds Information and Information feeds Truth.

Victimhood refuses any sense of Balance and throws off Truth by usurping Information.

I hope that the Artists soon realize the politicians and, actually, the enemies they truly do hate (the 1% of the 1%) are using them to brainwash their fans and fake interest groups.

Betrayal

Two Russians were introduced to the Hollywood version of betrayal and exclaimed it is worse than how the Russian mafias work.

A business dilemma sparked my lifelong fascination with a few rules of the human condition, namely Betrayal, Artistic Integrity, and Sophistry.

How it all went from one of the most perfect combinations of creative and financial forces to a messy deal and production is easy. A woman presented a story that was her own true-life story. On first look it was fantastic. She was trying to make millions as a single woman. The Mafia wouldn't allow it. Miraculously, she survived. Over ten years of beating the Mafia at its own game.

I brought the deal to a Producer I had known since she was in her first years in Hollywood. I trusted her. She knew in a minute it was as super-hit series. She had me do far more creative writing in building the Bible for the series. Everyone thought this was a true story for two months. Then I found out it was a lie, and she stole from the Mafia. I called a friend whose family knows of such things and he confirmed if the problem isn't settled someone could die. I related it to my friend the Producer. She and her peers decided to bury the truth to cash in on the highest level and not to worry about someone being murdered. I refused to be part of this because the solution was so simple: negotiate a payback with a Vig on top of the debt.

I'll cut straight to the consequences. I would not agree to do the deal. They could not go forward without me but did not know it yet.

Greed and ego drove her and the others she convinced to go along with ridiculous strategies that are best described by the old quote from Confucius: "It is the clever man's fate to outsmart himself."

They all used Sophistry to get to a variety of explanations that would prove them justified or innocent in the future if someone is murdered. Sophistry is the building of an argument you do not believe in so that you can get to a preferable conclusion that sways the minds of other people you need. Using lies to create a new truth. To get to that truth, the initiator might just need one person swayed, or a slow building mass of people swayed, such as in politics.

Facts dismantle Sophistry.

Betrayal can never be justified.

I told her that her quest is like sour milk. She soured the milk. It's in the fridge but not in the dinner. She's insisting two things: she didn't sour the milk, and we should all taste it to see how sour it is. That ingredient, sour milk/ a.k.a. Trust and Integrity, can never be used in any recipe. Like an egg, the values of it being a part of a long menu cannot be quantified as so many dishes must be avoided.

Betrayal is the worst of all social and deeply personal crimes. A great movie starring John Travolta "The General's Daughter" illustrates this fact so well.

It is agreed upon by the Brain Brain, but it is interpreted and inflicts pain upon the other four Brains more. It activates reactions from all four of them and requires the Brain Brain not to act on them. It's severely damaging to Truth, art, and survival. And the connection, through Ether or not, should not ever be repaired to a bond between the betrayer and victims, even if it could be.

This is not about forgiveness or empathy; it is about Humanity.

The poisoning of the Five Brains with Betrayal closes all portals to sublime inspiration and the sharing of Information from the Universe.
That is why the theft of Original Content and a creative process resulting in Art is unforgivable.
NOW... FOR THE ULIMATE BETRAYAL

The poisoning of the Five Brains with Betrayal closes all portals to sublime inspiration and the sharing of Information from the universe. That is why the theft of Original Content and censoring a creative process resulting in Art is unforgivable.

Extrapolate Responsibility – To an Artist Today

Artists are the messengers to the universe.
Leaders into higher elegance.

Of all the qualities that Artists share the most common one is Responsibility (with a capital R). Passion, self-destruction, self-awareness, vision, Artistic Integrity and more qualities would all crack under pressure without a personal air of Responsibility. While that word might not be used by Artists much, if pressed into conversation, they'd see what a critical element Responsibility is to them and their works.

Responsibility appears in nuance, as a foundation, in the Brand of the Artist regardless of which field of Art.

Responsibility becomes noticeable in, for example, the discipline a ballerina must maintain – she (or he) is Responsible for diet, practice, clarity, selflessness, and generosity ultimately when an audience views this Art.

For a sculptor, the choice of clay or marble, the Truth this work must tell, all fall into his or her Responsibility.

Photographers use light and lenses to capture the Truth in a moment. The popularity of it depends on who relates to the Truth he or she captured. Yet, his Responsibility ties him to this for decades or more if it becomes truly popular. The photographs of Martin Luther King Jr., of the burning woman in Vietnam after a napalm drop … of the jubilation of children playing in an inner-city fire hydrant spray … of the emotional cost of both war and opposing war. Even the Truth in the Smiley Face "Have A Happy Day" carries both consequences and Responsibility in the message. These pictures stayed with me long after reading the articles supporting them.

Writers are the first to be killed by totalitarian despots because they held rulers responsible. Marx, Hitler, Mao, Castro and other monstrous leaders knew writers could tell the truth in ways that capture the emotions of the masses, amassing them in the light of truth to overthrow them. How did they know? They used writers to overthrow the previous Czars, Kaisers, and Presidents. The propaganda machines they developed kept them in power by saturating the minds of the young as well as adults with their mantras. This illustrates the incredible level of Responsibility writers must accept.

I had a dream once. It took place in Beverly Hills. A throng of fans paraded down the street with me. In dream-thought, I realized that I'd made it! I'm famous. Fans! I have fans! They led me into a church. The Minister didn't read from a Bible, instead he found his sermon in one of my screenplays. Talk about feeding my ego! My writing would lift the masses! But when he began reading, he twisted my words to chain my fans and his congregation to damnation, raising the power of evil behind him. Hollywood evil as the Devil's evil. I freaked out and tried to silence him but it was too late. Thousands had chosen dark, negative "Lies" masquerading as "Truth." I woke from the dream at 6 a.m. in a sweat! An impending sense of doom and guilt hung over me. Hopped into my car at 10 a.m. I drove to the Catholic Church on Santa Monica Blvd. only to find the doors locked. I thought, what kind of religion is this, to lock out those in need?

My hypocrisy tickled me as much as the Truth of this being a future possibility haunted me. And began my study of Responsibility among varied Arts and Artists.

The assumption of Responsibility for Artists remains impotent until you (use I.Q.) and apply "One Concept To Many" … and then bring "the many back to the one – you, as artist, or you, as audience, for Responsibility is contagious.

Let's see how many philosophical and empirical statutes can be applied.

The Seven Great Arts from Greek consensus were chosen by the way that they advanced humanity. Architecture, Sculpture, Music, Poetry, Dance, Painting, and Performing Arts. You can see how each of these Arts swayed and elevated societies in pockets around the world. Eventually these pockets shared Truths that allowed their civilization to thrive until meeting other great civilizations. The Responsibility to both tell the Truth through the Art and choose to use it as an Excelsior becomes more obvious. The freedom to enjoy all these Arts became in part a reason for social evolution and revolution.

Bringing that back to filmmaking and writing, even the technology platforms of today, it becomes obvious that those presenting the world their art bear the Responsibility for this action. This perhaps is why I saw some value in Xi's "Four Comprehensives" and "25-Point Plan." He knows that the only way for a Truth through Arts to survive is to appreciate the core Culture of the nation. Look at the Terra-Cotta Army/Soldiers. As incredible an artistic achievement as is imaginable, the Truth sculpted into them can only be appreciated and made immortal when the Culture that extrapolates a socially advantageous message from them remains intact. A jade sculpture, Ming vase, tea service choreography, and song carry no less Responsibility.

Let us move down the Arts levels to the Seven Liberal Arts. Grammar, Logic, Rhetoric, Arithmetic, Geometry, Music, and Astronomy. Together or separate, notice the untapped "power" to persuade in each of them. Lives dominated by Astronomy and Astrology study. Geometry with its control over the shapes of buildings we live in. Grammar, logic, and rhetoric combine for the powerful orator, writer, or politician.

See basic Arithmetic now as a "Great Math" that seems to be the new way to interpret our entire universe and its Fields.

The old and new eras of songs, and how they turn into chants and mantras for the young people that are still too interpersonally inarticulate to form their own ways of "how to think" rather than follow "what to think." Each of these is used to control people. Yet few artists think of this while composing the Art.

When delivering a screenplay or a film, I try to apply its early and final Vision to these Seven Liberal Arts to measure its Truth and impact, its "passion element force." Why? Because with my Gift comes a Responsibility, with the Fame comes a Whip meant solely for self-flagellation.

Finally, the Seven Elements of Art: Line, Shape, Color, Texture, Form, Value, and Space.

These are simultaneously methods of choice of how you tell your Truth to the world, and due to the mimic characteristic of humanity, they wane immortal on how contagious you make them. Your choice of color inspires the fashion designer making the dress that inspires a nation when worn by royalty or even a pop star.

The line creates images (such as in a fine drawing) that shape pop culture and last in the national culture. The space used in dance theory carries down to the rave mosh pit, as an expression (or illusion) of freedom for youths that exhaust themselves because they are free enough to do so at that moment.

Truth begets freedom begets Responsibility. Because free people drive to goals that, if they last, elevate society. The importance of saturating both the youths that want inspiration, and the older politicians that want power over them, with Truth, bears great Responsibility.

Hollywood's stars engage in the most powerful Art form as far as how it shapes the world now. Everyone eventually sees movies. The themes of these movies only prove effective when the many art forms making the film – Directing, Acting, Art Direction, Music, Cinematography, Choreography – are delivered with a Truth that penetrates the inherent self-absorbed nature of people today.

Actors love playing true life characters that shaped humanity. Bad guys, good guys, doesn't matter. An overwhelming number of Academy Awards are given to true life character films. In these films, the real person(s) had to accept the great Responsibility of their actions and philosophies. This is great. This touches the audience and reminds them of the fact that the human spirit is simultaneously fragile and invincible. And that the viewer has his or her own battles, quests, and demons to overcome to mirror the Truth the star played out. This elevates humanity.

For this reason, I'm confused when Hollywood embraces so many causes that essentially abandon Responsibility, and often Truth. Reasons are rampant. Crowd mentality. Guilt for being overpaid and over-worshipped for playing who you are not (that is called a Con in any other arena.) Habit. Echo chamber. But I think the biggest is being gullible. Especially among politicians. Doesn't matter what Party. Politicians use Actors. They have since Chaplin and "Potemkin." To sway thought and push propaganda. Somewhere along the way, Actors became blind to this. All others in Hollywood, too.

Here is where I divide the line between film and politics, and why I stay away from them, especially in the United States, the heartland of Entertainment/Hollywood.

All aspects of making a film in the arts departments revolves around telling the Truth even if the movie is about liars. The lighting must look True to be effective. The words, actions, and dances must balance in a Truth of gravity. All players in the project seek a union of the Truth in the Director's Vision.

The exact opposite of this is Politics. Few in Hollywood know that when setting up our government, the one excluded group from political office were attorneys. This comes from France, which also excluded attorneys for as long as they could. They argued that the attorney (popularly named because they "turn the truth") cannot be trusted, only the law can be trusted. Not the argument. Attorneys took over our legal system. That is half the equation of this problem. They need the artists to hook the public through passion and performance, not through Truths. I find this incompatible. My Truth loses potential as the lies of politics bleed my energy and attention.

The other half of why Hollywood embraces so many flawed causes plays an even closer role to Responsibility.

I was born at just the right era to grow up as the American family had the relationship to personal responsibility challenged when it comes to child raising, or parenting. It started with Doctor Spock, telling parents they cannot hold the child responsible for their actions, that it must be the parenting at fault, or overwhelming pressure to improve amidst peers to blame. This idea, if you look at it, means that any personal failure can be blamed on someone else. "It's not my fault, it's my parent's fault," Or my country's fault. Or lack of wealth. Drugs. An addiction. Now obesity is classified as a disease, and any attempt to point the finger at over-eating or laziness is called Fat-Shaming and is punishable.

You can be fired, you can have your children taken away, and more. The problem is that, of all the methods used to have a fat child trim and be healthier and happier, making them feel bad or limited about it worked the best. It worked on me. The embarrassment of being skinny pushed me into the gym at 14. And due to that, my whole life was lived healthier. The kids that pointed their fingers at my skinny legs did me a favor, and I guarantee I'm in better shape and health than any of them.

Take as a less personal example the demonizing of oil. There is no other commodity in our history that has done more to save lives, evolve modern civilization, improve health, fairly distribute wealth, open up opportunity, capture the potential of an individual, free women to more purposeful lives, and even allow for the art of filmmaking to happen. Yet Hollywood treats it like a demon, a great evil.

There are only a few sources of power. Water, Coal, Oil, Wind, Geothermal, Nuclear, Solar. Of these, only one is truly portable. You can easily take gasoline to any spot in a small portable container and use it. Due to the laws of thermodynamics and the greater law of Entropy, all use of anything on Earth results in a wasted or less organized byproduct. Each form of energy embraced by mankind eventually resulted in a glut of waste. Horses were dying so fast in New York that their mountains of bodies bled toxins so much they feared that if gasoline didn't advance fast enough, the horses would pollute the Hudson River and all water sources.

Add to that the idea that people by the billions strap themselves into a moving bomb of machines and gasoline every day, and even the most ignorant of us never get harmed by gasoline or its engines, especially in cars.

While the Tesla is cheered, most celebrities forget that 60% of electricity in this country still comes from coal. To drive the car that politicians twisted the truth about, they feel like they are avoiding polluting. Incredible transfers of money from the hard-working people to failed anti-oil and anti-coal plans or plants could have been used to feed and educate the poor.

The cost of Stimulus jobs dedicated to Green Energy ended up costing up to $2 million per part time job. Yet it was cheered as useful. How about using that money to pay for the education of 200 inner-city kids? It takes too long, is not as shiny, and requires thought. But most of all, it does not enrich or empower the politician. And the entire profit margin of 2015 was due to taxes and pollution credits.

Tesla has regularly missed promises and deadlines, and is still sticking taxpayers with And, the Entertainment Industry played a part in making a god out of Musk. The better example of Integrity displayed by leaders in electric cars is China's investment into Fisker. Great choice.

Most people do not know the real story as to why Fisker failed. It was due to politics of the Left in the United States. Do I admire Elon Musk? Yes, he's brilliant and delivered something of meaning. And his new ventures into space kicked the ass of NASA and most of the military industrial complex related to flights and space exploration.

I drove one of the first Fiskers. Wow, what a terrific car, and a better drive than a Tesla. The Obama Administration got votes and lobby money from the Green Agenda people. So through Joe Biden, they approached Fisker, declaring it the great future proving the Green Agenda works even on cars. They offered Fisker a huge loan they needed at almost zero interest. Fisker agreed, and the details were being worked out. The public read this and their stock rose up dramatically. Faith was behind Fisker and it should have been there, they deserved it.

Then Vice President Joe Biden confronted the Fisker company. Knowing that it was too late for Fisker to pull out of the deal without a stock collapse, Biden made two demands. The first demand was that the car must be built in Delaware, his old home state.

The second demand was that it had to be on the road by the next elections in 2 years. Fisker immediately said both should be seen as impossible requests if Biden expects the car Fisker planned. All the manufacturing plants and raw materials were in Detroit, where Fisker planned to build the car.

Moving it to Delaware, building the right assembly lines, would greatly increase costs, and hit uncountable problems. And the company would not have the extra year they needed to work out all the bugs in the new car, mainly in the batteries and solar panel, as it works with the dashboard. But Biden demanded! So, Fisker had to move. The initial bad news was that due to the higher cost of moving to Delaware, the promise it would be as affordable as they promised could not be met, they were $15,000 higher in cost. When the cars came out, at first, Biden and Obama bragged how they brought the future to us! Hurray! But then the batteries caught fire! It was not just Fisker's blame, the new batteries even in small phones caught fire years later. But the news was out! Too late!

Fisker recalled all the cars, but new buyers would not want a car like this now. That quickly led to the failing of the car company. They tried to explain why, but the government made sure that truth never got back to hurting the President and Vice President. It "was not our fault, it was Fisker's fault" even though the reason was 100% on the government.

Yes, coal and oil have terrible byproducts. We've elevated the science of limiting them at incredible pace, and need more of it. "Clean Coal" is not the same as the new era of coal, in which the open space between atoms of coal is squeezed out, leaving a more clean coal.

Nuclear has its terrible consequences, yet is the only source of power that, barring a leak, results in zero emissions. So few people have been killed by nuclear power compared to all the other sources of old power that it is not measurable.

Let's now compare two companies – one demonized and one championed and even idolized. Exxon vs. Apple. Exxon has saved hundreds of millions of lives, evolved society and cities, and made modern life possible. Apple is a great company to be sure. It brought the touch screen, smart phone, mouse, and superior computers. Exxon is painted as the epitome of greed and abuse of people. Apple is embraced as the best Santa Clause ever! Now let's look at the greed factor first.

Oil profits are 1-2%. Drilling it puts Exxon in war zones, pits them against a dozen special interest groups, forces them off-shore to keep delivering what the fans and stars and artist demand – cheap oil and plenty of it. The engineering jobs and even the drilling jobs are some of the highest paying jobs on the planet, with great insurance and stock share options, as well as employing an unnecessarily high number of women and minorities.

They open free schools and hospitals in many countries the drill in. Exxon executives are often targeted for kidnapping in half the world they drill in. And they always pay their share of domestic taxes, in part because so much of Congress hates them. If oil stopped flowing, within 30 days civilization on the planet would almost completely stop, and all world finances would crash.

The profit margin for Apple is over 700%. Their product requires massive amounts of oil and other energy to make. They price it for rich people. They are so greedy they moved their factory to China, where wages are so low employees commit suicide, and injury due to over work is out of control. They hide their money in offshore accounts that the oil companies could never consider.

Steve Jobs was one of the biggest betraying assholes in big business. They crushed competing innovators. They need such a huge volume of rare earth metals that they are behind many government subplots of insurrection. They bribe leaders to get these metals. Aside from projects that better their future, they have never opened schools or hospitals. If the Apple products were suddenly taken from us, life would go on, with competing tech. Yet, they are embraced by Hollywood. Go figure.

Before I digress and get off track, the point is that in all arts, there is a personal Responsibility. Many of them usually. And society now is championing, and glorifying, countless causes that do the opposite. The causes champion lack of responsibility. Now why would politicians want this? Because people that are confused and unable to think or help themselves are easier to fool and control.

The confusion that permeates most of our new political causes can be boiled down to distorting Truth or ignoring facts. All of this happens only when a group is convinced it is not their fault, and second, that a politician can deliver an alternative to Truth and responsibility that give something for free to the voters and celebrities.

A campaign to encourage self-pride, regardless of look, ethnicity, background, or intellect is an admirable goal. The world seems to embrace taking a small segment of society and turning it into a political issue. As if politics has cured any major social woe. It barely has, but largely has failed, in all nations, in all eras.

The idea that a politicized goal cures a social malady is misleading. But leaving that idea for now, let's look at the campaign from a universal I.Q. perspective, with or without Entertainers involved.

If self-love is a thing of beauty the individual's creation (how he or she looks, acts, and thinks) should be valued as a creation of Art that has passion and value even if it is only in the eye of the individual. If an actor or actress does more for the sake of beauty it can also have a positive impact on what roles they get. Extreme make-up for monsters or devils is expected to convince the audience the scenes are believable. Likewise, if someone uses beauty or a nice body to convince the audience this should also be allowed. The cinematic devices have to be True to the story.

The two fundamental elements of Art are Truth and Information. If both Truth and Information from the backstory are withheld from the Individual that we should glorify as a talented actor, then how can it succeed in suspending disbelief? It cannot.

If the facts of civilization are withheld and the "Truth of Responsibility" is not allowed, then the person has no chance of finding inner peace. Unless politics promises it and gives hope. Do politicians ever want these splinter groups to reach Nirvana, that total one-ness with the world? Hell no. They insist on chaos and despair to control the people and get votes and money to make the very people they pretend to lift ups do the heavy lifting.

By turning all facts into emotional issues they bypass inspection of the facts. This does not mean that facts and emotions do not matter to actors and actresses. It only means that the system they set up encourages manipulation because their face, their Twitter, is so popular that what they say carries more than "one citizen, one vote." All it takes are a few top actors to start a political stance and Hollywood buys into it.

Look at what they call Wage Disparity. There is no group in the world more guilty of the two sins of being the 1%, and also demanding incredible wage disparity than celebrities. They earn tens of thousands of times more for just one film or tv show than the people on their crews. They are the ultimate hypocrites. Look at the argument that women are not paid as much as men when they star in a role. They're arguing if they should get 1,000 times the wage of some man or woman in the crew, or get 1,250 times that figure.

Then the argument for more women in top positions is valid but it did not pay off as promised when the top 5 green-lighters were women. Here are some facts. The majority of television viewers female, so more shows have to make sure they appeal to females. There is an abundance of female-only shows. And most talk shows are geared for women or hosted by women.

When it comes to film, women might not be the Director on as many big films. But who decides this? The people with Green Light power at the studios were: Disney – Kathleen Kennedy; Warner Brothers – Veronica Quan and Sue Kroll. Fox – Stacy Snyder and Elizabeth Gabler. Universal is Donna Langley. Paramount has Shari Lansing at the top. The other studios allow great power to the top women. So why don't these women force more films to have a female Director? Because of Truth and responsibility.

The responsibility is to the shareholders and audience. They must deliver hits, not social agendas. Almost all films doing the opposite fail. And then what happens? These female execs lose their jobs.

The truth is that to be a Director on a big film usually requires many mid-sized or small films, and the stars themselves, including female stars, demand high power Directors that are mostly men.

The fact is that 10% of the people wanting to be Director are women. We don't force them out, they simply are not drawn to stay on course to be Director films of that size. Is there any force holding them back? Yes. Consensus. But there are also forces holding back older Directors and movie stars, shorter stars, ugly Stars, Black Stars, Asian Stars, Latino Stars, gentile filmmakers, and Artists that have too much passion and not enough business sense. BUT, only a few categories can be politicized. For that, we have these battles in Hollywood over discrimination. What the most vocal should do is study the actual stats for minorities in films and Oscars. That would be a start. Get the facts that support the Truth and then add in Responsibility.

I was/am well known for writing screenplays with strong female leads in big films because I like to, and it is good business. Not for an agenda. The same can be said for films I write and produce that have as a core demographic, African American audiences with cross-over appeal to white audiences. Because ... it's good business. It's smart. I do not agree with artificially forcing any Artist to change the composition of his or her work to follow an agenda, because that forces them to refuse the laws of the Seven Elements of Art and my Seven Elements of Responsibility of Artists. Not to mention the Information these punitive groups demand be ignored.

And finally, Hollywood is a bully. One of the worst. Once a campaign is presented by an organization that essentially "politicizes" a topic, and Hollywood embraces it, then the star power can destroy businesses, careers, families, and anything they want. For this reason, I stay away from politics. Hollywood like other media outlets for the last 200 years is being used and conned.

Ask for the celebrities to face actual facts about who or what they demonize of force on us, and they either won't face them or know little about them. Actors are people that are expert at faking reality, they do it for a living. Not one celebrity I know is aware that all the U.N. Climate Change graphs and facts were given to MIT to decode and present to the U.N. and world. The first year it delivered the models and information posed by Al Gore and others in about 2004. Their models showed that yes, Climate Change is real. From then on, global politicians and greedy aristocrats used this to make fortunes and gain more power.

MIT was ordered to deliver a comprehensive report on existing Climate Change and review all data delivered by scientists and explorers in this field. The Second report, also relying on models presented by scientists, again declared Climate Change is real. But, in 2017, there was finally an independent body leading all research. So much evidence came out disproving every single model they had relied on from politicized experts for its report to the U.N. and world they reversed their decision. They came out and declared that Climate Change has not altered the world in any significant way or any more than normal climate events.

Their report declared that 90% of all Models presented to them failed to prove to be true once the Model was put in the field for actual observation or other work. They told the U.N. that while pollution made by man is indeed harmful to the environment, that all other predictions were false or faked. And all rises in temperature they recorded and used turned out to be due to errors made by the scientists that got paid huge sums to get to the proof that Climate Change is real. Now, who is the bad guy in this? Follow the money, as usual.

Al Gore was the single greatest recipient of financial benefits from or if Climate Change. Michael Moore made a fortune on his movies about it. The U.N. transferred trillions of dollars into new organizations that basically funneled the money and power to rich elites that are the greatest abusers of the planet and energy you could imagine.

Yet, Hollywood has not gotten the bulletin. Soon I expect they will move on to a different cause.

This leads me to the most important part of this entire topic. What I call the Seven Elements of Responsibility of Artists.

TRUTH
CULTURE
FREEDOM
CONVICTION
CONSIQUENCE
SACRIFICE
JOY

I write and produce films for women or minority audiences because they honor all the above. Telling me I cannot honor them all is a Lie. A Political move on art. That is the start of tyranny -- the opposite of freedom, hope, individual pride and power, and a better future.

PATRIOTISM TRUMPS TRIBALISM

(Patriotism sustains culture, lifestyle, and safety while Tribalism lasts only as long as it is fashionable or useful to politicians.)

Tribalism leads to the Balkanization of a society. Yet it is being glorified in media, film, news, politics, and schools. Why? The only obvious conclusion is that those in power want the American Dream destroyed from within. Make people believe, repeat, and live lies that are devised to make them hate their nation and they are not united as citizens and will not fight to preserve the nation they live in. The tribes turn against each other. With that comes violence and the influence peddlers, the NGOs with financial gain coming from destroying families, religion, and education.

Patriotism becomes a sin. This has been seen in dozens of nations that turned into dictatorships run by tyrants and military gangs. [This was written years before Trump became President, just for the record.]

Our democratic republic built its constitution on the premise of protecting the "small" from the tyranny of the "big." Diversity in religions, countries of origin, and personal goals reigned supreme in a truly remarkable history of a nation. The Constitution ensured the best political arena was kept in check, knowing that for people to get together to serve a greater cause was the nation's secret strength, and secret weakness if it ever failed. People came here to better their lives, contribute to a free society, make it one that embraced them as individuals. The knowledge, skills, sources of foreign finance, family structures, and so many other elements that came to our country built a unique World Leader. However diverse the dreams or origins, the crucial element of patriotism guided their actions.

In the United States, more than any other country, patriotism was a "choice" that could openly be challenged in public arenas. Due to the remarkable state of the nation, the freedoms at stake, and the building of nearly perfect family foundation over generations, the idea of acting against patriotism was nearly unthinkable.

While this might be more evident in the United States the factor of patriotism is the foundation of all successful, strong nations. Exactly how patriotism is nurtured greatly changes the future of the country, its politics, it's creative force in the innovative world, its international esteem, and spirit of its people.

Dictatorships, totalitarian regimes, Communist (and to some extent Socialist) countries, and now "fundamentalist nations" use patriotism to the nation country or its religion to erase hopes in any direction other than blind faith to a leadership. Patriotism for the military strength, sovereignty, political agenda, and even the very soil of the country was force fed into all the people, regardless of personal belief. Anyone resisting usually disappeared. Look at what is lacking in the areas that the citizens were demanded to patriotically defend. There is an obvious elimination of anything created by an individual with a free will: no patriotic worship of fine art, fiction, or design other than military statues. Why? Because the leaders of brutal and mentally-closed regimes know that the first duty of an artist is to "tell the truth." And the truth in their politics or country is hopeless, cruel, and ultimately terminal.

North Korea provides the clearest example of what unhealthy patriotism can do to a nation, its economy, people, progress, cultural achievements, and coexistence with the rest of the world. Patriotism fueled by distortions on politics, within and without the country, leads to fear and fanaticism – the two best ways to force your citizens to do unthinkably violent or world-wasting actions. The deduction, then, is that there are forms of "unjust" or "unhealthy" patriotism that simple cannot be argued away by demanding freedom of expression or personal views. In the hands of tyrants and evil-doers, patriotism is a poison pill force-fed down the throats of the people.

Patriotism in free nations, in contrast, shows how wonderfully the practice builds a nation. Patriotism in Japan after its defeat in World War II was muted, but it was a driving force behind the people establishing themselves as one of the great production and creative markets in the world. And, patriotism allowed them to maintain their identity during the very trying occupation by Allied Forces for decades after their defeat.

Exclaiming "My country, right or wrong" in some senses is like saying, "My mom, drunk or sober." Yes, there is a love, and yes, mistakes or weaknesses are forgiven. Yet, loving to the point of endorsing a drunkenness in your mother hardly makes for the most functional family. Patriotism requires steady (not fanatic) examination of hundreds of factors that might pull a country off a path of greatness and future security. Open forum has been the bedrock of most all the "great" civilizations/nations in history -- Rome, Greece, and the United States being the most dominant examples. The discourse has been captured in movies, Shakespeare, and many important illustrated texts to save this moment for future generations to learn from and use.

Yet the current trend within the United States can both erode and embolden "healthy" patriotism. Literally hundreds of special interest groups or business are demanding to be supported based on their rights in The Constitution. If Patriotism for the country trumps their personal or political beliefs, then, the Big Government is evil and must be changed or dismantled. The dreams and rights of smaller groups should be protected, but, not to the point that it decays the strength of the country that protects their right to demand. If a war protecting a nation under obvious threat of terrorist attack goes against smaller interest, then, the war can now be openly challenged. This stance was unthinkable in World War II. Has life changed so much? No, only the public allowance for rhetoric against patriotic duties has been slowly embraced as a politically correct way to protect our nation – which, upon examination, is absurd in the new world of global terrorism.

Watching the news inspires the question of whether the Constitution should also be applied equally to defend the "big" against the tyranny of the "small." But the better question is, "How did it become fashionable to demand the rights of the small over the safety of the big?"

A simple factor allows this: the belief that it is more important to protect every small culture within the United States above protecting the safety of the environment that allows the cultures to coexist.

No country in the history of civilization has survived long when attempting to be "multi-cultural." By that it is meant that the culture is more important than the patriotic adherence to one government, one peoples of that nation, one identity to the world, and one brotherhood citizens. By making the culture of a minority more important than the national government or identity is the first step in having a citizenship that refuses to fight for its country, believes short term gains are more important than long term futures, and creates many factions who will not follow the laws that allowed the country to form in the first place into a healthy political and economic nation.

Patriotism cannot be blind. It must be challenged. Yet, it must be also nurtured by morality and virtues that keep the nation united despite being under attack for the right reasons. Individual groups demanding socialism or justifying criminal behavior are now mostly funded by forces that want the nation to collapse. The people spouting the rhetoric do not have "patriotism." They have customs, languages, and goals that all must fall subservient to the greater good of the country. They might be at a protest to just make a paycheck. They might be there because they are anarchists. At the same time, the nation's leaders must always recognize that without healthy sub-cultures, people are not as productive, secure, or happy. Roots mean as much as religion to many people. How to balance the demands of the "Small" while ensuring the strength of the "Big" in America is the greatest unknown in the future of leading the people to a safe future.

The Modern-Day "Robber Barons"

The men who led the world in industries were damned.
The men who lead tech damn US.

Oh, and how we do worship them.

First, you must know what the term "Robber Baron" means. They were the "Captains of Industry" that made the nation what it became. But for popularity's sake, they were renamed to The Robber Barons. This is a derogatory term for the Titans of American expansion and ingenuity that lived from the late 1800's. It is used mostly by groups that want to blame success and innovation for problems and disparity. But a closer look proves they did more to increase social evolution than anything in our history.

Who were they? J.P. Morgan, Banking. Andrew Carnegie, Steel Industry. Andrew W. Mellon, Banking and oil. John D. Rockefeller, Oil.

The list grew to include anyone that profited highly in wealth and or power form their vision and resourcefulness. Those also associated with Robber Barons, somewhat like Mini-Robber-Barons include Cornelius Vanderbilt, Mark Hopkins, John William Gates, John Jacob Astor, John D. Spreckels, and Marshall Field.

Consider what they brought to our nation first, and the world copied it to lift their lower societies up.

J.P. Morgan loaned the U.S. Government enough money to save tens of millions of people from starving to death during financial meltdowns – not once but twice.

Carnegie donated the largest fortune to free education for the poor in the history of this nation.

Spreckels offered the only safe river and rail passage through murderous territories.

Vanderbilt went on to fund one of our finest colleges and gave enormous grants to the poor.

Mark Hopkins donated enough money to fund medical research and hospitals across the nation, saving millions.

And although he is demonized, Rockefeller in his quest for global dominance in oil brought that portable energy source to every nation in the world, allowing them to build electrical plants, roads, schools, and exports.

It is easy for us to point a finger at them and create a negative image for eternity. But at the time, they allowed our nation to feed and educate itself.

The only way to properly see this anomaly is to apply it to today.

Today's two biggest Robber Barons are Steve Jobs and Jeff Bezos. Close behind is Bill Gates. You say, "But wait! They brought us cheaper deals on things we buy online. They brought us PayPal. And Smart Phones and Software cheap enough for anyone! And Gates donates a lot of money. And The Cloud allows us to store unlimited selfies! And Google is just freakin' awesome! Facebook lights up my life, I get most of my news from it! And I stay connected through them."

Yes, and they ruthlessly put competition out of business if they could not buy them out long before their profits were maximized.

I am not accusing these modern Robber Barons of being evil. Quite the contrary, they make life more functional, more livable, and push us into the future.

But do not fool yourself. There will come a time when civilization looks back and sees these modern idols of wealth and invention as Robber Barons. Or it will, if that is allowed.

The modern Robber Baron doesn't just maximize profits and destroy competition. Every Robber Baron today wants something no Captain of Industry could imagine.

They want your Mind, your Imagination, Your Heart – they want what makes you an individual to turn into one of two things: a drone in a predictable hive making them richer, or a commodity based on your ability to imagine and create.

In no field will this have greater consequence than in the Arts. In all things creative.

Silicon Valley has its Unicorns, that feed into the Robber Baron culture, that like the ones of old, are idolized by parents pointing out their genius to their children. That mantra is drilled in not just by schools but by daily reliance on them to manage your life.

Is Time Travel Possible?

Quantum has proven that what we do now
can change the past.
Is our concept of time a Human Agreement
we can deconstruct?
How high of an I.Q. and how
much Will driving us to higher Elegance
would it require to step outside
Time as we accept it today?

There are countless movies and television shows that use "Time Travel" as its theme, or a device into it.

Because of this, it's more in vogue for scientists to talk about it. To try to explain it to us.

They've got one thing right, in that, you cannot talk about Time Travel without Space (or what we call Matter) linked to it, so what you are actually talking about could be called "Space-Time Travel."

[It must be mentioned that Time and Space-Time Travel relates to gravitational forces or scientists that know far more than the author of this book will go ballistic. There is so much about Time this book cannot cover. All we can spotlight is what might give us an Accelerating Philosophy to apply to execution of a plan toward a goal of the Qualia Avatar realized through Art.]

However, I don't think they're going about it in the right way.

First, they discount what they "just discovered and announced" – that what they for a hundred years said – that space is a vacuum – is actually holding the majority of measurable mass in the universe. They now call it Dark Matter, something I call Ether. I won't even go into all the relationships they need to factor in on this part of their faulty approach.

And I won't talk about the absurdity that we will travel in something we build here.

I'll just focus on one part.

They assume that Space-Time Travel will first be experienced outside our body. The opposite will prove to be true. And it seems so laughable to me that they haven't jumped this shark yet.

First problem with going outside the body is that we are still fish in a fishbowl talking about how cold the water is – regardless of what the elements and distances are. We cannot view ourselves, we cannot view our Universe, not from the insides of it.

What can we observe, what universe do we play the utmost Macro role in, to solve the Micro mystery, and has this experience.

We are the universe of our atoms. The atoms that make up our body.

Once they switch viewpoints and see the first Space-Time travel will be the journey inside our own atomic and genetic structures, there we will step out of our concept of Time. The atoms in a cell follow the Circadian clocks of the universe, but un-quantified aspects of it still exist. And I argue these are flexible, but not measurable, in our Rules of Physics.

Time travel will not be realized or related to others through words. And it will not happen with us securely set inside a machine.

What will travel is the "Essence of Us" in form of Information that alters as it merges with other Rules of Physics that define the other dimension or Universe.

The communication needed for the merge will source from our Chi/Gut Brain on impulses other than words, the knowledge moved through "Inner Ether" (which I'll explain in another chapter) that translates all our plans, allowing Intuition to steer the Journey without plans made up in our Brain Brain with Words, and the form we take is pure Information. As we pass through Information is frictionless, it defies gravity, exists in multiple Time frames simultaneously, so we won't lose years here from space travel we'll simply alter the basis of our Information upon what can be visualized as a merge return to this Dimension.

If that sounds cool to you ... remember this ... we cannot return without Information from other Dimensions stuck like Velcro to our Information.

The decoding of this Velcro will finally be the proof, method, and key to a whole new form of communication, that speaks the language of genetic wisdom, which exists in all time since our existence within the forces of Ether that structure the incubator we call Space-Time. That's all this is after all, an Incubator for Information.

With this Information we shall finally be an active player within Ether on its expanding scientific category "Orders," instead of a bug clinging to a log whipped up into a tornado.

The beautiful aspect is, that the Laws of Ether will dwarf our ego and reason, and balance shall be sublime within our Minds.

What does this have to do with Art? With Creation? My belief is that Art is the secret ingredient missing from advanced Physics and its approach to Time or Time Travel. What will inspire the cosmic Will to curve the edges of the universe toward us, or warp a bubble universe to allow us to witness one of the secrets of our existence? Are we ready for the secrets revealed when we engineer a transmission from a decentralized source? Meaning, from every individual human mind after a breakthrough "tears a hole in the fabric of humanity" and we witness what changes literally everything.

By viewing the energy and Information in various entities we observe or can't observe, the "Spark of Genius" speaks to us. This Spark ignites the unlimited potential of the being we call US, humans in this case, run by a Brain Brain (supposedly) and this body will slip beyond what we call limits to our creativity.

Imagine cancer being cured by a certain dance, with your body moving to the rhythms of a beat from another Dimension to a sound that we don't hear in ours, and you get some glimpse of its potential. Or a painting you draw wearing a blindfold with a brush you cannot feel on a canvas that is separated atoms that your dog can see and it knows today is your birthday.

How freakin' cool is THAT!?

THE "PLIABLE" STELLA OCTANGULA

Velcro to "Raw Universal Information"

What is a Star Double Tetrahedron?

It is one pyramid stuck into another pyramid of equal size, the intersection points exactly in the middle through one side. (Also called a stella octangular.)

Why is this important?

The pyramid is the perfect shape to build anything upon that is limitless (but only in this physically 3-Dimensional world. The opposite is the circle, which has no support for so much of its space—and expanded to three dimensions it looks like a bowling ball, but a dot does the same in two dimensions, as mentioned above.

On Earth, in our Space-Time visuals, a pyramid looks perfect. Building a long wall or a building out of a series of them would be a strong architecture. There is no wasted space. It is solid.

But in other dimensions, or rather in dimensions when measurement is more than in 3 dimensions and the reality is more than 5 dimensions, a pyramid does not hold up a wall.

I am about to dive into a realm of Physics, Astrophysics, Mathematics, and Humanity that uses accepted, known terms in those fields, yet presents new "words" to carry the concept of what I've seen as the reality of our future. While everyone accepts the term "Double Tetrahedron" usually call a stella octangular and added the word "Pliable" because it simplifies many terms related to the concept that Time is not an arrow and we can know a future event if we link Pliable with plasticity of the mind. (I know most scientists or experts will say this is meaningless or just junky terms but it helps get the point across to people considering this topic for the first time.)

The genesis root of my asserting has been proven already, I am adding a cool way to look at what is about to happen, in a way that is measurable – something that has been happening for a long time. There is nothing supernatural or Sci-Fi about it. Words are incredibly inadequate in explaining the leap into a higher level of interpreting our Dimension. Open your mind, let go of Human Agreements, and simply "sense" through your Intuitive Brain if this sounds right to you.

[Spoiler Alert: The Qualia Imagii Avatar will be navigating the Multiverse created by Event Horizon Studio. The Avatar size will change often as new "planets" in our Multiverse are entered. Information from other players and the games or environments will "stick to" the Avatar if in advance it is pliable (in a Will and Concept aspect) and designed for this insight into the soul of the Avatar.]

Only a double tetrahedron can do that and not lose space to small voids, and still have all angles of pressure supported. Yet, people cannot fathom this. Why? I'll get to that in a few paragraphs, first, back to its original conceiver, as seen by an outsider to the study of physics (on an academics scale) a role that is essential to getting past their bound reasoning.

Nickolai Tesla believed the Double Tetrahedron was a unique shape in our dimension as it interacted with our Laws of Physics in this Universe/Reality. The ancient Egyptians build the pyramids to last forever. Its walls do not flex, it is rigid.

Look at a fishbowl. The fish are unaware of the water. They treat it the same way as we do air.

In the last decade, Astronomers have announced that Space is not a Great Void dotted with planets. Instead, it is planets in a sea of "Dark Matter." I of course call this Ether. And that in this Dark Matter, most of the "mass" of the universe exists. We just cannot measure it. Usually, if you cannot measure it, it can't be proven – in Science. But they're going all out on this theory. The very elements they use in equations are their barrier, though, like the form of air, they do not see it. Why? Again, soon to be answered.

I have always thought that this blackness of space was in fact Ether. And that Ether had certain properties.

Einstein focused all of his work on breakthroughs that exist within this dimension and according to our Laws of Physics. Tesla on the opposite end, did not take for granted the absolute truth in the "Human Agreements." One of these Agreements was that all we know or could do must obey the laws of Physics, and the example of existence, of this dimension.

My mind works more like Tesla's.

Here is what I believe.

Air and Space, according to Scientists, is made up of a substance that to us feels like it is without resistance, is totally clear, can reflect light, and then once out of the stratosphere, it becomes a void of blackness. Fair enough.

Astrophysicists and Astronomers recently announced the proof that I was right all along when they announced that up to 70% of the universe is made up of "Dark Energy." Not just Dark Matter which now they estimate accounts for 25% of all that is in our Universe – but Dark Energy. Energy does not disappear, it transforms if it changes at all in our Fields or Laws of Science.

I think Air and Space is a Medium that holds Mass, Memory, Information, and Spiritual Energy. It goes on forever, and never has a gap. What could this possibly be made up of?

It's shape must therefore be in a Star Tetrahedron. But there is another characteristic that must be considered that I deeply believe is true.

Ether is formed out of Pliable Star Tetrahedron-shaped substances. Think of it like Jello, but more elastic. This Pliable substance is a conduit for many things: radio waves, love, Chi, Plant language, and so much more.

We cannot see it because … well, for the same reason we cannot figure out how sand or crystal turns to glass. Most people do not know that science knows how to "make glass" but not how it turns from solid into something pretty much invisible.

They've tried all sorts of hypothesis. But none hold up.

My Intuition tells me that this substance in the shape of a Pliable Star Tetrahedron that can appear at any scale and Information around it sticks in patterns as the event unfolds quickly turns transparent because the pattern emerges and humans connect to the pattern as a whole just like we don't focus on the granules of a crayon when it draws on paper. It turns from the rigid separate element we call glass, into something so pliable we cannot feel it when we are in it. However, and this answers the above questions, the reason why scientists keep running in circles is they refuse to accept that properties of Information do not just exist in this Dimension. If it transcends Time-constrained Dimensions then it cannot be made of what we call matter. And it is scalable, which means that our concept of its size, which depends on our antiquated idea of matter, also does not function as a part of any equation.

Why they can't see this could be ego, or could be that they cannot get grants for it, or don't want to admit the idea can be imagined better by a pet. A pet dog assumes different things. He does not "think" of you as his or her best friend, it defines "friend" by you and you alone. Other animals like it or unlike it they use different Brains in their bodies to react to, and that reaction itself determines their emotions toward the animal. Is there a correlation in human life that has been proven? Yes.

In Down Syndrome Children. I mentioned this in previous chapters. They have proven that Downs Children do not do one important thing in infancy. Normal healthy children look at the faces of the mother and father and others around it all the time, and it recognizes smiles, the closing of the mouth, the movement of the forehead when frowning, and all the other facial movements. They do this long before their minds shape words to explain things. The movement in the face they see then makes their cells (their Brain Five, fully mature at birth though coordination lags) contract to mimic the movement. The movement itself then helps develop the emotion associated with it. That is why no kids smile when crying or frown when tickled. So the reaction of our cells to what we look at shape the mind, and the mind shapes our immediate and distant environment and how we perceive it (though which senses processed by which brain.)

Bring this back to the Scientists and human studies so the I.Q. experiment is useful.

Looking at the Universe, it's Dark Matter, the way we do, makes us like Downs Children. We do not form from the start the cellular reaction to seeing the true substance of Dark Matter or Space. What side effect does this have? Our bodies never learn how to react to the passageways to both other Dimensions (where cats often go in mind only) and hold on fast to the only form (hard body form) taught by parents or agreed to by scientists and teachers. To make this simpler, in dreams you might fly or feel out of body, or better said, with "no body." This detachment from cells is actually how our Mind relates to the body and brain, but that is another chapter.

Information is both bordered and not bordered because it is scalable, which means like Tara's Brains, it is infinitely small as it is infinitely large or limitless. Our idea that Information has any border is what we do not see as Downs people when we approach it with the Brain Brain. The Brain Brain cannot let go of this because it is actually what its existence makes as an "Agreement" in this 5-by-3 Dimension. The process of acceptance is similar to the virus around the jellyfish explained in previous chapters. Or the hive mentality of pattern making that Starlings make in their murmurings. But as advanced as we think we are as a species, we have not yet jumped that shark to that acceptance. Is it because this Truth is held back from us? Is it from the over reliance on the Brain Brain? Is it because other brains like the Heart Brain and Chi Brain need to be synchronized to be part of it. Rain does not imagine it makes a rainbow as it fades in the sunlight, yet it exists in waves our eyes see but dogs probably do not see. So we see what we want to see in part, and are kept from seeing what we desperately have sought to see since the evolution of our Minds: Who are we, how were me made, what is our relationship to all else of existence?

The answer might be found in the Pliable Double Tetrahedron, not so much as it passes through Dimensions, but in our very ability to let go of our 5-Dimensional Universe and allow the probably 6 or 7-Dimensional reality allow us to merge into it. This is the release of our Agreement that Information has any borders. Like Love, it is a concept. You can only whisper of it, for the idea, chasing it, scatters it like fish in a bowl. Fish are curious. Extension of our Minds form this hold on 5 Dimensions will bring the "curious of the higher dimensions" closer, and they sometimes touch us.

The hair standing up on your neck when in primal fear processing lets us let go of our Brain Brain's hold on 5 Dimensions just enough for the static charge of crossing dimensions to merge in the information (processed by the Intuitive Brain) to "read and integrated down into 5 Dimensions" so it is felt. Information on pure senses carries static charge not of electricity but of the ripple wave effect of breaking the Feedback Loop, mentioned in previous chapters. "You cannot go somewhere without leaving something behind, you cannot come from somewhere without the stella octangulas picking up "sticky" energy. Or, more accurately, sticky Information.

This plays heavily in two things: Time and "creating something out of nothing. "

This leads to the ultimate answerable question. "During the moment of spark of genius in the artistic and unbound creating something out of nothing, has the Velcro effect brought that from "All-Time" and parallel or accessible Dimensions? Have Artists in that moment accessed the Ether unbound by our laws of physics, not seeking data coordinated in Words based on Mathematical observation of our Universe, but simply seeking Information, succeeded, and is their form of Art (dance, music composition, painting) the personification of it … and the ultimate question therefore being are other forms in other Dimensions waiting for us to jump that abyss of unknown to contact us?"

Let's go back to glass. There are many scientific goals that think that if you transform one element from solid to gas or another element it becomes transparent or translucent. Think of how water turns to steam, right? But it is still H2O.

What if the fundamental building block of Matter, which on Earth is entirely rigid, becomes supremely flexible?

Transparent Aluminum, Transparent Silica – these have both been tried. Some success. But they still work within the confines of our idea of Matter having a cell wall or made up of elements that are rigid enough to give resistance.

In another Dimension, a form in the shape of a stella octangula does not have to obey our laws of Gravity, of Matter, of magnetic pull, or of sharing communication. When the original stella octangula is released from the limits of this Dimension and Laws, it becomes what I refer to as pliable. What happens then? As someone passes into it, though it has no resistance, it does have a sticky element.

I am pretty sure that Tesla discovered Pliable fundamentals (requiring plasticity of the Mind" because it is fundamental to frequencies. Discovering and exploring this would allow for many things he did: breaking through the walls between dimensions, creating matter out of air, sending energy without wires. All would be possible with Pliable Star Tetrahedrons. My moving through Ether, Pliable Star Tetrahedrons would either stick to Tesla like lint, or, it would pass through him like a Gamma ray, and touch points of Memory, later to be retrieved and experiments on.

How do we imagine this? Well, like most things, our far away ideas come close and into focus in Science Fiction and the Sci-Fi movies.

Remember Interstellar? At the end, a Tesseract appears What is a Tesseract? It is a formation that can be infinitely repeated. The star McConaughey jumps into a perfect Tesseract. It is like an infinite number of hallways, that observe every moment in his life. Gravity and love can pass into it and still hold energy. A Tesseract is just a form that exists easily in our Dimension.

What if it was Ether? Same thing but there would not be walls. What if it was a Double Tetrahedron shaped as a stella octangula? It would exist in all Dimensions. So let us say that the same thing could appear and exist in all Dimensions if it was a stella octangula. In other words, the rooms would not be pulled away from each other by stars, or pulled into the center to collapse by its own gravity.

But as a thought, as a Principle, what was the Tesseract supposed to be?

It was the imagery example of what is called the Akasian Library (or Akasa). Limitless moments and wisdoms of our past and future – seconds when messages were attempted but few noticed. What could house this? The shape of a stella octangula. But ancient Seers said that it was invisible until accessed. Aha! That is what I call a Pliable Tetrahedron.

What my Secret Project does is … it takes the idea of invisible force and structure that is Pliable, and it transforms it into a mode of Information sharing. Einstein was seeking the answers "within" this Dimension … Tesla sought answers "outside" this Dimension.

The conduit, the carrier of secrets, was invisible and limitless, almost godlike in powers, and it was in my words the Pliable Double Tetrahedron form used in the Akasa. Limitless. But, all one unit. An organism of sorts.

Farming Imagination can only happen when Information is centralized in a solid form. The Pliable Double Tetrahedron frees it. (No matter how hard you might try to keep a new realization secret, the Ether immediately converts it for the Akasa, but that is too much to explain here.) It is where its Spark of Genius comes from. What the Secret Project does it is Uber-izes the access to Pliable Tetrahedron Secrets, History, and Energy when it is in the form of Imagination or Information. These both have no mass, can pass through anything, can be held invisibly in anything, yet contain all forces under pressure. Since all in Nature must be balanced, we find that on the other side of the Feedback loop. For better or worse, we call it God.

Imagine all our realities and dimensions are the inside of a balloon. With the constant breath of Information in stella octangulas, it is limitless yet particles of stuff that then becomes energy or Mathematics or even Faith and Love. God is the wind. The wind that blows up the balloon with energy or call it breath, putting all this into our realm, and when God inhales, the balloon deflates (but really it does not, the countless and yet singular as a stella octagula simply scaled.)

When God breaths in, it/he/she immediately turns into energy (we might call it an idea or a Law of Physics) and reads it, and blows back in ... what? ... Our Dreams. Our Hope, essential as is Corruption. All in all, it is the limitless power of our Imagination, which is the vault in which whatever made us invented to see what we pull out of our Imagination, with God watching why, when, and what for. God is the trust factor between all that we know and all that could be known. Art, the Creative Process, is in what form and what method God measures it, meaning, God is a shared yet singular form of Nature in our current Dimension.

What is our peril? What shark do we need to jump? It is what do those in power do with Imagination once the Scientists discover this. (They killed Tesla before he could reveal it.)

How would Farming Imagination feel or be? Like Uber. You would contact an Imagination Driver. For a fee, he would pick you up, and take you for a ride through an expanding Tesseract, picking up Information like lint.

Why is this important to Artists?

The Spark of Genius, the moment of Creation, uses the Balance, Energy, and Wisdom held within the Flexible Tetrahedron which is simultaneously infinitely small and infinitely large.

Why am I so sure you can come and go through this theoretical Pliable stella octangula canvas/avatar skin and the proof is Art?

Simple.

Entropy.

Scientists all agree that the universe moves from "Order" to "Disorder." But wait, how do many loose colors of oil paint become a painting? Did you just go from Disorder to Order? The argument that scientists make is that the energy to make the art was greater than the energy held within the art. They refuse to see that the missing energy was absorbed by Pliable stella octangulas.

I pose the idea that the transference of Information from the universe via pliable stella octangulas could explain the existence or purpose of man. While this is covered in more detail in other chapters, it is important to mention here that the entire universe is moving into a state of higher chaos or disorder. Within Mankind, in our DNA and all our actions, we are driven to exploit the highest potential. And that brings about a primary urge to make order out of chaos – to explain the inexplicable, and to break out of our simplest existence to mimic Nature in a way that mirrors as close as possible Nature in all its forms. The stella octangulas, sticky with Information, shares it in Binary, instantly almost, so that inspiration, Art, and genius bombards our brains. Some brains are more attuned to pick it up or to absorb the Information – then through I.Q. bring it back to the One so it spread order.

Deeply creative people I know all agree that genius comes to them from without. This entire chapter came to me from Ether. The visuals in a stella octangular form materialized in front of my eyes. The memories of being taught otherwise were easily accessible.

I feel its truth, but it requires I give up a Human Agreement. That Space is a Void, that something we cannot sense, feel, or see or measure doesn't exist.

It does. In the form of Pliable Star Tetrahedrons, which will last forever.

DISTANCE FROM POLITICS

There are three rules to politics in The Age of Imagination.
The first rule is we do not talk about politics.
The second rule is we do not talk about politics.
The third rule is if this is your first time
joining The Age of Imagination
you MUST imagine without it.

One of the most (if not the ultimate) paths for an Artist is to tell the truth though his or her medium to empower people that experience what they created from within.

The modus operandi of politicians is to twist the truth for their own personal power.

Artists distribute power and nurture interpersonal balance.

Politicians centralize power and play to imbalance a person to control them.

Artists distribute power by filling fans with joy, which is an innate human goal from the beginning of our existence.

Politicians take your power and leverage your emotions usually devoted to seeking goal, and that leverage is often negative emotion, such as fear or hate.

Artists break into new wonder for the world through their chosen medium, thereby exhausting the unnatural fear or hate towards life in an individual.

Politicians keep you contained in their arena to get your votes and money, and they want consensus without inspection or introspection.

Artists take their first clues for "Real Truth" from Nature, family, and pets resulting in a learning experience that often transcends the Brain's learning systems.

Politicians (especially now) try to limit what you decide is Truth by controlling the media, message, and facts. And to do this they pay fortunes to experts in manipulating emotions and minds.

For the reasons above, I shun politics as much as possible.

This is not to say I avoid absorbing "news" about events. I study it, discount a lot of it as political propaganda from all and any sides, but don't let it control my actions, thoughts, or goals anymore. One reason why is because there simply are no news organizations that are not biased and self-serving.

At the beginning of the Internet, every single legacy news channel or paper shouted that the Internet news cannot be trusted because it is not vetted or put through stages of verification in editing. Internet news was the only news "held harmless" by spreading lies or ruining lives. Now they all capitalize on the money they get by feeding viewers what they want to hear, which is not usually what they need to hear. CNN executives even got caught on camera admitting their news is a lie, the content and perspective is chosen just to get ratings and followers on "click throughs" or "click baiting."

There were periods in which I was extremely invested both mentally and emotionally in politics. While I do love my country and know my mind was opened by its policies on personal achievement and responsibility, I no longer devote my creative mind to political thought or discussion. Why? Because of three main reasons.

First, the biggest reason. Politics is a black hole of hunger that eats up your time, thoughts, creativity, independence, attention to Nature, Balance, Truth, and open-mindedness.

Second, a good portion of what I thought was true on first viewing turned out to be either made up or exaggerated. And, we're never given the full story on anything from policy, to currency to warfare to scary diseases to government control.

Third, I swear it made me dumb – it killed my creativity. It made me impotent in ways to personally impact people in a positive way.

All my creative imagination went into decoding half-truths or partial information. Then it hit me.

This is where they want to keep you and me.

Not individually, but overall, creative minds need to either be controlled from political perspectives. They understand the incredible responsibility all Artists must take on when using their talents. The ability for Artists to motivate other people through emotions hits a magnitude that propaganda cannot. Young people especially eat up the message with the emotions driving it past reasoning.

I read a statement long ago: if my children are not more liberal than I am I'd be worried, and if they don't grow up turning more conservative I'd be worried. The curious mind is easy to manipulate. The person responsible for children or leading a company of employees tends to be cautious.

Both the United States and China reign supremely over the world now. And Artists within each country do need to understand the responsibility that comes with creating something out of nothing. Especially when their Art is rooted in Truth and in genetic understandings of what brings us joy or gives us a state of Awe.

My talents should prove over the long run to benefit humanity and the stability of both nations. My gift should be a way for both nations to communicate between their leaders and governments, to bring things that inspire their citizens in positive ways. My impact should not be on or for me, but for the people that are drawn to my true belief that your Culture is your foundation. Be proud of it. Root your goals in ways that nurture it. And realize that you might wish for all things you want at once, but life is a process, and most of us are not ready for the responsibility that comes with getting what we want.

I might have an impact on you – but listen most to your own genetic truths, all rooted in your great cultures in both nations. You are here because for generation after generation you kept evolving within one single great national Culture. But what does this really mean?

I'll explain it like this. As an animal, no matter what era, you woke up each day facing both dangers and rewards. Dangers were other animals, other people, famine, disease, and war. Rewards were food, health, love, peace, and pride in the people you called friends. You communicated with them in less fear because you shared long-standing beliefs. This is what Culture is. In any single moment, you respond to an event or idea from your Cultural past. Is this thing a danger or a reward? Does it offend my Virtues or embolden them?

It is your perspective that determines the angle into your beliefs. You know this most of the time immediately without having to think about it much. If you did have to think about everything that happened around you that you would never evolve. Culture fed and protected you so that you could make better food, better parents, better citizens sharing common goals and beliefs.

As an Artist anything I create reflects my own Culture… I hope in its best ways. You do, too, whether you knew it or not. You have the blessing and freedom now to explore new frontiers in imagination, freedom, wealth, travel, and creation of new things. If all your government or my government asks us as individuals is to remember to let your genetic memory based in Culture help you go to what is good for your country, then that is a very small favor to ask of you.

In every creation of mine, in all my publicity and speeches, I do all I can to stay away from politics so that I can do the most to impact you in positive ways. To make you more free to explore yourselves and bring joy and pride to those you love. I can't do this if I get trapped in politics. Besides, I know what I am not good at, and I'm no good at politics. And it consumes my mind, which should be doing other things.

In the United States or in any other county, abiding by what the government wants is simple because I assume the responsibility of my gift of creating things. I also see the deep value of the roots of Culture.

Be free to explore and make you and your nation truly amazing. Break through to futures so amazing you can't imagine them yet. But understand that if you did not have Culture to stand on, you would sink, you wouldn't rise as high. You might do well but why make it so much harder for yourself. I can say this without ever getting into the politics of it all. I won't comment on specific events in either nation. I will say now that the chaos and sorrow in the United States is the result of an attempt to refuse the Culture upon which we are based. This is all I will say. This should be enough warning to anyone to please enjoy and take pride in your Culture.

Go out and explore the world but come home and realize your own nation and city is the foundation of your strength and reason when in other nations. Explore the nature outside of the cities. Go to monasteries and wilderness and absorb the Truth in these places. Reflect on Truths and see how your Culture preserved them.

NET NEUTRALITY: A GOOD THING?

Net Neutrality was voted on in the years
this chapter was written.
Now we deal with the consequences
because as this chapter hinted it
delivered the opposite of what was promised.
It is worth remembering in
any Accelerating Philosophy
taking on AI's immoral conquering
of human gifts, joys, and Free Will.

In question was Net Neutrality. Why is this such a difficult topic to correctly describe, much less hold a solid opinion upon?

The answer in part lies in the origin of the words shared on the very beast itself – the Internet. We now rely upon Google, Facebook, and social media more than Nature, family, spirituality, and personal exploration to "give us our opinions." Dig deep enough, and the foundation sits upon "Likes" and Search Ranking.

Already, Google has proven itself biased on what news to promote, as well as Facebook. And the trolls scare off those seeking independent answers. What does this then require of us? To have proven sources of Truth relatable to our epigenetic code, as well as our emotions rooted in family and Nature, to qualify what you read.

Then I suggest as always two processes: Derrida – which is believing the opposite is true and see if more Truth appears from that view; and IQ – applying many ideas to one, and bring it back from the many to the one to monetize or take use of it.

Consensus is the anesthesia of the curious and just mind. Remember that, and flow away from consensus, in this case with Derrida steering you. With so little weight put on faith aside from the rising religion of "Faith In Popularity" and since the Bible was written so long ago, it is not the best tool to use in forming any argument to stick with on this topic. Net Neutrality does not seem to adhere to or protest many of the lessons of the Bible.

Let us then use Derrida to take a temporary stand that goes against consensus, which means you must support your argument with facts traceable in history. But to have either conclusion on this topic effect mankind, it must have "meaning" or go against a "perceived meaning" created out of nothing but consensus.

The best example is the perceived meaning of guilt which is dampened through Virtue Signaling. Can we find then examples of consensus trying to give Meaning, in other words, replacing a Faith or God with itself? Ah now, there is a great start on this journey seeking both Truth (Christian and empirical) and Balance (Buddhism and Commerce?)

Should the Government hold the power of what we would call Net Neutrality, giving it ultimate say over what is Fair Play and "Good" (or at least allowed) on the Internet, and then, should they be thought of as the gatekeepers on the power or access that both corporations and individuals have on the Internet?

Does the government have as an outstanding a track record of policing, innovating, and fairness on the level of individual providers on the Internet?

If I believed the government oversight or intervention would give more freedom, privacy, individual pursuit of wealth, health, and information I would be all for it. But, the United States Government would be the "oversight" on Net Neutrality. These same leaders and governors of things that blew our education and health systems to hell, abused power in all forms, has an approval rating from 11-19%, can be bought by lobbyists, abused both its monopoly and anti-monopoly power, used Internet as a social and political weapon already, and spend ten times to one hundred times more per service than private companies wants to back what sounds good.

Have the companies or government let you down? And, once given this power, what company is above the law or is excluded from the very law it says others must follow? The politicians' record in Insider Trading glaringly states they will abuse this, and that analogy is very fitting. Most people do not know that only Congress and the Senate cannot be prosecuted for Insider Trading, and even fewer know that they cannot be held liable for either breaking any Commerce or Utilities laws. In other words, they can use Net Neutrality to silence opposition and make a profit.

I remember when air traffic controllers went on strike. They swore only they could watch over our skies. Within one week, the President replaced them with private companies, that did the job better, cheaper, and with more accountability. It cut the budget and took abusive power away from lobbyists and Congressmen.

Look at TSA. The majority of major airports contracted the government-run TSA, newly formed after terror attacks. They have an abysmal failure record, in addition to their record for poor service, fraud, and crimes that if this were a private company, it would be out of business. If you fly into small airports that refused TSA, like Portland, Tucson, you are met with very friendly, fast, service authorities, that cost the taxpayer nothing. These companies have high ratings and pass the tests.

Are there more examples of when the government screwed up a good thing? When the government decided under Johnson in 1966 to get involved with making sure college was more affordable to the poor and all people could access it, and made sure of it by loans, this is how the college tuition went up 40 times, or about 10 times in cost every ten years.

When it comes to privacy, the government "claim" that only the Internet companies or service providers would get to spy on what you say or your business model. Wait – they admitted now that multiple companies would spy on you? Well, that's bad enough. But the government has never stopped short of spying this way, as was found out when Apple phones needed to be hacked. Now they openly admit they go in back doors to spy on the spies. This is what the stress test was all about, and why Jaime Dimon was fined billions – because he told the truth about the Stress test, it was extortion by the Justice Department, that fined billions but kept it, until it disappeared.

This leads to the question: When the government can announce arbitrary fines against violators of Net Neutrality, it will be the big companies that can pay, and the small ones that go out of business. It is sort of like forcing a Netscape or Blackberry failure upon companies without the finances to innovate as fast as, say, Amazon or Netflix.

Next you have the fact that the USA and world is being taken over by the great "Duopolies" (altogether FAANG.) Or multiple monopolies creating oligarchs. Best recent example are the thugs that took over the Russian oil business. But we have "friendly" Duopolies we love now, right?

Google wants to centralize all information and knowledge. Makes our life easier and searching easier.

Facebook wants to introduce people to other people like them and products picked for them. Cool. Makes it easier to get by without being challenged.

And Amazon wants to make us stay at home more, buy more online, and put businesses out of business that hire us when we are old, or young, or uneducated. Bezos follows the "flywheel" model for business. Is this good or bad for them to have so much power? A look at what they did when they took over Whole Foods proves what happens. Whole Foods was the store for local growers that delivered more wholesome and safe foods. Whole Foods shut out those suppliers and this lowered their standards. Whole Foods is dramatically worse for shoppers. So, should we trust them with a Net Neutrality influence?

Facebook (Meta) recently got caught in a series of felonies involving taking money from countries and companies spreading lies that they let appear as truth. Then Meta sold off the data on who responded to them, creating cults of followers to liars.

Google got caught constantly scrubbing information they don't want people to know.

Both Facebook and Google got tricked into supporting the Arab Uprising in Egypt, because they got duped. Yet we all bought into it as a great idea because it came from the gods of the Internet.

Are there examples of a type of Net Neutrality in the world that can show us examples of what happens when the government becomes the arbitrator of what is "good" or "bad" or "fair" on the Internet? Yes, in Spain and France.

Spain ruled that a person guilty of terrible crimes can demand to have all this erased from the Internet. And if the facts of him being a child molester or date raper come back online he can sue people or companies for violating what is called the "Right To Be Left Alone" law. What has resulted? Rapists, child predators, and drug dealers moved to Spain and into safe areas and committed crimes because people didn't know any better about them.

In France, fathers running out on childcare and husbands that are serial cheaters were given protection. In a few countries, they sued Google and Facebook for being too impressive with winning converts to their ideas. Yet, there is no replacement for these services in Europe. So people that previously could gain income and employment were out of work. All because the governments run by officials wanting votes aligned with lobbyists and unions instead of citizens.

In China and Asian nations Tencent and Alibaba control up to 70% of all Internet traffic. To no bad effects yet. However, you must ask, what differences exist in these nations? They are not free, they answer to a very controlling government body, the Core Communist Party. This Core has actually been extremely successful in making sure all people and small companies get Internet access. Yet, once they become popular, these companies are eaten up by Tencent and Alibaba because there is no anti-monopoly law there. There is one other reason why the Core lets this happen. They trust that major companies including JD, Baidu, and WeChat will do a better job of maintaining equality for all because they are private businesses, and cutting off a major fanbase is Cultural suicide. Yet, these companies cannot conceal their data or reword Truth to fit their narrative. So you must now ask, "Does Silicon Valley, the source of our information now, have something to gain by keeping Net Neutrality in power or making the Internet a utility?" Yes.

What Net Neutrality wanted to do was define the Internet the same as the old phone companies from the 1930s. This definition is what made Ma Bell a monopoly so big it never innovated or served well. It took 20 years to break it up. Then came Sprint and iPhones and all their innovations. Good deal.

But on what basis was the first utility law passed in the USA. A utility is something that the average citizen cannot live without, such as commerce, electricity, water, and radio. Ways to organize and protect, as well as feed and find shelter. Basics. The Internet is essential to improving your odds but is not a Bonafide utility. It is not utilitarian to existence. And is far less of this than is religious Faith, yet you don't see Faith Neutrality suggested to make sure all Faiths are equally honored.

Internet is not like water or electricity. They are not the old phone companies, that stretched phone lines from one coast to the other for 50 years without anyone able to compete. All the US system did was prevent new companies. Net Neutrality was getting to the same thing for FAANG but without the costs.

Then there is the bandwidth argument. Charging everyone the same price is not fair once you include who uses the most of it. Data caps will be high enough that individuals won't be affected unless they are in some shadow business like Bitcoin mining. Who will have to pay more? Netflix, Amazon, Facebook, and big companies that offer free wi-fi as a drawing to customers. When up to $1/5^{th}$ of Internet traffic can be shown as just due to Netflix, well then, if you don't use Netflix, should you pay for their extra bandwidth? No.

Now, points in favor of it not being under Net Neutrality are that it will allow ISPs to innovate. They were prohibited until now. What would they do? They have a laundry list of experiments they want to try out. It would or could greatly reduce cybercrime. But the US government won't allow it. How about if they wanted to build a backup Internet to their clients in case the Net ever collapses? They are not allowed to do that under current rules. Only the US Gov can, and they are dubious.

Silicon Valley profits when there are fewer companies able to get bandwidth, and that would happen as rules force newcomers that cannot afford policing their own fanbase or material. The government did this with the 'good sounding" international finance laws, a part of FinReg. They claimed this was to prevent black market money or illegal money from transferring. But, what did it really effect? They have measured that less than 6% of all illegal money was even considerable, and that less than a percent or two of the big criminal operations was caught. Far more is lost in other ways. They ignored that the biggest movers of money around the world are Foundations and Charities because Congressmen run so many of them. Oprah bragged about her Foundation for the girls in Africa to get an education. She moved $500 million to South Africa, untaxed and unmonitored. Less than $5 million has been spent. Only one school now exists. There is no attempt to investigate her.

But who did this financial law really hurt? Drug dealers? No. Foreign banks, especially in developing countries, and even in major cities. Singapore had to hire 6-12 full time employees per branch to do all the research the USA requires. Basically that money and time goes nowhere in helping the citizens. It in fact empowers the USA to bankrupt foreign banks by freezing or seizing all their USA assets. This brainchild was dreamed up in the Clinton years, and was pushed through in the Obama years. Yet, these two are thought of as being hard on Wall Street and banks. Not so. While Bush and Clinton both prosecuted approximately 1,100-1,400 bankers, Obama had only prosecuted 2 by almost his last years, instead fining them via the Justice department and Fed, and keeping the money until it was lost in space, untraceable. But, he got Wall Street's votes because he only fined the bankers, he did not trial them.

So you see, once the government gets a financial punishment or regulation tool like Net Neutrality, they abuse it as it matures. The same will happen with Net Neutrality.

Now with the Internet so powerful we look at Facebook for clues to the question: does Net Neutrality give Facebook so much power it can change the Truth and also replace natural sources of common sense? Look no further than Zuckerberg's announcement that the world does not need religion, especially Christianity, because the world has "him."

Hypocrisy is the homage Vice pays to Virtues.

This points back to a previous point. Meaning. Give the people Meaning, and they will follow it through hellfire. Net Neutrality poses to allow the most powerful hubs like Facebook, Amazon, and Google to define Truths according to their desires.

Here is a good way of assessing the Integrity and Virtues of the FAANG.

Facebook wants to automate away your capacity to think by replacing it with Likes and Popularity. It limits your ways to choose. It formulates "meaning" by consensus to forge an Internet homogony without Faith values.

Amazon wants to minimize choice in shopping to their portal.

Apple wants to put you in the center of the universe, feeding your soul seeking with App solutions.

Google wants to hijack the gift given mankind to explore on his own using multiple senses to evolve his ideas of right and wrong by stockpiling information in artificial minds that cannot sense higher feelings.

In Silicon Valley they can hide their propaganda within native advertising, excluding competing ideas. Try looking up books with strong legal cases against Google, you can't find them.

What to fear with Net Neutrality is the Hive Mind that cannot appreciate Nature or Spirituality, both are anti-FAANG. Algorithms make our ability to think for ourselves weaker and minimized the right and wrong genetic coding within us.

A.I. that will soon be used by all these companies will put this on steroids.

The Codifying of the Chinese Consumers

Regardless of what you think of China the way they advanced consumerism teaches valuable lessons.

Learning from any foreign nation that makes a positive game-changing social system is a good thing, even if it is a nation we are not politically aligned with such as China. My point is that if other nations clarified their consumers this well it improves trade and design. This explains what I call the "Gravity" that appears to be growing in power. It's growing due to social media profiles, and the companies in charge can seduce entire slices of the pie and determine what motivates them.

One of the most significant facts about China's immense market for luxury goods, entertainment, and media is that the market is evolving faster and with more impact than any market ever before in the history of commerce and communication. While at first the new liberty with responsibilities, along with new wealth, created a high-earning group that loved showing off, this did not last long. (A comparison might be Russia, where their new-rich and rich have always liked flashy "bling" clothes and products.)

Chinese youths and new affluent classes reversed their trend of trying to see a Virtuous and exciting reflection of themselves in American style, art, and goals in 2016. Looking inward to their own nation, their own Culture, for symbols of success or worthy-of-souls, is transformative. It is a key to their Creativity. They reject terms in favor of creating their own to identify the countrymen from a distance. This is a genetic trait in all surviving humans – the ability to assess someone approaching on multiple levels to start the Probability Machine in their Minds. This launches the ways that a person shares Information with another person in hopes of advancing both of their goals the fastest and highest.

On a micro, individual level, it is doing two things: It is proving the importance of China's 5,000 year Culture, thereby proving President Xi's push for Culture in all his speeches and hopes … and it "Distributes the best concepts of Communism down to personal relationships" so that the cost and weight of these best concepts shift from the State to the individual – essentially Xi is Uberizing Communism, and it is working. It is also part of what will get the young and most advanced groups of Chinese enthusiastic about his remaining in office.

You will see that there are different stages of consumerism in China. First phase might be called the Bling phase. This led to the next stage of more nuanced buyers, more savvy buyers, and buyers that valued the internal benefits of what they bought: in other words, they bought things that developed them as individuals and wiser people.

While this second stage created the next list breakdown of consumers and fans of entertainment, the entire nation evolved its buying power yet again, in a third phase. This third phase will be presented after the second phase. This is essential for not just selling things to them, but in keeping them united, proud, and growing as individuals. Overall, they want to be involved directly with the expansion of their culture, identity, and joys.

The U.S. hoping to profit from the numbers of people in China expected them to happily follow in line in our language. We blew that chance by not being honorable and respectful. Now, we cannot even speak the same language of Trends, in which we originally invented the Tech Code Words or Slang of the Digital Age. This is especially important to the Brand that any American force or business (such as a movie, studio, or bank) has hopes to win the loyalty of Chinese, even as it evolves. To achieve this, it needs to do two things:

1. Speak its language as it evolves
2. Have a "Meaning" behind the Brand that reflects the best of Chinese that shows off this language as a sign of where they are in their social and intellectual evolution. In other words, the Morality Merit Badges shown or spoken in Trending Codes (words, appearance, or skill.)

To give an example, I'll present the most significant group of words that have made a shortcut in how Chinese identify with purchases or style. First, there was MAC's and KOL's.

I must credit as a group influence online Chinese news including "The Sixth Tone" and "The Beijinger" and "Jing Daily" for this as a lot of the list is duplicated from articles written a decade ago and it is impossible to find the original articles.

Phase Two of the Consumer Groups in China

This codifies the next Group Brands that will lead trade and revenue with and within China.

I like starting with this word because it's close to the previously "Trendy Do or Die" or "Funny or Die."

1. Fu'er Dai (Foo-er-Die for you English speakers) – These are the 2nd Generation Rich that show off more status symbols visible at a distance. They wear logo clothes and give splashy weddings and Trump Gold. Born mostly in the 80's or early 90's, often to wealthy Chinese the rose to the top with the new economic freedom. Their parents are industrious and brave, and found Virtues as they grew up. Relevance. The core luxury consuming group of the future. To win these over, you flatter their egos. They want the red carpet treatment, a public display on small scales, like VIP treatment by stores. Sicong Wang gifting his dog with bling and iPhones is a cute example. The good aspect of these groups is if you can befriend them, they do help fund new ideas from poorer or smarter Chinese. The KOL's are on pillar under them, the other is the power that comes from wealth. This is important, because if the KOL's turn against them,

and they are a fickle group, things get iffy with their ego, and they need to alter what they do or go into hibernation until the storm blows over. It's sort of like backing Tiger Woods ... before his meltdown – he never recovered. Fans of Fu'er Dai switch allegiance fast. Chanel is doing a great job giving meaning to Chinese, I'll explain how they're doing it some other time, but it's something East West Bank should do with/through me.

2. Daigou. A type of personal shopper, or the purchasing commuters, for Chinese wanting to save money on luxury brands or any brand, really. They come in a few forms: professionals, students, KOL friends. They work with two things: currency differences and price differences. Brands are learning to woo them. Yet, they are both good and bad for China and its economy. They circumvent taxes in some cases, smuggle contraband, might be duped into buying counterfeit, and reduce the sales out of brick-and-mortar stores that really need revenue from MAC's. There is one other factor that few realize: they can either import bad ideas from abroad, or they can promote the idea that circumventing the Chinese law is a good thing. I'm iffy on whether or not to be involved with any Daigou. To openly state I appreciate their efforts and applaud their creativity is okay, but, the future will be better as they become less relevant. How do help this happen? Simply make the yuan stronger, and hire a local young successful start up in China to be key creators of our Brand logo merchandise and movie goodies, keeping the money in China, and making a new strong business, based on following the laws.
3. Shanzhai (Shi-an-zye) – These companies see a hot new style, and slightly alter it to be original. This gives the buyer something that looks trendy but is cheaper. The best of these are very high-quality and they move super fast. There are examples in the USA. A company in Santa Monica watches every Oscar or Music Award or idol, and quickly does a close original. The smart ones know how to avoid being accused of stealing a design. These are bought by MAC's, but, where they are really valuable is the class below them, rising up. These companies can innovate faster and alter designs to better suit Chinese weather, Trends, etc. and do not use slow-moving company Think Tanks, they move within hours or days on personal studies of all aspects of Internet (they tend to avoid Big Data analysis, giving it more of a soul of China, which is paying off – a Data can't pinpoint a soul.) We cannot avoid a Shanzhai copy, so we should embrace one or two early. And our currency of trade? NeoCoin, which enriches the Chinese Treasury, and gives Social Cred to the NeoCoin, making President Xi seem Trend-Friendly and a seer of the futures that are good for the rising classes.
4. Tuhao (Two-How) – The people identified first at a distance and later close up are the super-MACs that grew into wealth seemingly overnight. They did not have time to adapt the manners and culture of longer-term wealthy, and are often seen as crass and their purchases are looked down upon as uneducated or not based on anything but copying what the older money and status symbols do or did. A good example comes in the booming business of wine. They don't really appreciate its fine points, buy bottles to show off, and more. They're like the "Ugly American Tourist" in many ways. Most of Hollywood is actually a Tuhao, but they do come from a general sense of politeness. Still, the crassness of Actors, the need for idolatry, blind

their statements in public. Not good, but not bad, just part of life. Tuhao can propel a business for a short time but not a long time. It's also too high-profile, the Government has already made a few steps or condemnations against over-extravagant display. They came down hardest on the Entertainment people in late 2016 and early 2017. Announcements that are not hard laws came out stating the government will not enable and might even curb the rise of a star if too much Tuhao is shown off. This resulted in some stars being blackballed for more than 4 months. But, the up side is, they want in on the locations that are trending, and this could help with the popularity of Jing-Jin-Ji as it develops, and most of all Qingdao. More on that later.

5. Singles Day – As so many Chinese remain single in pursuit of better lives or simply because they can afford to be picky, Singles Day honors them like Valentine's Day honors couples. Alibaba gave it Meaning – in other words, their brilliant Brand campaign for it let Singles find a value in their being alone, and that value is paid forward some day, enriching others. The idea of a childless family or woman still prevails, so, this is important to the Culture, too. President Xi recognizes the need for two-child families, he did so openly in the last 6 months more than once. He wants the free minds of singles, but also family values. This Singles Day is a temporary cure. What needs to be done is a fast track from this status to solid Chinese married couple status. Core Policy in a Xi announcement has reminded citizens of it in a monetary/Social Security sense. The government social security system there will only support families that stay together. This is where we might find a card to play, to help Xi help Singles be proud but also just in a transition stage that will help China if they eventually move to marriage. Does this pay off? Yes. Alibaba did $20 billion in business on the first 5 minutes of the event. The event happens in the end of the year, allowing for companies to show off a big bump in the first quarter profit report, sometimes just in time for Chinese New Year luck! JP.com launched their version called 618. Amazon followed. We can't create one of these, but we can create a Brand item known for it. I know what that Brand item is. Note: this cuts out sales for brick-and-mortar for weeks.
6. Sheng Nv (Shung-nyu) – Chinese "left-over" women want a new name, image, spending habits, and … once again … Meaning – that is flattering. (Remember, give a meaning to what someone can do, and they follow it because it simultaneously gives sacrifice and indulgence justification and purpose.) Quick-to-notice International Brands have already profited off this new group that materialized in 2016. SK-II did it in make-up by creating a short movie (see how I can work in this arena?) The short documentary gave the Sheng Nv hope, honor for their sacrifice and situation, and went into Culture in a positive way. We can do this, and do it in not just movies, but in a Beauty deal, and in any segment of Xi's goals.
7. Duo Shou (Du-ah Show) – It's a funny term for a shop-a-holic. It comes from the funny promise they make they will cut off their hands if they need to if it is the only way to end their shopping addiction. Ecommerce made it too easy to be a Duo Shou from home. Bad credit card debts racked up. In an arena like WeChat, that builds a mirror of approval for many young and newly married women/people, it became not just easy to impulse shop, it added to Social Cred.

The new App that measures Social Credit will both help this get stronger and put a curb on someone that can't stop. Their online Credit Score will effect their Social Credit Ranking.

8. So, now is a good time to talk about these new Apps. WeChat and Alibaba will end up being the biggest and most powerful. While these Apps do indeed measure Credit Score history, they also will show what class listed on this page someone is in, if they cheat on their partners, if they are playboys, if they are hiding some problem, if they troll or stalk girls, if they are "haters", and much more. While we do want to enjoy a flash of Duo Shou when it comes to our Merchandise, Movies, Beauty items – we can enjoy this while also helping Virtue, the Sheng Nv, Singles, and Xi. Short films, that make funny messages about Duo Shous, work, but only if the outcome is it pays forward a Morality Merit Badge, domestic usually, but in Coin, Internationally. In other words, it is like a Carbon Credit. If you Duo Shou, you can admit it, and you can buy an item, or Brand rank, or Coin, that at least puts a little bit of your extra money into something good for China. Namely, the "Left Behinds" … the children left behind in villages when both parents go into major hubs to earn more for their education. The two tiers of helping the Left Behinds, will be 1-12, and 12-16 when they need help going to a college for a better life. With up to 70% of some family wealth going for this, to pick these forgotten kids is both incredibly humane, and also a brilliant Brand move for not just Creativity First Films, but for KOL's, solid MAC's, and most important, it helps President Xi in an agenda item – which, if a source of this is a Merit Badge (NeoCoin or our Coin from outside China) then the Duo Shou can show this off. How do they show this off? In a brilliant connection. The two Coins that China is inventing have different specialties. One is a commerce-oriented Coin. The other is a ledger history specifically outstanding in Charity and Loyalty tracking. Charity is helping these Left Behind kids. Loyalty is like supporting your favorite team (both physical team or eGame players, helping Wang indirectly.) Give Merit Badeges in Morality, Loyalty, Culture, the Yuan strength (making import buying cheaper) and Xi and it's a great Brand Indentifier for your company and mine. And it helps the Duo Shou either break their habit or find a role in Culture.
9. Zhong Cao Machine (Jong-ztow) Figuratively it means "planting grass." To pick words with emotions, memories, and future worth is what I do in my book One Fish Says To Another Fish – so this is an important group Brand in China to recognize. A Zhong Cao Machine is not a machine per say, it is someone that generates seeds so wide it is like a machine. KOL's are a category, idols, stars, and also hobbies the centralize groups (like gaming and movies.) Actually, anything that can bypass language to send a message for "What I want' or "What they want" or even "What I wish I could make them want," is the strongest type of Zhong Cao.
10. Why? Because this bypasses the Mind's normal process of evaluating the pros and cons of a purchase, alliance, loyalty, faith, or trend. What we want to do once we recognize this group is to provide the Meaning, the Bigger Reason to respect Zhong Caos. Why would we want to do this? Because as this happens, if we give KOL's a Meaning that is more than just being followed, we give them what is the duty of all "Artists that Create Something Out of Nothing"

(in their case Value and Popularity, which all Lux Brands chase.) But what we are enabling is something much more valuable than that chase. We are giving to them Meaning through Responsibility that will support Xi's Agendas: Culture, helping in poverty, good moral values, advancement in Tech, and more. What we do is say that the KOL's "Last Kilometer" in Xi's Point Plan is to give Trend Purchases of Luxury Items a Meaning based on Uber-izing both Charity to the Left Behinds, and also Art development at the special new branch of University North of Shanghai, it will play into Hendu Theme Park, and the value of Qingdao's Resort area. If you don't participate, you're still a good Chinese, but one of the scores on your Social Credit Score will be blank. This reflects the "good heart" of an individual Chinese Man, who wants a Bride of high rank and intelligence. You get this through ... NeoCoin or our Coin if abroad.

11. Xiao Xianrou (Se-ow si-an-row) Little Fresh Meat and all their fans. Good looking, tall men with nice skin – the greatest category of all the MAM's, and it fulfills the vision of the women. There is no accident that one of the biggest categories of rising Online or Live shows is Xiao Xianrou telling women how to show off their inner and outer beauty while still getting ahead in work or school. While the actual Little Fresh Meat are on stage, all MAM's are on stage Online. What is interesting is the spending habits of the ones following Xiao Xianrou's. Unlike in America or Europe, as the Male grows up, the fans don't forget them or go to other young meats, they evolve their tastes as the MAM's and Fresh Young Meats grow up. This means a lot to both International Brands and Domestic Lux Brands and Trends. A good example is how the hot trend of Sports Wear, Lux Gyms, and Lux Gamer Clubs cluster and flourish, and then International and Domestic Lux Brands will fill the shops that malls need to be successful. While yes sales do happen, the Lux Brands want to hitch a ride with the MAM's and Young Fresh Meats. We can appeal to them in Media, Beauty (helping Sicong again) Merchandise, and movies with sports in them like Fighter With Heart (an American version of the super hit Dangal, it is the story of the only woman that disguised herself as a man to prove women are tough enough to box in the Olympics and in opening games for big bouts.)

The listing written by several bloggers on Jing Daily above proves that the MAC's are finding specialty categories to recognize themselves in – something the USA youth do but as they grow out of cliques like Nerds or Jocks, they are not encouraged to carry this with them. In China the opposite is happening, because they realized they needed to not look to the West for an honorable and true mirror, but in to themselves. By dividing, into groups and enjoying their class in it (bragging rights) they find Meaning in their sacrifices more than their accomplishments.

This is a basic tenant of Xi's desire to Uberize the popularity trends within Communism: Responsibility, Virtue, Honestly, Generosity, Faith, Balance, Nature, Environment, and the longer history of China Culture rising from a long slumber in the DNA of all Chinese, especially strong as it is nurtured in the people under 35 Xi wants to win over before his next 5-Year term as President gets obvious. The clock is ticking ... those domestic and even more important foreign businesses and Minds that help him do this, will be singled out for more favorable deals, allowances, and transfers of money.

Phase Three

Experts within China have determined that consumer groups are driven by three forces that define them on a higher level culturally. The three forces driving them are: value consciousness (seeking the best value for a given price) susceptibility to normative influence (conforming to social needs like saving the environment or inclusion or cultural vectors), and the need for uniqueness.

At the top of the consumer list is the "Luxuriant" which makes up 22% of buyers. They live mostly in Tier 1 cities, which have a higher percentage of women in them. They are fashion leaders. Their decisions are also based on more humane or virtuous reasons than just showing off, such as for health, environment, and family safety.

The second classification is known as the "Demanding" group, which makes up 13% of users. This is a hard group to please. They favor products and events that make them stand out. Yet, they demand a deal when they can get it.

"Flashy" is the third grouping. Big spenders on brand names. But because they usually have less annual earnings or family wealth, they seek out cheaper prices, go for sales, and they spend more percentage of money on flashy clothing, accessories, and gadgets.

The "Urbane" group making up 14% have less money than the Luxuriant, are brand-loyal, and care more about quality and long-lasting use. They are not so driven to show off their purchases.

The social "Climber" is found more in tier 2 and tier 3 cities. They are heavily influenced by sub-groups and try hard to identify and socialize with more wealthy people. They need to be more price-conscious, and actively seek out cheaper options. The "Down To Earth" buyers are younger and place more value on family than social life. They do not buy to show off, they buy things for overall usefulness and reflection of family values. The "Enthusiast" are mostly in teir-2 cities. They may not have as much disposable income, but they love buying items that make them stand out and to show off.

What city they live in effects their entertainment decisions. These are the HNWI's of China, making up 70% of the population of consumers. The one thing they have in common is that they like to feel like they are playing a role in the rise of China's culture at the same time they are playing a role in the luxury and entertainment foreign providers. In other words, stage events that engage them, and let them live out to the fullest their value system. This means understanding what your product will do to enhance their identification and values, as much as keep them on the path of rising worth. Harrods, Chanel, and other brands created pop ups or store events that made Chinese shoppers feel part of the history of the Brand and store.

CULTURE

Once again we can advance ourselves by studying a nation labeled the enemy to "American Values."

China has the largest population in the world. Their culture is 5,000 years old. It has learned some things and now the leader is increasing his pressure and expanding his influence. Learning about the man that leads the Culture will add to your I.Q. because you have more reference points.

Any system that organizes people (government or religious) will not last without two things.

The first is Balance, which Xi learned in the monastery (most people don't know that he was stationed at one after his term in the military, and this does not put him on a pedestal it simply is a somewhat extreme and relatively unknown example of how politics seeks balance, too.) Viewing Nature changes a person. All things durable in Nature must be in Balance. Anything out of Balance tears itself apart from the inside, which is part of why we see a Culture War in the great U.S. unfolding right now.

The worst thing Chinese Youths need to do now is look to the West for Balance and confirmation of their reason for cultural balance because it would not be innate, it would not be natural.

The Youth of China stand on a point in Time. They look left and see the infinite history of their Great Culture. To the right they see their infinite future. They unconsciously found a pivoting and accelerating Meaning that keeps pace with their absorption of knowledge in all forms, and art in all forms.

The second essential element of durability of systems is this "Meaning" mutated for the technology age. In the United States, adaptation occurs, but it is not so noticed. "The American Way" epitomized advancement based upon merit in a free society.

If any leader or thing can infuse Meaning into the minds of any individual, this person will unconsciously align with the goals and message. They do this because they "feel" freed. In the era of mass bombardment of Information, we want to make sure that the message is based on the one thing that is real and undeniable in China: their Culture. The philosophers and monks that instilled evolutions of understanding Virtue made it to them.

Artists have a supreme responsibility that goes with the explosive power of their Creative Works. The West refuses to see this as more and more artists take political sides. They forget that Artists were often the first fooled by tyrants and the first killed once tyrants got into power. The new era distributes information so fast. Yet, is it so hard to come up with fantastic Entertainment and Tech that supports visions and hopes of Chinese Youth groups? No, not at all.

The fact is, that Hollywood wants two things: Chinese Money, and Chinese dependence on them. Hollywood is broke. And the distribution industry cannot survive without China's theaters. How do we keep things in balance in this arena of art we know as film making?

I have written my second book, One Fish, to show how easy it is to trio-up Culture, Innovation, and Virtue. How easy it is to give to China the material that naturally builds pride within each Chinese person. Chinese can smell propaganda a mile off. This Entertainment/Writing must be artful.

Two Oscars ago, a global message was spoken by Jack Black and Gweneth Paltrow. The first message was "Save Our Art Form Called Movies." The second message was, "If China is our last source of finance, someone, please, step up and guide or protect them." I was in Shanghai watching the Oscars in the morning, and two hours later was drinking tea with a top appointment in a China government body dictating over film and media to do just that. I'll show the card in our meeting.

A famous film industry leader said that the problem is not just that some are sick of the system of Hollywood, the problem is that the system is sick of itself. There is not a cure. Another said that if he were tasked with making Hollywood healthy again, he would start by bringing in a motorcycle gang with chains to rip apart what's standing.

There is only a better alternative, not a worse one really, because a worse one brings quite simply the end of film.

Much of this is in my book. I shorten the message to a few minutes. Yet, it applies now to all arenas of investment.

Whether someone likes President Xi or not is irrelevant to the forces. Understand that Xi governs over the single greatest and most challenging political transformation in the history of the world. The safe transference of intelligence to the youths at the same time as the transference of new leadership to a nation over 1 billion people.

It is my duty as a master of certain arts to dedicate myself to the safest possible transformation for the good of all people. The gift to create things that circumvent the limitations of words and reaches the hearts and souls of people must be remembered today more than any day. Yet, it must be set in a foundation of fundamental truths and sources of longevity. One pillar is Culture. The other is Originality. The other is Nature. And several more pillars. All my messages in my book take great extremes to not sound Political in motivation. This can be done. It can be enjoyed.

This can be done while still being an American citizen and loving my nation and culture first. Xi is not flawless, but he is indispensable and irreplaceable for at least 5 more years. Why do I say this? If China collapses, 200 million people will starve. It will spread disease 100 times more than COVID.

Any institution or system recognized as doing this of their own free will and faith in Art and Chinese potential/Culture can help immeasurably. My loyalty is to the United States, yet my duty is to humanity. If I can ease and improve life through my art and visions, that is why I was given the gift.

Almost the first question anyone asked in my 2014 visit to China, whether high up or street workers, was, "Hey what do you think of our President, pretty amazing, huh?" This is when I realized how different he was, and how much, even though people know some moves increase his power, they saw in him a gift of leadership they need.

I am convinced of his sincerity regarding the environment and awakening the free will and creativity in youths is real, and it comes from his time in the Monastery. In this stint after his post in the Peoples' Army, he was introduced to the one eternal source of Truth – Nature – and its requirements – Balance, Virtue, and more.

He took great risks in opening the young minds to Western influences, and I feel it is my responsibility as an Artist to help him help them. It is my true belief that as mankind moves toward "Humanism" and "Data-ism" that we will not advance to sublime understandings unless we clear the path for Chinese to raise themselves up with their own heart, soul and minds to perceive the great contribution their ancient philosophies bring to a nearly crippled and at least bloated and lazy American way of seeing the future. We need them also because they interpret Truth through Intuition and Chi. Balancing Brains.

On October 18th, the 19th 5-Year Congress convenes. Most people missed the subtle change in words used to describe Xi. This is not to set himself up as a new Mao, in fact, his goal lies in his lessons in the Monastery. They are to free the minds of the young and bring them into the Tech of tomorrow as their own leaders … to fix the environmental problems in their nation … to stabilize the currency while lifting poverty as much as anyone can … and to -- this is supremely important -- for China to finally receive what I call an International Morality Merit Badge (one that transcends the unfair accusations of doing good things in the world just to spread Communism 2014-2021.)

All Xi's Points prove the opposite. At its fundamental level, why did Communism (which sounds superior) fail and Capitalism flourish? For the simple reason that Communism Centralized Information (of all types.) All of Xi's moves are to de-centralize Information and he is doing this at a great risk.

The last time Information was poorly used was in around 1958 and it resulted in the great starvation. We cannot know all things Xi thinks about, but we can be sure that each day he asks for guidance in how to evolve Chinas' minds and empower the youths - without such a tragedy recurring if there is a breakdown in supply of food. We need to and can help him, simply by opening doors for Chinese youths to see, to find ways within, in Cultural Pride.

We are entering a new era in which Xi is asking that the best of Communism be put on the individual. In a sense, he is "Uberizing Communism's best elements" especially the humane part of it.

For this to happen, he cannot rely just on the replacement of voting members of his cabinet to elect him another 5 years. In 2021 it is the 100-year anniversary of the founding of the Chinese Communist Party. He is not going out of office before that happens and should not be replaced, I think due to his rare time in the Monastery (which explains why it is so important to him, and therefore us, to do some things to make Jing-Jin-Ji a reality.) He must "Balance" the Core voters with the mass of China's population endorsement and faith.

All aspects of investment in all foreign investments and even the domestic deals or expansions must help to support the strength of the yuan.

There are several major moves led by Xi in his Presidency that people are not connecting. (I will keep one of the biggest keys secret until the meeting.) They are in order:

1. The secret establishment of a PetroDollar with Russia, India, Kuwait, and a few others, defying the US, due to the monetary policy of Obama/Fed/Bernanke
2. The accumulation of gold to back the yuan
3. The establishment of their World Bank which got approval as the 3rd Reserve Currency behind the Pound
4. The announcement they will include the NeoCoin in the Treasury and treat it the same as gold.

There is a big way we can help in #4. I will explain it in the meeting. In Xi's today quests to make a Coin as valuable as gold, it requires both a Perceived Value and a Moral Value. The Moral Value can be achieved with our help in one of his main goals in his Oct. 18 Speech. I will explain this in the meeting, it is easy and fascinating. But it requires full understanding of the origins of our idea of what "Money" truly is, how it relates to gold, how gold relates to Coin, and a few other points. Coin has a foundation of quicksand (an implied value) until it incorporates in its definition (per Coin debut) of what I call "The Six Pillars of Encryption". I will explain them.

Xi knows the power of the NeoCoin is that it decentralizes Monetary Policy and yuan value. How do you increase this beyond doubts on its value?

The greatest value to China is in trade. The second greatest is reducing shadow banking, money travel, and online gambling. I expect them to make that CBDC essential to Gamers.

Those are special steps you will enjoy learning about in our meeting. Why did he allow Coin to be invested in, then pull it in? That too I will explain without a shadow of a doubt.

How Do you Know What I Know and What You Don't Know?

Knowing what another person knows. Knowing what someone on the other side of the world knows without meeting him or visiting him. Knowing what most animals know.

This could be the most significant innate skill that elevates humans above all other animals. It allows our minds to act more like superior probability machines. By doing this, we ushered in the era of using animals to improve civilization. While small societies warred among themselves over common ground, greater civilizations more often worked out agreements for trade and avoided warfare. How could they do this if the people came from across the world and did not look or talk like them?

It started with "knowing what they know" and what stage of mechanical evolution the foreigners were in. Knowing what they know gave people common emotional and spiritual ground, common food likes, and appreciation for foreign music and dance. Customs might seem odd but were respected. By leading with their best shared desires, they could debate on the lesser goals such as trade routes, taxes, and avoiding war.

Yet this did backfire a few times. When the Gaul first met the Romans, both parties brought their best customs and foods. Both knew that they represented the most advanced societies in their regions. Both had empires. Yet when they met, the Romans made the mistake of serving cheese. And the Gaul wanted to toast the Roman leaders, so they pulled out the skulls of their parents, drank wine from them, and offered the stranger a chance to drink wine from a family member, which was the utmost offering of respect. The Romans left thinking the Gaul were barbarians, the Gaul said they had to annihilate any rulers so sub-human they drink the secretions from lower animals. In this case, both parties did not know what the other knew.

But all ancient philosophies from the eras based the texts that evolved humanity upon what we know all living humans know. What are these?

Family comes first. That there is some spiritual higher being even if it is a human called a god. That Nature is a mysterious all-powerful force that must be recognized. That disease is bad and sharing stories to memorialize epochs is good.

Dance is a release of energy that transcends generations and should be respected. Sharing technology brought prosperity.

Keeping some secrets like how the Chinese made fireworks and weaved silk from silkworms gave them trade superiority. Discovering that one distant civilization believed in the same ancient stories and myths gave them avenues for learning based on similarities.

In this way, knowing just a few things led to great advancements that no other animal considered important. A few domesticated animals, especially dogs, found that their weaker dogs could find a better life outside the pack protecting humans. But they never knew what we knew, they could just mimic approved behavior.

Yes, some animals form a bond with humans and know what they are thinking at a specific moment. And elephants, dolphins, primates, dogs, and cats share understandings or communicate on various levels. Elephants even paint oil paintings, using their trunk to hold the brush, and the drawings are better than I could draw. And birds create amazing nests feature advanced "beauty." These facts leads to a few primary questions.

1. Is a human-level mind necessary as the antennae for creating something out of nothing?
2. "Experts" claim this awareness resides in our Soul, and possibly in our Colostrum in the Brain Brain, so does this prove animals have Souls?
3. What skill or inborn trait do these animals use to tap into creating Art or even appreciating it?
4. Does this prove that Artistic vision or inspiration comes through the Ether or from other Dimensions?
5. Do spiders spinning glorious webs, or birds building architecturally impressive nests tap into the Euclidean Geometry in Nature, and if so, do they do this to show off, proving they know what impresses other spiders or birds or even humans?

This is not so far off a concept. Whales send messages so complex over hundreds of miles that they can teach a song to a pod of whales a continent away. The songs seem to be lamentations on the past or charting new discoveries. Regardless of what is in the message, they found a song to be the easiest way to remember the concept. Or is it the other way around. Did the music or Vibration first haunt the whale until it picked up the Circadian timing related to magnetic poles in flux?

It's not important to know the answers to these questions in this chapter. They simply prove the point that only humans ponder not just what someone else knows, but how deep that knowledge goes, and where that knowledge came from.

Here is one thing that they DO know. Elephants look at humans with the same mental state that humans look at puppies. They think we're cute. Fun to watch scampering around and trying to interact with them. Which begs the question: do they know we find them fascinating? Do they know some people look at them as the key to understanding communication in this dimension and others?

Do whales, who are so gentle with humans they find lost at sea, know of our awareness that death waits for us all? Are whales observing us, waiting for us to jump the shark in communication, so that they can send the message out into the Ether to Aliens waiting for us to be ready?

Humans know that almost every human they cross paths within a day knows basic survival skills. And one of those skills is self-preservation. Which should give us pause. But stronger social skills push us for more organized works, that elevate mankind, such as cathedrals. Before that happens, different people must be curious about each other.

Spiders as artists designing a more glorious web do not necessarily refute the Darwinian Evolutionary models. What pattern the spider weaves might be an automatic exercise. Yet, how do you explain other insects like the Praying Mantis? The female will fall over and play dead when an unsuitable male tries to romance her into mating. This must mean that the female knows ahead of time what the male is thinking.

Shamans swear that plants talk to them and tell them what they can cure. This must mean that plants know what we know. Where does their knowledge reside if they do not have an identifiable central brain?

I mean, outside of their individual root or seed. This leads back to Ether holding an external brain, that networks with the systems of objects held within it. It means that the Cosmic Net is "aware" unto itself, and might have one giant sprawling Soul.

In our case, Mother Earth is the central brain. Mother Earth knows what we know. The plant is just a messenger system. Again, this comes back to distributing information as opposed to centralizing information. Being "part of the Cosmic Net," the Cosmic Net does not have to wait for it to know how Mother Earth and her creatures are doing, any more than you have to wait to know which organ is making you feel ill – you simply feel ill due to an imbalance of forces (fever, energy, etc.)

Now let us consider this on a more opinionated scale.

Does man know what animals and other humans know because of the network of Mother Earth? What does this mean?

If Ether transmits all creative information from other Dimensions and brain functions exist in Ether when Matter and Time lose rigid structure, then we know what other things know based on what we accept and/or absorb from the Ether.

As each Dimension can look down onto the lower Dimension and interact with it with known results, it tests us constantly to see what we can pick up at Information. At times, Will in the form of a stella octangular might pass through Dimensions with Information clinging to it, that drops off or sticks in lower Dimensions. When we respond in the right way, perhaps more Information of a higher level can be sent to us.

Higher Dimensions know almost all we know, and it appears incredibly simplistic to them. We might be little more than an ant farm to them. This does not mean that concentrations of Information or entities look down on us. They simply know our state is fragile compared to theirs, and our limitations are overwhelming.

Let's look just at vibrations, which are the touchstone of the next level up, Dimension 6. We have worked on raising vibrations for centuries.

Look at India and all the meditation there based on Vibrations and then expand it to Tibet and many regions in South America. It is as if higher Dimensions visited us, gave us the methods for piercing higher Dimensions, and somewhere along the way we lost the map or systems.

Maybe that was the point. To make it a challenge. To see what ways we developed without higher intervention. Some people believe the dinosaurs were wiped out because they could not evolve as groups. In other words, they could not use what they know other species knew.

Altruistic goals prove to higher Dimensions we are capable of evolving at a speed that makes knowing more about them meaningful and useful to whatever game the higher Dimensions are putting us through. And games is a good word.

We are almost in a game simulation in which the general two rules are: what are your observation skills, and how much do you want to get to the higher level? It takes repetition. Playing and playing and playing. Playing. Not working on getting to a higher level.

This is why I think that Artists and not scientists hold the secret to escaping this Dimension, entering a higher Dimension, and coming back to be re-assembled into Matter observing a Time order not based on Circadian Clocks. But on Memory. So, the goal is not just to take us into higher Dimensions, but to allow us Memories that can be unlocked in a Time continuum that is linear.

The reality that Humans can know what other being or energies know is a telltale signal for us to exploit this higher than the Dimension we live within. We have so much experience in this Dimension, who are we not exploring it as a system to rise. Tesla was doing this exact thing. And it scared the government of the USA so much they killed him and took his secrets.

There is not a need for faster computers, then, for this to be possible, but simply for just one capable Mind to be open to using what other people know as an acceleration vehicle, instead of a behavior control vehicle. This requires very diverse people working in a Vibration rhythm so Information flows. Art is the convergence. We're talking about plasticity of the mind and deconstructing Human Agreements again.

Art allows for sharing of Information in ways other than words, which require observing linear time, the byproduct stored in individual minds or machines. The external Mind seen in Nature is there to give us clues. A whole room of young people dancing to music without a thought in their Minds is a blank computer waiting for software, or simply shared desires for things outside of this Dimension.

Kids dancing don't know what all the other Minds around them know. and they don't have to. They just need to sync to the Vibration - and the energy builds. It removes barriers. The addition of drugs breaks down the hold this Dimension has on us. I'm not advocating drugs, I'm saying that there are ways to break free and truly use this gift of knowing what others know.

Monks performing a ritual to tones might be making more slow and graceful moves, yet it is the union of Sound and Movement, over Time – or more simply stated: it is Frequency when compounded.

Once again, the key to this all is observing Artistic impulse at that moment of "creating something out of nothing."

UNFORSEEN CONSIQUENCES

"More tears were shed over answered prayers than unanswered ones." Capote
Or how about this?
"No good deed goes unpunished." And finally …
"The only law is the law of unforeseen consequences."

Dismiss the human ability to distill logic, and then watch unforeseen consequences predominantly turn a good idea "bad." Sometimes tragic.

Replacing truth with what sounds good might make politicians, economists, and activists popular, but it leads to suffering.

Why would I consider this important to the future world of Art, Media, and Technology? Simple. These three expressions of human creativity hold incredible powers to persuade the minds of fans. What seems like an honest, innocent, or even good idea to include in a story will have some effect on the emotions of viewers and users. Emotions drive the sense of purpose or meaning for an individual as much as genetics, family, and Culture.

Once "the genie is out of the bottle" you cannot "put the genie back into the bottle." And one of the key problems is that the people often responsible for forcing a new agenda never have to suffer the consequences for being wrong. A good example is Sean Penn, who flew to Venezuela to be best friends with and promote Hugo Chavez. Chavez destroyed a great nation and now has left tens of millions of people starving, without electricity or water, or even basic hospitalization. Sean's push helped President Obama favor Chavez, giving him political cover through the Arts. But where is Sean Penn now? Is he donating ten million dollars of his own money to feed the starving Venezuela people? No. He's in hiding.

Outrage is prejudiced and aimed only at people who insist that the individual must take responsibility.

A few more examples help prove that when it comes to invention, innovation, and media, people need to accept that many ideas that sound good socially end up with bad consequences. Who should pay for them?

Environmentalism began in the late 60s and really took off in the 70s. It acquired significant political power in the 80s. Due to extremism, environmental issues became the second strongest driver for political elections, second behind sexual identification – at least for one party, the Democrats. However, it should be stated that many Republicans are environmentalists. I'll put that aside for now.

The hotbeds of environmental activism were Oregon and California, especially the universities. I was an avid camper, into zero impact, so naturally I joined one of the first environmental groups called Project Intercept. We would pioneer the idea now called recycling. At that time there was no such thing.

Within a decade, Recycling became a huge industry. So, the investors behind recycling got their way. Or did they? The plastic and metal from recycling was inferior in many ways, so they made demands on companies to use inferior material. This led to more frequent repairs, itself a contributor to environmental damage. After some time, recycled products found their role in garbage bags and lower quality goods.

The NGOs and trash companies do a good job hiding the environmental cost of recycling. It manifests in the trucks they must use, separate bins for garbage, taxation that removes money from the economy, toxic waste from it, and more.

When recycling became common it produced so much recyclable material that a large part has to simply be incinerated or put into ordinary dumps.

The availability of cheap recycled plastic bottles enabled the boom in bottled water, which is now the biggest source of plastic pollution. China and other 3rd world countries had to be used to dispose of the plastic. How does it get there? On giant freighters that use diesel. And then, it is driven in old trucks that have no smog control to dumps. There, people rake through trash so they can buy things in their country made without environmental control.

Environmentalists demanded an alternative to crude oil as an energy source. They used every tactic to ensure this shaming, taxation, violence, courts, fines, and political weight. Did they get the result they wanted?

Yes, if you are asking if they got major funding and someone raked in a fortune. The alternatives that are clean compared to coal we not championed.

Natural Gas emits no pollution and it is cheaper than oil. Far more can be transported in a single container. It does not spoil over time. Extreme Climate Warriors pushed incentives, not simply to get Natural Gas, but to penalize and shame the major oil companies. These companies were driven to invent new ways to acquire Natural Gas. Entrepreneurs invented fracking to release trapped Natural Gas. This led to an energy sufficiency boom that lowered all energy costs and resulted in an incredible decrease in heavy particles emitted mostly from automobiles, and then also by power plants, ships, and public transportation. You would think the people that demanded this would be ecstatic.

They're not.

First, the point never was to make the environment cleaner. It was 1) to punish oil companies, 2) feel better about themselves, 3) demonize political enemies. But more than anything the people got into Climate Change to give meaning to their lives. These groups only find meaning if they are angry against something.

All of their demands led to other problems that they can now complain about – because they have no purpose unless they have a cause and victim to demonize.

These same groups also demanded recycling.

Now they cry that Natural Gas has made plastics so cheap to make that there will be an exponential increase in plastic products. And they are made so well they do not degrade.

Let's look at one of the alternatives that plastic displaced. Iron, Steel, and Aluminum. To mine these metals ruins large expanses of land. Transporting it causes pollution. And it rusts so it must be replaced. It is heavy, so it requires more dense building material to support it. All around these metals cause more pollution.

Look simply at the automobile. It is hundreds of pounds lighter while offering ten times more gadgets – almost all made of light-weight plastics. One of the biggest automobile pollution problems was in radiators, made of metal, that would rust. The money spent repairing them could have gone for food for the poor. So, by replacing a metal radiator with a plastic one, less money is spent on gas, repairs, lost time, water, and oil. In addition, most plastic radiators now last the lifetime of the car. Then look at the oil itself. It lasts longer, and some is synthetic. Again, from recycled plastic, made cheaply with Natural Gas.

Another crowning achievement for environmentalists is the electric car. After decades, they finally delivered one that is popular in the Tesla. But how good is this for the environment? Let's take a look so we can predict that the tears soon to be shed over this answered prayer.

First, Elon Musk's greatest achievement and skill is not in making an outstanding car. While the Tesla is very nice, there is almost nothing in it that other car makers do better. All he brought was a more reliable battery. A better transference of energy system was already invented by Porsche, BMW, and was pioneered by Fisker.

A half-educated public voted to subsidize his company losses. For every Tesla sold (mostly to rich buyers since it was a $90,000 car) taxpayers paid $7,300 even if they did not buy one or could not buy one. The entire profit of the company came from Emission Tax Credits, often paid by states that at the same time are greatly increasing their homeless or alien populations, which also demand more money.

But this is not Tesla's ultimate two unforeseen bad consequences. Though bad enough to warrant I think legal investigation, the failed plant and subsidized losses are nothing compared to toxic pollution his car creates.

The two greatest pollutants come in the form of coal burning emissions, and toxic batteries. Most Americans proud to own a Tesla crow about saving the environment. They refuse to accept that (depending on what part of the country you live in) 60-90% of the electricity they need for their car comes from burning coal. The next source is oil, then Natural Gas. Only 2-3% comes from solar power.

Bigger than this problem? Batteries. Batteries so toxic they kill the children and garbage miners of the world by the tens of thousands now. The majority of batteries were sent to India, Pakistan, or China for disposal. Now all three countries are beginning to refuse to accept toxic batteries. This leaves Bangladesh.

Yet, all the movie stars promote the Clean Energy initiatives regardless of what they require and these people never pay the consequences.

To show another car example of unforeseen consequences, consider the Fisker. The pioneer of great electric cars. Fisker was on track to be Tesla before Tesla. All going great. I drove a test model. But then politics got involved. Joe Biden led the electric car popularity train. He offered Fisker an almost zero percent interest loan of $500 million. Fisker agreed and their stock rose up dramatically. Fisker was going to be a great success. But then, Biden put two demands onto Fisker, purely for his and Obama's political gain. First, Fisker had to move the plant from Michigan, where all the suppliers and test tracks were, to Delaware, a state Biden wanted to favor. This created chaos. Then Biden demanded the car be on the road before the mid-term elections! This was impossible. Fisker needed 9 more months of testing. But Biden threatened to pull the loan he promised, and this would crash their stock. So Fisker had no choice.

What happened as a result of Biden's stubborn demands? The batteries were not tested long enough. The same fires Apple suffered with their iPads happened. All companies making these lithium batteries had to go back and add in more safeguards. But for Fisker, it was too late. Batteries caught fire, people died. Did Biden pay the price? No. Fisker went out of business, thousands lost their lives. Did Biden or Obama pay a price? No.

Finally, Fisker was recognized by China as a great company, so they bought it. This led to other designs off theirs, and this will allow China to make their own electric cars. And they can achieve their goal to replace 90% of all delivery trucks with Electric by 2020. Because Biden demanded Fisker to change procedure, he indirectly gave China the means to build a superior electric car, blocking future US companies from being able to compete selling cars inside of China. Biden crushed future jobs in this division and allowed China to buy Fisker and all its technology for a fraction of the cost.

The Snowflake faction of the Millennials was designed to be extremely weak in logical prediction, setting them up to be victims of unforeseen consequences without them ever having the logic to trace it back to make the ones responsible pay for doing this to them. And most of this is done to them to gain power over them, and to get votes from them. A perfect example is the latest college loan offering from the Democratic Party. "Students should be able to pay their student loan off with their Social Security earnings." Now, let's think about this. Students sign up for loans to take classes they aren't sure will result in a job, that may or may not be there in the future. They sign up not knowing how long they will live and make a deal not knowing if Interest or hyperinflation will appear before they die, or if the dollar still holds value.

The politician that increases loans and forgives student debt gets elected. Students have been trained to think this is not a fair debt already. And many are fatalists thinking they won't live that long. Why not sign up for this?

Let's do the math on this, math over the life of the student. Assume that "Free college" became a reality, and the politicians voted to forgive all student debt. The student does not have to pay $25-50,000, which was the average student debt up until 2000 but it is now far more than this figure. To pay for this, they would need to increase taxes. Not just once but keep increasing them. So, the student graduates and gets a job that soon pays $100,000.

Taxes were 30% before this new free college. But after, taxes went to 50%, then to 60%. This is realistic, because Sweden has taxation that high. Each year, this college grad pays an additional $25,000 approximately. Within two years he or she paid back the college loan. But now must pay it for the rest of their lives. No one points that out to them. Why?

A better question is to ask why the politicians suggested this. Politicians expand their base by offering something for nothing – freebies, get out of jail free tokens – and do not answer for the laws they vote in. Their party believes government should run most everything, and that taxation should be very high. When students sign up for this plan, they are not reminded that at its current taxation, Social Security will be bankrupt long before these students use it. What does that require? A huge increase in Social Security tax put upon these earners. So instead of having to pay off a loan, the student then simply has the government demand it out of tax on their earnings.

They are forced to pay a tax they did not sign up for, which is taxation without representation. Sure, the party will announce they must keep Social Security financed, so taxes must go up. If these students live long enough, they will retire with far less or no money in Social Security, allowing the government to greatly overcharge them for a student loan now. But these students are not taught logic and civics anymore, so they cannot see this coming. They just go with what sounds good.

Education is probably the one industry suffering the worst unintended consequences. Ebonics sounded good, but had the opposite end result, in lowering black student scores. Bi-Lingual California attempts in classrooms also failed. Free iPads failed. And most of all, throwing more money at schools failed because it just bloated high earners and politicians who benefit from only one thing: increasing spending.

Bringing all this back to film and media, and technology… we gain greatly from invention. The Internet and personal computer took humanity to a higher level – but created cybercrimes. "Mass Shooter" video games made fortunes for companies, but now children are not doing physical sports as much, and some are suffering health and anti-society results.

The one thing all this has in common is "blind consensus." The people that want their idea to make them rich usually frame the proposal in a way that demonizes those that question it. The blind consensus won't allow opposing ideas. The greatest one being that each time a bad idea goes forward it creates a larger body of people that naturally go with blind consensus. This is not accidental.

Social Media does one thing that already has profound effects: it introduces people of like mind with shared goals to each other. This sounds great, right? But … but … for all of humanity, societies improved, and individuals improved in logic, health, and politics, by meeting people "not" like them. Forcing different types to listen, learn, adapt, and find common goals is what sets humans above animals. And it is one thing that the Art world pivots on – but in the opposite ways, if the Art world is healthy.

I'm not saying this because I'm older than the young people. I'm basing this on real life consequences; the type politicians and celebrities never have to take responsibility for.

By making it too easy for people to surround themselves and fill their days with "people that think just like them" their minds do not get challenged or expanded. And they can become easily fooled, led, and turned into fundamentalists.

Facebook is the worst on this. They get data and feed information to subscribers that make them click for mor and have set up algorithms to make this addictive and self- circling. The goal is to make Facebook the meaning for life. Already Facebook is caught guilty of providing news or ideas that do one of two things: help Facebook get to its goal of being a god to people, and limiting news to that which feeds their financial and political agendas.

DREAM AMPLIFICATION

Eastern Philosophies place a lot of divine knowledge within the realms of dreams and any Accelerating Philosophy should include Eastern respect for dreams.

Dreams have been recognized for thousands of years in civilizations in every continent as a passageway to secrets of the universe, and truth about your own being.

Dreams are highly symbolic and determining what you saw and what it meant is often up to you. Ask 10 experts and they will probably give you 10 different explanations for what you dreamed and what it means. This is an exercise that is at the very least amusing and even fascinating at times. People are more subconsciously looking for "meaning" and "patterns" every moment … much more than they think. And dreams can remove your ego from your thoughts and show you truth about life and yourself. Many people also believe that the Cosmos tries to give us lessons or tell us about the future through our dreams.

Some of the most brilliant people in our history as human beings admitted that some of their greatest insights and discoveries came to them in dreams. These include Einstein, Tesla, Socrates, Plato, Homer, Disraeli, Kant, Freud, and countless great thinkers from China and India.

Since obsession with dreams seems to be a human trait, dreams can also be useful in our attempt to remain human while computers try to replace our souls and epigenetic truth values with "data-ism." So, let's explore dreams. While these are my dreams, you have your own dreams that you should start to study.

Lucid dreams. Dre am wherein you realize you are dreaming. Often multi-sensory and acute in physical feelings. Smell and hearing are often heightened. An inability to monitor Time despite trying to track time happens. A realization you are more aware of your environment than others in it also is part of Lucid Dreaming. There are many other factors.

A book of very few pages I read at the old Bodhi Tree explained additional realization like the "Silver Thread." This is a tether that attaches to you wherever you go, and the other end is secured back to your soul. The Silver Thread usually yanks you back into the body before you can take control of the dream. This small book on how to control your dreams presented a method to remain in the Lucid Dream at that moment you tell yourself or realize "I'm dreaming."

If you can master this, the Silver Tread does not yank you back. The author said to look at your hands for five minutes each night. Time five minutes each for both sides of your hands. Studying them.

And then go to sleep. When in a dream at that moment you think you might be dreaming, look at your hands, notice details in them if you can. If you can do this, your mind accepts it did something on purpose, and gradually you explore your dream, and you'll be in control what you do in the dream. It is possible Lucid dream without this, of course, and most people do.

One other way is to recognize a common element in these dreams. This works for me, but it took me a long time to figure that out. Practice this trick, it works and is worth it.

Once I got better at not getting afraid in my dreams, I realized that some part of my dreams seems to be that one single fact – meaning, the moment I see something extremely bizarre or a threat to what I am comfortable knowing, if I remain calm, usually the next thing in the dream is fantastic to behold. If dreams are a window to greater truth, some of that might be very confusing or upsetting.

I like the idea that dreams test a person to see if and how much new truth they can handle. This book is about stretching your mind so that you expand its boundaries, so that you can stay far ahead of computers, even A.I.

The human ability to discover the secrets of our existence and the cosmos will lead to opening new areas of our mind to higher activity. There is a lot written about how most of our experiences travel to trusted parts of the brain for storage. And there is a lot of speculation about how the brain, mind, and soul interact or relate to each other.

Dreams could be trying to show us, or trying to see who correctly interprets the message, which usually is delivered in symbolism. I do find that my dreams offer far more Cosmic wisdom and new ways to look at things since I began focusing on these books and my purpose in China.

When studying Chinese literature and history my mind repeats some of what I studied all day, and now it appears in my dreams. For example, a sound I heard in a dream reminded me of a common Chinese phrase I learned. This simply told me that I am on the right path to learning more and making it usable.

One symbolic factor that repeats in my dream is water. It might appear in many forms, such as in rivers, hoses, and raining, or huge building waves of water. I'll explain the dreams, all five of them. They are very brief, so I hope it won't bore you.

The first Lucid dream that was so strong it has stayed as a big part of my self-awareness occurred in about 1977. I explain it in a previous chapter but the longer version here is better. The dream was so clear in my mind I can see it perfectly after all these years. I worked and lived in San Francisco, age about 22. In this dream, I was transported into the water of the San Francisco Bay. I swam in the Bay, just under the Bay Bridge, about a quarter mile past the towers that hold up the bridge.

I could feel the cold and hear screams of fear. The bay was covered with small white capped waves that got wilder and bigger every second. I found what looked like a 2x3 meter wooden hatch for an old wooden ship, floating easily on top of the ocean water. It had ropes that were attached to it. I climbed onto it, water splashing my hands and knees as I stabilized on it. But I was safe. Then I looked to the Golden Gate Bridge. A gigantic wave almost a kilometer tall raced at the Golden Gate. I watched it tower over and crash through the San Francisco Golden Gate bridge, with the bridge falling over in slow motion. Every boat in the water was crushed under it. The wave rose up the side of Telegraph Hill to the top. The city would be buried under this wave. The wave was coming to me in about 30 seconds.

All around me I began noticing people I knew well. Guys and girls I knew, co-workers, business contacts, high school friends. I looked to them as they were screaming for help. I was beckoning them to join me. They all wanted up on the wooden hatch with me. But as the water rose and got more violent, waves pulled them away, making the distance greater.

When the tidal wave hit me, I wasn't crushed under it. I rose up it and to the top. Then it went past me like a wave you miss while surfing. I was alone up on top of it, realizing I am in a dream and I was there to see this from a sort of an "Eye in the Sky" perspective. The water level went down. I was one of the few survivors. I heard someone or something watching me and looked around but saw nothing but water. Then felt I wanted to go back to my life and snapped back into bed. I woke with full memory of it. I thought I was witnessing a future catastrophe.

This also was the year that San Francisco broke big in Silicon Valley Technology. I did not relate the two facts together. I only thought of a tsunami burying San Francisco and me witnessing it from a safe place.

It was many years until I had my next dream related to this wave. I was in Hong Kong, scouting locations for the film "The Quest' starring Jean Claud Van Damme. I was learning about what will happen when it returns to be part of China. I was there and had a dream that a huge wave was coming, just like the San Francisco tidal wave. I was on top of a skyscraper watching it and the roof lifted up in the air. I watched the wave pass under me. So, this was in 1998.

I had another dream about a huge wave of water. This was in 2014, February. I went to Shanghai the first time and stayed in Pudong in the Grand Hyatt. The city was still mostly riding bicycles, that is how long ago it was. I remember strangers all over the city asking me the same question: "What do you think of our new President and how he's going after corruption, and opening up the Internet?" WeChat was becoming a big thing.

On my last night there I again had a second dream of a wave, this time it was coming to Shanghai. I'd already had the dream with a wave in it that inspired me to write about "The Last Ship" now came a second wave dream. I was on the floor where the pool is at the Grand Hyatt. I was looking down on the Satellite skyscraper everyone knows is on the riverbanks on the Pudong side. I saw the wave coming in from the right, in the river, covering up all the boats and the many cranes that were rebuilding along the river. I was so high up that the wave went to my feet but not any higher, and the Grand Hyatt stood all right. A lot of the buildings on The Bund side were drowned in the wave. Again, I felt confused and sad why I could not lift up the people I saw to protect them. The important thing was what the water symbolized, that the people put in my mind. This was the year that President Xi was going after corruption and allowing more freedom in the Internet.

In Singapore in 2015 this time the next evolution water dream started telling me something more. I was transported to fly in the air over Singapore, feeling like I was in a balloon. But I moved very fast up high. Then suddenly I was watching Hong Kong from about a mile off the shore. The massive wave hit. I realized I had someone next to me. It was sort of human, but a little bit like an Alien from space, tall and wearing a simple grey long robe. He had skinny fingers. And I said, "Why am I here? You showed this to me before." And he said, "Because you could not see what it really is." I looked back and all the white on the wave was not water. It was … a virus.

I finally realized that what I was watching was the moment when Information on a higher level were washing over the people. Giving them the freedom to rise up as good, smarter, people guided by Virtues as much as the quest for a better life. If I could describe the alien now, he was made of pure Virtue, and was more like a Tao Monk, with a long white beard.

The second I looked down I saw the city of Hong Kong was not destroyed and there was only one ship, a huge ship, offshore. It was the last ship. Suddenly the words hit me at the same moment I woke up. I said the words out loud, "I am the last ship." I sat up in bed when I said it, so it was the moment I snapped back from the dream. And right after that, the rest of the meaningful words came out of my mouth. I will keep those words to myself for now. The important aspect is that it clarified that I had to make a choice. To live with a purpose, meaning being willing to sacrifice everything to achieve a greater purpose. To hang the value of my life on it.

Since that day, I knew my purpose. It has defeated every obstacle given to me. It is what I speak when I do my morning yoga.

Then I had one final tsunami wave dream of this type. It was on February 9th, 2018.

In this dream, I am not at the shore of the ocean. I am in a sort of bar that is part indoor and part outdoor, looking over the Chicago River. Like being on the walkway along the Shanghai River, or the marina in Singapore. Everyone was dancing to music. I look out the window and the water on the river was getting very big, with many whitewater caps. And it was rising fast. I told everyone we had to go outside and see why. I began to realize I was in a dream, and not just any dream. In a dream like the previous dreams. There to give me a message.

Indeed, the water was going crazy, very violent like a fast river going downstream. Lots of whitewater. We all walked into water that was up to our ankles. I told them we had to go into the water where it is not so rough. We walked into water that went up to our knees, then our waists. They were afraid and it was cold to them, but not to me. A wave came at us. So I said to swim out to be safe. But instead my friends got afraid and ran away from it. I swam to it and was surprised, it was not cold, it was not scary. In fact, it was warm and welcoming. At that moment I knew I was in a dream for sure.

At first< I thought the wave would break over me and take me with it. But it did not crest where I was because I swam out at it alone. I rose to the top of it and could see how it hit down the river. Everyone was buried under a huge wave. And in the dream I knew then it really was like the other dreams. And the message was clear. I was the Last Ship and would be alone.

This is the burden. That I must accept I will be alone in what I present to the billion people I lift up like a giant wave of water. They will rise from the rising Information wave.

I have been tested now five times, tested to see if I would understand that my gift of thought had to be alone. And that is okay. I knew the people with me in Chicago would be fine, just like the people of Hong Kong were fine.

I had one more dream where water played a role. I was in a very nice garden, full of flowers, that went over the hills and through rocky walls. I noticed a giant beetle, so far away I could barely see it. It was bigger than a bird. It flew right at me, with a very scary pincher nose and four wings buzzing loudly. My first instinct was to run away. But I said out loud, "This is a test." So I stood there and held out my hand, and this giant bug that could have killed me grew smaller and smaller until it was the size of a hummingbird. It landed on my finger, and it wanted the water that was in the rock bowl right next to me. So I lowered my hand and it drank water … and then it smiled to me. And we were friends. Something scary was my friend because I let water do for it what it does for all things: it gives a gift very humbly. And I felt humility, and adored this bug, and is smiled. Then it flew off.

So this is what I realized about each dream.

The San Francisco dream was at the moment when a new era of Information in the Internet age would hit mankind. The first Hong Kong wave was when new Information would be given to Hong Kong as it returned to China rule. The third wave for Shanghai was when President Xi opened Information and freedom to be successful to the people of China. The fourth wave was when I was helping leaders in film in China access the new Information level of Media. And the fifth wave was when the finances from Chicago would free me to sail alone as The Last Ship to deliver my message in these books. The last water with the giant flying beetle was to show me that I was not afraid of Information in different forms. I had passed the test.

The new wave of Information is not just waves on their own. They are waves that build upon each other. They are getting in sync. As they do, the wave will amplify by double each wave that joins the water, the endless pool of Information.

Now I know what most people will think – all you did was guess at what the dreams mean through symbolism and maybe the dreams led to nothing. I could think that same thing but once I realized them as guidance to write the books then it became what I needed it to be or thought it was after I dreamed it.

Here is how it all ended up.

The waves are named. They are known. They are merging. They are:

The first wave of the Computer Tech era.
The second wave of innovation on a global scale.
The third wave of youths worldwide engaging the Tech Era.
The fourth wave of Social Media and Entertainment dominating Information.
The fifth wave:
The Intersection of Media platforms and the Internet of Things.
This wave has an energy, a realism pulling people into it
as it covertly makes youths comfortable with spying via IoT.
The time is now to balance this wave with Human Virtues.
An Accelerating Philosophy to match an Accelerating Intelligence
presents the ways the Virtues are measured and recognized.
And I must do this journey as if

I am The Last Ship and so are you,
giving what is pure like gold
and unbreakable like Integrity.
Even if this journey is only in my Mind …

It is all becoming synchronized. The waves are not connected yet. I am the only one that sees they are coming. It is my gift to bridge the sine wave so the wave of pure Information arrives people have a bridge. The burden I carry is solitude, and the secret realization that what I give in this will ultimately kill me. Yet, the knowledge in me will be shared, and the wave will lift a billion people.

Knowing this, the final wave, in Chicago, has meaning. This is where my money comes that allowed me to sail as The Last Ship. The time is now. No more do I need confirmation from those around me. Explaining is wasted energy. They will only know it when the wave passes them and they are in a sea of the Intersection of Media, AI, and the Intersection of Things.

Will someone come along to add to the novel with a book on Physics that moves us closer to solving "The Three Body Problem" this era? I think so. But maybe the dream should be taken more literally. The Last Ship was a real Red Cross medical ship. Does this foreshadow a medical storm akin to what was predicted in "The Broom Beach Disaster" film?

In the novel, "The Three Body Problem" was solved by two concepts. A person that can notice details others do not see and therefore cannot use to solve an unsolvable problem; and a person that possesses the Planck Unit made of unbreakable connection – the only spool of nano wiring connecting it all. Together they maximized the potential of what is most unique about human beings: we have an accelerating intelligence.

Will is the essence of all our realities, known and unknown.

Information is the Planck Unit in this 4-dimensional universe with Space-Time.

The connection is nano, unbreakable, and it is the thread of Virtue.

Information signals the Cosmos when it is ready to transform its contents into Intelligence in people. But they must meet the challenge and be unafraid to expand their minds and Virtues to hold it and use it.

That means they must enjoy the Will in each cell, in their every thought, action, and emotion. We will achieve in its "Pillars of Encryption" the "Planck Variables" that respect the simplicity of the five Planck Constants. Essentially, this will lead to standardized Planck Variable Tests for companies and their Ledgers or cryptocurrency, much like standardized Electronics led to advancements in televisions.

Ontological Truth Trumps Social Vectors

The Internet linking us all is driven by consensus. After all, isn't that all that Google is? A consensus search?

Throughout humanity, a person grew up in a protected environment of the family, where interpretation of the world was governed by the parents, and to a lesser degree, schooling that centered around the basics. These basics emphasized Mathematics, language skills including vocabulary and penmanship, and the sciences.

Mathematics and the sciences dealt with mastering the Ontological Truths that cannot be disputed, such as the observable laws of Physics (gravity, rotation of the Earth, numbers.)

Mastering Language involved mastering logic, which will now be shown as essential in the new age of Data-ism. Penmanship is mentioned because the process of writing characters in Chinese or cursive in the Romance languages and English requires attention to detail and appreciation for Culture. What was not widely known back then is that after thousands of years, the process of writing calligraphy and turning pages to read actually anchor and propel knowledge.

They have found that the cadence of Shakespeare and the complex musical arrangements of classical music in both Eastern and Western Cultures evolved us to the point that humans are drawn to hear it play out. All this, while slowing down thought to capture the optimum word.

With these skills for interpreting truth in the world a young person was ready to leave the nurturing family cell and meet people vastly different from them. Through meeting people vastly different, the use of logic merged with whatever level and ideals of Virtues this individual upheld. The individual gradually learned "who he or she was" by overcoming obstacles that gradually formed higher levels of self-awareness. He or she became unique by testing it against what "was different."

Logic emerges from the greater source of Ontological Truths that surround us in Nature and the cosmos. "What do I believe in?" "What is important to me?" "How do I cultivate joy?" What gives my life meaning?" "What is precious to me, what am I entitled to call my own?" Challenging what you might know as Truth strengthens Truth if your logic proves to be stronger than untruths.

All of these essential steps in self-development made humanity and the Cultures within nations what they are today. And all these are being undermined by the Internet and Big Data. Realizing this offers a way to preserve and respect this process of finding "your true and better self."

Culture and Truths must be upheld before the Technology is apply it to society is part of the starting process .

Key to this evolution is properly identifying the Social Vector of each region, nation, or national Cultures. In past generations only one or at most two major social changes would arise. And these would challenge Cultures and parental guidance. Over the last 4 decades, there seems to be a new major social change every year. Every year our core beliefs are attacked to allow these social changes.

As they all line up they drive society in the direction of the most energy and impact: this is the definition of a Vector. In Science, Vectors must be proven true. In society, most Social Vectors are not proven, and some are later proven to be very bad. Yet, no one pays the price for taking positive goals and energy away from people easily convinced to change directions and their Vector. This convincing includes using guilt, shame, blacklisting, prescription medication, scary viruses, confusion, lies, and other unethical techniques.

Within a nation the Youth Culture tests Truths now in different ways due to technology. Yet this should cede weight of opinion to Culture that both the old and young share. Culture cannot be fragmented for this. It must be appreciated in its whole. Just like you cannot have "some Integrity" you cannot respect "some Culture." Integrity like Culture is one of the few "Absolutes" of human society in that it must be considered as a whole, not as an action reflecting appreciation for one moment of culture. Culture and Integrity are "indivisible" just as your body and soul are indivisible.

You cannot shave off a portion of your soul and apply only that part to your personal evolution. You cannot use only part of the Culture of your people in your survival and growth as a person – you must appreciate it all. That does not mean to use it all equally. It simply means that you acknowledge that your Culture is a vital ingredient to how your body, mind, and soul integrate to go through your life's journey.

This is where Social Vector becomes so important. Because the Culture of the United States is under attack based on the manipulation of our national history it is useful to see what China did right.

If a nation like China has a youth class that treasures the entire Chinese Culture that goes back centuries it can also know that each phase of Culture does allow changes they need. The processes through which a person seeks Virtues and joy align in all generations and serves the nation as a whole.

The point of Culture was to protect, increase health, advance Art, evolve the soul, and unite even extremely different people toward a common goal. If someone refuses this idea then usually this means they want to take only a fragment of their Culture to serve only a fragment of their future. Civilization has never honored this. It simply rolls over it. But, if all someone hears is that fragmenting Culture is good and not doing so is evil, then they will make the wrong move out of ignorance.

To make sure the long-term Culture was valued and used as a whole in Western countries the wise men and leaders aimed the visions of people, especially the young, toward the Ontological Truths of the world we live in. Growing up, they could see these Truths as clearly as they learned about the forces of gravity. Then as the mind developed Logical Truths advanced living situations by enabling better laws, wider learning opportunity, and philosophical studies.

But if the studies that enrich reasoning and Logic vanish, a young person is witless and powerless to articulate Truth. We do not need to be told all the Truths of existence – we know many of them from within our being, reflected in humanity and expressed through empathetic actions and laws. We are in part talking about Critical Thinking.

Facebook and other social sites draw you in then make it easy to find people that seem on the surface to share a lot and think alike. This then is "monetized" and sub-groups of people are created in their advertising algorithms. These algorithms already know what to hit the young people with so that they feel like they belong and come back often. The algorithms reinforce this with Likes and punish it with social attacks.

This culls the herd of free-thinking intelligent youths, that use fewer and fewer words of wisdom. They do not need to develop logic or base new ideals on Truths or Logic. What this does is it "centralizes Consensus."

Facebook or Google would love it if you just gave them all of this. They want to hold your life hostage by increases the fragmentation of our Culture by dividing citizens. This means when anyone looks around they think they are reading what everyone else accepts as being true. It is very subtle censorship of options and perspectives. Demonizing people makes it easier to feel good about limiting your opinions.

This prevents you from escaping to personal responsibility. They especially do not want you to find any meaning for living and thinking outside of their company. This was proven in 2017 when Mark Zuckerberg announced Facebook users do not need religion or ancient philosophy like I Ching – they only need him because he is a "savior." He will determine what is Truth. He will then use it to define how you find joy and love. And how to raise your children. With the upcoming era of A.I. Facebook will monitor a profile user to make sure this happens all day long.

In clear words, the Social Vector of Facebook is to make you rely on them for joy, attention, approval, wisdom, a sense of belonging, your self-esteem, your opinions, your friends, and your image.

The current trend in the United States is to replace Truths with "what sounds good" and what makes someone feel good. Replace it with what gets approval. This is called Consensus, or Consensus Rule, and Virtue Signaling.

All advertising, news, web design, and profits extend out of what Big Data now tells companies is the greater Consensus for their product or service. Why is this important to Art and creating anything, and in a separate universe of computer causality – why is this important to us in relation to A.I. when it appears and takes over?

Truth – for most humans – is determined through words. Words can be twisted. They can be overvalued or undervalued depending on Consensus. This consequence expands without Cultural appreciation, unity, and display of action. The simple act of returning home during Lunar New Year or sweeping the graves of our ancestors illustrates how this unites even up to a billion people.

Yet, dismissal of a long-term national Culture becomes automatic due to the new Tribe fracturing of Western Civilization. This is also called the "Digital Re-Tribalization of the West." The book "The Strange Death of Europe" presents how this is happening in Europe. Why could this be happening? Because the computer minimizes the two things that prove a Truth: Nature and Logic. We interpret both individually through our senses, and this develops our inner drive to follow what gives each life meaning.

To be loyal to a national Culture has now become illegal and proof of this can be seen in the limits and punishments given out to citizens that want to describe the Truth. In England they are expanding laws to appease millions of immigrants. In many nations it is a crime to state the nationality of a murderer or rapist. This hides the real dangers of immigration. Why would the E.U. do this? Simple: for more control. Fear is a method of control.

Philosophy that upholds Integrity and gives value to the genetic Truths get replaced by "Likes" and "Re-Tweets." If this isn't enough, the bullying of the "Consensus Power Mongers" intimidates individual recognition of Truths. The damage this does to an expanding national society compound. The opposite tactics are being used in China.

Nations must remain competitive and that requires opening up of Information that holds Truths. Competitive technology giants like Tencent, JD, Alibaba, and Baidu need to simultaneously lead, free, inspire, and protect the youth groups.

The unifying goal tool is appreciation of Culture, acknowledgement of Ontological Truths, and development of personal skills in Logical Truths.

Nature presents Truth on an undeniable, gargantuan level in an unavoidable environment. The saturation of Truth by comparison makes the human ability to determine Truth by all other means insignificantly small. This Truth that Nature bombards us with is the greatest source of what is called Ontological Truth. Examples are the balance of Nature, the Planck Units, Laws of Physics, Order of Species, positioning of the cosmos, mating, and existence as we understand it and interact with it. None of this is spoken. Ontological Truth is what really exists in the world and what has "real being."

The "The 5 Brains of Tara" chapter explains how we absorb this saturating Truth, and so much of it without words.

Genetics is the next level down in presenting the human mind with Truths. Its impact relative to Nature's impact grows as we learn more about genetics and the human genome. Genetics follows a very slow curve in adaptation. Epigenetic Memory, or memory that adapts faster due to social constructs, also holds Ontological Truths. For example, we cannot deny that genetics passes down traits from generation to generation. We cannot deny that a mutating cancer cell appears out of nowhere in our bodies.

We cannot deny that in our genes lives the fight or flight impulse just as it also holds appreciation of Fire. Words do not tell us to be in awe of the greatest examples of Nature, whether it is an eclipse, tornado, Half Dome, or the birth of a child. It is Culture that tells us how to calm a crying baby, or defend Philosophies, rulers, and the cycle of the harvesting of crops.

Culture provides the next level down as a source of Truth. Customs that gave humans in a society a common goal allowed their energy to share a direction long enough to accomplish great feats, such as building the Pyramids and Great Wall, holding off hostile enemies, surviving famine, organizing crop harvesting and selling, spiritual guidance, and protected the family unit. Culture for a nation continued if it was just adaptive enough to keep up with innovation. World travel and major wars are both good examples of what impacted Cultures deeply yet allowed people of a nation to sacrifice for each other and adapt to one another.

The I Ching and Tao hold the largest reflection of "human interpretations of Truth" for Asian people in all categories: Ontological, genetic, and cultural. And in the I Ching stones were etched the first examples of Binary Code. Unlike Internet sites, the I Ching goal was giving this Truth of tool to anyone to find their own wisdom and Balance that reflects the greater Ontological Truths.

The Bible holds Logical Truths expressed in stories to define Virtues and the penalties for ignoring them. But it offers little on a philosophy to appreciate the Balance found in Nature. Perhaps this is why in the age of Data-ism fewer and fewer people find meaning in the Bible. Conversely, more Asian youths find value in Nature and ancient writings. Their emphasis on balance and observance of Nature for cultural and human Virtues and motivations proves this again and again. Yet, both works understand at their core that words become a trap when Truth is involved.

The most common category of Truth for centuries was based on Logic in the Western World. This is odd because the Fathers of Logic – Socrates, Aristotle, Plato – gained most of their words to express Truth and their Philosophies by observing Nature first, genetics second, and what worked for advancing Cultures third. In pursuit of power, Nature got left behind. Culture remained strong but now is fragmented, or Tribalized.

Logic requires a minimal level of appreciation for Virtues because without Virtue a lie and a truth could have equal value. Only one of these serves humanity in the long term.

How did the West lose its lead in Logical Truths? It simply became more popular to dismiss the role of taking responsibility for one's own actions. Power over people can be gained by removing their skills in Logical process. In place of this the Tribal leaders offered a free way out of suffering the consequences of unethical, selfish, and or weak action.

A coward is now a victim, even if he or she did horrible acts. Victimhood is championed. Nature and Natural Laws, along with natural laws of humanity, are dismissed because they develop free thought and the sense of responsibility that must go with liberty to expand the flow of Information to anyone. One key proof of this is the announcement when certain words or moments in history are forbidden. Without the correct word, history cannot honestly be studied for improvements in the future.

A great factor in the rising arrow or vector of humanity is valuing the long road of a Culture: asking honestly, "What got us here?" The answer does not always serve those seeking power. And they minimize the individual, his or her rights, and access to Information to gain more power over them.

Truth matches words with ideas. If you don't know the source of your words then you can fall for any idea, especially if you do not spend much time exercising your individual skills in Logic. In other words, you are easily manipulated. And often defenseless against arguments or laws and rules that seem unnatural. Consensus insists you follow them.

In the Technology Age, no one is held accountable for lies and misleading people. The news now is all about driving people by their emotions to click on an ad, link, story, or picture. The greatest sources of Information in the form of words are Google and Facebook in the West. Both are completely compromised when it comes to distinguishing Truth from lies propagated by Consensus of their users and fans.

They find the most manipulative methods and use them on the people easiest manipulated. These also happen to be people easily driven to outbursts or violence. Such as mob mentality, cult actions, and low-education groups which effects all immigrants to a new language and culture.

Logic gets thrown out to ensure more followers, to get laws passed by lobbyists appealing to politicians that need votes, and to control the masses.

The United States increasingly damns greater sources of Logic and refuses to hold them up against Ontological Truths. For all the campaigning for saving the planet and Climate Change the same people are demanding their Tribe ignore Ontological Truths.

Some of these relate to the Truths evident in Culture that relate to mankind. A good example is the family. Popular causes demand the eroding of Cultural Truth in the family. Groups champion their "ideas" as superior yet refuse to subject them to either Logic or Ontological Truth examination. They base it all on feelings, emotions, and some vague idea of what is "fair" in the world. No one allowed to question the origin of the Idea, or who benefits from it in the long run. A good example is degrading women who prefer to not work and raise a family: humanity did not get this far by forcing women to work through guilt trips. A second example is demonizing strong men and natural leaders, labeling them as toxic. Destroying the image of the free-thinking father or individual man is an example of a "Tribal Un-Truth." Every source of Truth far greater than that proves otherwise: that strong men and fathers are essential to society and family.

The Tribe bends the will of the individual and incentivizes the majority of users to make Social Media popularity more important than anything. Computers allow the majority to rise without facing Logic or any consequences for spreading a lie. They risk nothing, and live in luxury, often paid for by others. The very people damning the upper 1% maximize the power of money to avoid Truths and replace them with Tribal Virtue Signaling.

This fails humanity fast.

One key question can be answered by this: Does it give you true joy or did it simply give you a message and a divisive meaning to follow? Evil will do all it can to act out its potential. This is a universal law from the individual cell to a Tribal political movement. They are not responsible for building up something new, they only care about tearing down something. As in a Nihilist. The governments cannot allow this. The United States is going through its first and – if it loses – its only Culture War.

THE NORTHERN HEMISPHERE: ANTENNAE TO THE UNIVERSE

Almost all invention and advancements in early civilization originated in the Northern Hemisphere: Agriculture, gunpowder, aqueducts, the Bronze Age, Iron Age, spoke wheel, scratch plow. Ninety percent of the manmade wonders of the world are above the 33rd parallel. The great computations of Egyptian mathematicians, the designers of gothic architecture, Pi, large ship building, steam power, the computer, airplane, car, spaceships, and smart phone are just a few life-changing inventions.

All the great thinkers of humanity and their creations whether Eastern or Western lived in the Northern Hemisphere: Confucius, the Tao, I Ching, Aristotle, Plato, Galileo, Copernicus, Newton, Einstein, Hawkings. These names give a clue. What existed since Egyptian greatness up to now, since the I Ching? The stars above us.

The Southern Hemisphere sees a different sky at night. Observing our rotation around the Sun allowed civilizations around the globe on the Northern Hemisphere to advance at almost the same time in harvesting, sophisticated music, warfare, and complex philosophies. Observation of Nature resulted in deciphering the balancing act needed for surviving famines and wars. Above that these geniuses led us to complex mathematical equations that began to explain our Fields and Laws of Physics. This led to studies of Biology. Again, all in the North spreading horizontally.

Could it be that cosmos engineering and Field creation beam Information thought Dark Energy onto the Northern Hemisphere?

Neuroscience is mapping the connections and firings within the brain and discover patterns that copy star formations in our cosmos. This happens to the point that images of the brain and universe can be laid over each-other and bright spots correlate to brain functions. But only in the Northern Hemisphere.

There are a few isolated civilizations of greatness in the Southern Hemisphere or bordering it, namely the Aztecs, Mayans, and Incas. Yet, their method of measuring time outpaced their mechanical innovation as we understand it today.

It was only in the Northern Hemisphere that the great thinkers of societies were driven to compare involuntary action with evolutionary action. Only there, did the concept of a human gift for individual potential get tested and extrapolated. The West was always "involuntary." The East Evolutionary. Both must be based on Virtues or there is ethical darkness. The source of life is the source of joy … it springs from within.

Art is the opposite of a subconscious being told it is powerless. Art is achieved through realizing you have power and reservoirs of truth being stimulated invisibly. The balance of our actions comes from these three parts of the mind working together as a tripod to support the journey into the outer realms of experience. This brings about outer realms of vision and then application, that results in "creating something out of nothing." The explosive energy potential of the Information comes to us in chaos we turn into inventions.

Free Play is the childish nurturing of interacting physically with "what is" and the triggering of "Imagination to imagine imaginations." This is uniquely in the Northern Hemisphere.

Creativity is not an enigma that cannot be explained and children know this. It is a riddle, ever changing, that has a skeleton key for unlocking the rings of multiple riddles that spiral out of life. This reality of life happens to be presented to us in repeating examples in a Fibonacci Ratio pattern. Newborns "know" The Theory of Relativity, as proven in Peek-A-Boo. It does not take wisdom over years to "react to and relate to and within" a universe that does not obey The Theory of Relativity. Over a matter of months, an infant loses access to inter-dimensional Truths and reality. Adults chase after it for the ultimate questions of our existence.

As Information moves between the higher Dimensions through the entangled and binary shared communication, it interprets the Will of Fields within Dark Energy. This Information then is given the freedom to show itself, as White Energy, or what we know as our see-able universe. We explain what we imagine about this in words, which is not the real language of Dark Energy. We realize more each year that Mathematics, vibration, and the Science of music or sounds provide us clues to unlock secret.

Were we closer to understanding our reality centuries ago when words played a lesser role in interpreting the world and our senses as well as epigenetic impulses of knowledge were stronger? In the eras of glyphs?

With this understanding, the ancient philosophers gave a singular approach … how we drink from the cup of forgetfulness. By this I mean we discard the humane systems and the deep thought of our ancestors at an accelerating rate to appease a group of "accelerationists."

This is in part due to the increase in social norms within a generation due to the computer age.

And here is where Media and specifically fast-moving Games tech shows its essential purpose in our evolution. First, in testing our bodies to work with others without words but more like fingers on a hand – as seen in Gaming Olympics. Second, in recognizing the "magic trick" of mass entertainment. Considering so much of the evolution of Games came from Asia a comparison is valid.

While the United States making movies had everyone believing in the magic of the Ruby Slippers – China searched for "no place like home" in Honor, Virtue, Culture, and Face. We offered the heart to those that already had a heart, a diploma to those that had a brain, and a medal to those that already had them. And this is the honor of the "New Censorship" if you want to call it that. To refuse the Western Identity Politics is not the same as refusing to honor whatever Tribal memories and rites a person might follow and honor. It is a defining element (and perhaps Identity Politics' most damaging rule declares) if you are a victim… "you belong to us." This spreads both fear and human impotence. It tells us we are not in control of our lives and not responsible beings -- the group or Identity is in control. This ensures a downward Vector of Society. This requires controlling education because Identity Politics demands to be more important than the nation and long-term Culture. They demand they are the source of morality. How can you appreciate a country or yourself if you do not know your true history?

China offers citizens a trade: Liberty with Responsibility. Identity Politics demands that people take blame for things they did not do and that their ancestors did not do. It distorts what is truth and where to find it. It steals away a young person's ability to compete on his or her own merits. And in this process, it removes incentives to be a moral, ethical leader among friends. Once measurement and incentive are removed and replaced by mob approval the benefit from the true long-term Culture is voided. Because politicians are now the "intimidated puppet masters" of all this the laws they pass will increase the negative Social Vector of Americans. Tech Giants in Silicon Valley want this because it makes mindless fans.

A grander philosophy built off of core ideals set thousands of years ago ensures the upward Vector of the Culture where it can unlock secrets while in the flow of Yuen Sun. Then it can lift the United States and world and remind us all the Truths beamed to us in the Northern Hemisphere. To be given Free Will at the same time as consciousness powers the antennae for the raw Information of the Northern Hemisphere.

This ultimately must result in change of Frequency, to the point where unlocks the doors we so desperately need open to retain Free Will and Free Play.

Leaders in a New Era of Patent Protection for the Arts

[**Note:** Much of the ideas in this chapter were thought up in 2018 when Blockchain DLT showed so much promise. Anything related to Blockchain was ruined by the greed of people running the first companies. Now with A.I. the danger to patents and anything human is exploding. A solution is out there waiting for your accelerating mind.]

As technology advances, the traditional system for registering patents no longer works efficiently enough or fast enough to protect original creations in any art or creative form. This is especially true in the creation of software, and digitized entertainment (films, songs, videos, television series, ads, and many other forms.)

Can Blockchain DLT introduce a better level of patent protection? Will transparency help protect the individual singer or innovator in an era where FAANG seems to steal anything it wants without punishment.

The system that served inventors or businesses forging into a new service worked well for over 100 years. The most powerful and common patent protection was in the United States (and in some European nations.) Many emerging nations had trouble protecting "what is created by the mind" – known as I.P. Rights. Standard Essential Patents (SEP's) are recognized in nations. China's M.I.I.T. and the E.U. worked together recently to innovate superior, faster ways to ensure I.P. rights internationally.

Blockchain and new forms of Distributed Ledger Technology appeared more frequently since 2016. Two of the fastest companies to create patented systems were IBM and Alibaba. IBM's most famous success model is their logistics tracking with Maersk. Alibaba was pushed to develop additional methods of proving authenticity of luxury goods like Louis Vuitton because the deliver website had rampant problems with fakes of high-cost Brands. This was especially true with luxury goods from Europe, that were bought in record amounts by the rising Chinese luxury class called Mass Affluent Class.

After these companies, the idea of Blockchain providing a transparent and immutable history of any product or art spread to tech agencies. Already, the major global brands are teaming up with companies to create their own Blockchains that prove where the original products came from, and all steps in the creation of an item such as a purse or shoes.

China opened up five courts specifically for Internet fraud and I.P. theft, and have successfully prosecuted hundreds of thousands of cases, delivering harsh sentences for repeat or big violators in smuggling or the production of fakes. One reason these courts were so efficient was that they allowed the use of "Smart Contracts" as evidence. These Smart Contracts are self-executing and appear on screen as soon as an abuse occurs in media. In products if a fake QR code is scanned or otherwise recognized it alerts the parent system.

Alert systems are being promoted to full use in Europe, China, Singapore, Asian nations, and the United States. The top world court for I.P. protection is in Singapore: on last review, they had 7 judges – 2 from the United States, 2 from China, 2 from Singapore, and one that rotates (and they gradually will add more judges.)

Many companies realized the huge potential if they created I.P. protection Blockchains. Leaders included Corda/R3, Chainalysis, MUSE, ACCORD Project, BitFury, Accenture, The Chamber of Digital Commerce, Rimon, ACCESS (Association of Cryptocurrency Enterprises and Startups,) Coinplug, and other companies compete to see which one emerges as the global dominant forces for luxury goods, film and media, music, and software rights to protection. Several key changes in how the courts operate, honor, and then rule on cases helped speed this along.

One of the greatest challenges for any of the above listed companies is that they need to work hand-in-hand with those that create the Intellectual Property and deliver proof while innovating the system to higher standards. This is known as a "Patent Pool" or in Hollywood as a treasure chest of Original Content.

The tech giants like Amazon, Apple, Alibaba, Tencent, and others are investing tens of billions into Blockchain and need Original Content. A.I. will be able to copy patents and I.P. so fast a tidal wave of crime will crush artists.

The I.P. rights companies need to cover this growth this faster than their competition. Winning over large suppliers of I.P. Rights is a key to any one of these companies becoming recognized as the "Global Brand Leader" in I.P. rights protection. Anyone remembering the problems with the music industry when rogue companies offered free music can see how entangled the fight can be if all the artists are not represented together as one force.

In addition, working in the beta stage with White Hat Hackers like eIDAS must be done on all three stages: sharing the work created, the company protecting it, and the distribution leading to R.O.I. for all royalties, Patent use, and new revenue streams. In the future wherein Augmented Reality and Virtual Reality team up with IoT and Artificial Intelligence this is even more important.

If any I.P. Protection company also has dedicated a lot of time to the "ethical, moral, and virtuous use of I.A." it is ahead of the competition and should be put in a spotlight for making this integral to its achievements. This is how to win over the needed I.P. created by artists and software developers. And this time is not in the future, it is now! The Chinese company Nreal will be selling its "Mixed Reality" glasses that allow VR to appear in front of people, and for them to interact with it.

If a film, song, or other form of Media knows about this going in and plants the "Easter Eggs" into the Original Content/Production, then naturally the I.P. Rights need to include this and be able to work freely with companies like Nreal, regardless of what nations creates the hardware.

This is another reason why Singapore and Switzerland play an oversized role in the future of Blockchain and AI: we need neutrality in deals with these innovators. What this means is that some

U.S. companies are not free to protect rights to products made in China if new court orders or trade wars determine it is a violation for the U.S. base company. This means that European and Singaporean headquartered companies are favored when it comes to I.P. Rights.

This is known as creating a healthy eco-system of cooperation on the level that the industry understands it now. The Dynamic created by our companies was engineered to be the Organism-Environment that is explained over a hundred pages in the six primary books written to define and dominate the future of Blockchain in all industries.

It is important to remember some of the abuses by Facebook, Microsoft, and other big Silicon Valley companies – they either sued valid competitors out of business, stole the idea and outspent them in court, or swallowed up all ways to build a vertical business model.

Silicon Valley should not gain a global monopoly on all things related to innovation, art, and consumed Media. It would then bully European, Singaporean, Chinese, Japanese, and Korean innovators, artists, and musicians.

To make this even more complex, some industries are under a trade restriction in the war between the United States and China. This makes it even more important to make the national home for the I.P. Rights protection and distribution in a small, independent country famous for meeting regulatory standards. Most European nations that are members of the E.U. will have trouble creating an enforceable I.P. systems due to some rules in Europe such as the right to be forgotten, nations that specialize in stealing I.P. rights, and the ease of stealing from the Internet there.

As for the leading companies, there are some factors to consider.

Amazon is too big and their R3 alliance with Corda makes anyone entrusting them essentially their bitch. No one will have any loud voice in Amazon ventures. While Jeff Bezos is not a bad player in this arena, everyone simply has to recognize that Amazon will do whatever it wants to do. It will for sure grab a big share of I.P. to protect, but will it deliver the best protection and the best services? Will it listen to innovators and artists? By their history – no. They will maximize profits and squeeze out the little guy. And, Jeff Bezos admitted that once he sees a good thing he races ahead with all the money the company has and innovates off an I.P.

Chainalysis is a great company, but its relationship to some companies might make it biased. Microsoft's history of lawsuits makes this scary for artists.

This is one case where members of the Ethereum Alliance (E.E.A.) might have altogether monopolistic goals. If they get this, they have no incentive to improve services or play fairly.

China has the most experience and widest use in I.P. protection through Blockchain in courts and the use of Smart Contracts proves valid – but most nations won't trust China.

This reduces the number of companies to a shorter list. Accenture, MUSE, Bitfury, CoinPlug, and a few emerging companies in Singapore. Accenture has other key businesses. Which points out another problem no one is thinking about yet has been proven to be a huge problem for the near future. This has to do with how Cloud storage today functions. In Winter of 2020, Microsoft admitted that its Cloud slowed and almost froze due to overuse by gamers stuck home during COVID.

Crypto-Integrity-Tao realized in 2018 that the current Clouds will not be able to handle the explosion of Social Media energy and A.I. This is both a problem and a path to enormous riches for the bigger Clouds, especially the BAAS Clouds being promoted by Accenture, IBM, Microsoft, and Amazon. BAAS will simply be replaced by a new type of Cloud that holds vital data. What this will result in is a silk trap or golden cage for companies and eventually individuals.

By drawing companies into using Blockchain, these biggest Cloud companies can make the clients dependent upon Blockchain, just as they are dependent upon logistics data in real time today. If Blockchain is a rehearsal for A.I. then we apply this fact. The biggest Clouds will draw companies into A.I. in the early stages will do the same. The cost of Blockchain Cloud storage will skyrocket, then when energy prices go up accordingly, skyrocket again. This means huge profits for these giant Clouds – but is it fair to the companies and consumers.

And then there is the problem of Cloud failure, ransom attacks, or companies like Facebook selling the data of their Cloud users to companies like Cambridge Analytica, which specializes in government election perversion.

Seeing this problem years before it even hinted the chapters called "Dark Clouds" was written. And another chapter called "Why Personal and Public Data Must Be In Different Quadrants."

This will impact individual artists and the people that want their music, artwork, or media/film/Streaming content.

This reduces the choices to a few favored companies. But one factor that is important to know is this one: did the company write their own code for software or not? This is important because so many use Ethereum as their Smart Contract language protocol or base, and this has proven to be buggy and vulnerable to hackers.

A tangent hurdle to overcome is in getting the artists themselves to agree to participate in the art being protected. Remembering the first artists that agreed to go along with iTunes shows that this is the smart move. However, immediate incentives must be obvious.

The Responsibilities of Artists in Our Past Evolutions

With great gifts comes great responsibilities.

Many people ask why art is important, or question the roles and duties of artists, especially now with the computer replacing so many art duties like mechanical drawing.

So let me take you on a journey into our longer past in order to predict our near future (actually our current stasis.)

Hunter-gatherer homo sapiens observed their world so much more than we do now that even his primitive life proved more inventive and adaptive than our daily life now. They had to observe the food they ate and hunted. Women's thumbs opposed their fingers before men's thumbs did (if you believe the long-term evolution idea.) Yet men could throw objects and swing clubs harder than women.

So, the women were drawn to harvesting food in the ground and in trees. This centered women around homes or camps. Their creativity imagined countless things we take for granted, like how to spin yarn, how to preserve food, and more. Imagine yourself, along a river in India, washing clothes for years. You find out that the clothes you wash downstream from the lye on the river banks got cleaner, and soon began scooping up lye so your clothes always had this in the water when cleaning. Would you be able to observe this and make a change in human life?

Yet for all the inventions we introduced to the world, the grandest examples of our creativity were in the "Seven Arts" I mentioned before: Architecture, Dance, Painting, Music, Sculpture, Literature, and Stage. Up until The Renaissance these Arts mostly served one purpose: to make the supreme ruler into a god. Ancient Egypt provides the greatest proof of this.

Huge statues of kings so impressed slaves and conquered tribes they gave up their freedoms and lives willingly, seeing a hierarchy order to serve. These stories that cemented the powers of human gods got exaggerated until about 1,000 B.C.

The powers they gave to the gods usually revolved around food, harvesting, natural disasters, and wars. While they could use only their own logic, observation, and primitive calculations, they built the massive canals and dams that prevented floods and famine, that would be a challenge to build even today.

This, through art and observation. Art, presenting the Truth in the Fields (Gravity, Mass) ignored the passage of time. Generations died before the dams, canals, and pyramids completed. The human gods, called Pharaohs, held mastery over all animals and harvests. They needed the masses to believe this so they would not rebel. Artists conveyed this message in a way universally appealing so that slaves or traders from other worlds understood not to question a Pharaoh.

Now, after conquering most of the problems related to survival, we give the power of gods to corporations, computers, government bodies, media, and organized heads of social communities like Popes and mullahs. But really, we as a race are merging with the new machines giving us powers of gods.

The powers to remain young and healthy, the power to never be hungry, to travel around the world as fast as the sun crosses the sky, the ability to communicate without face-to-face settings.

As computers and medicine begin augmenting our human body with chips and new abilities like "The Bionic Man" our quest to be gods will be reached.

Yet, the powers of gods rests in the stories, the art, the humanity that binds large masses of people to work together. Reckless interpretation of art and storytelling has led to countless government overthrows.

Our strongest connection to the part of our "Being" that is not making choices based on an algorithm comes from the stories we believe in, and from them the stories we tell ourselves about our own identity. Our soul, or conscience, takes note and apply energy to evolve. Our moral decisions, our emotions that determine our mating and our generosity, our compassion, is what gives meaning to life.

The more life gets equated to computers or holograms, the more valuable artistic observation and execution become to societies.

Dividing Art into two categories helps understand this. impulsive art like Dance, Singing, Music, Painting, Sculpting … vs. Architecture, Computer Coding, Quantum, Fields, and more. All involve Artistic interpretation of our reality, yet one follows algorithms.

What is an algorithm? It is any process that results in a goal or answer. While the first category of Arts surely benefits from practice and study in the fundamentals of any art field, you do not have to start in an algorithm structure to begin or create.

Art inspiration falls into the huge category of things we cannot explain about our mind, brains, and existence. Child savants or geniuses can create music more sophisticated than they ever heard on the radio. Art comes from the part of the brain that tells us who we are. That values our humanity. That believes in what has yet to be created.

For this reason, Artists, or at least the origins of the symbolic stories of innate god-like gifts or qualities, will become the most powerful and important source for our technological evolution. It binds us to our animal instincts coded in DNA. Perhaps the evolution of Art has a threshold we soon cross that opens percentages of the other parts of our brain, or lets us self-identify and communicate with our DNA. What does this mean? What if you have a sequence in your DNA that allows you to perceive other dimensions or staves off cancer? Communicating with that part of you will most likely be through imagery or vibrations.

Now, what if we find out that the DNA sequence allowing cancer gets activated because the gene knows if we live longer a horrible future event happens, so removing us from the greater cosmic plan was ordered long ago? We already know that introducing cigarette smoke prompts cancer cells, so this is not science fiction. The only element in this deals with how we communicate with our own DNA. The virus in the jellyfish proved that introduction of one thought through many cells creates a brain wave of communication.

Do you really think that a computer hacker or scientist knows this in his brain or comes across a discovery through algorithm ahead of an Artist? Or a Free Play Thinker imagining or creating it with Art replacing long form code breakthroughs? The quantum leap in how we process information was shown in "The Matrix." By bypassing the internal dialog of the Brain Brain, or other Brains learn instantly how to fight, fly a helicopter, and ultimately become The One – a god over the code of existence.

As Asian Art and Philosophy merges with Western, quantum leaps we cannot imagine shall ambush us. This comes with responsibility to Cultural roots. Why? Because this is where the essential stories our who we think we are and come from. As we make ourselves more into gods, still by far our greatest mysteries reside in our Conscious Being.

Three ways to explain this are: cryogenics, inter-dimensional travel, and soul-merging. All shall happen I think within 10 years, because I ignore the slow process of algorithm discovery without the human element. All of these were imagined in stories centuries ago if not millennia ago. And what captured it was Art. Stories. Plays. Sculpting.

What held us back were the very things that allowed us to share ideals for cooperation: governments, religions, royal rulers, wars. Yet to say these elements should be fought off is exactly wrong.

Whatever or whomever provides more powerful storytelling (through emotional Art) that translates to communication and an ability to organize to cooperate en masse will ultimately hold rule over individuals. Think not? Look at the frenzy at a Justin Bieber concert and tell me what the girls think is more important? Wars or Justin, at that moment. Harnessing this power by responsible ruling parties will allow us faster breakthroughs as long as the culture from which the stories that shaped the personal foundation for our conscious is protected, respected, and seen as a tool to evolution.

Religion can be the greatest obstacle to discovery and improvement in life if we are not seeing it from a fair perspective. That is because they all demand that Time (and our evolution) doesn't evolve, and cannot be altered to allow innovation. Look at the Middle East. Once the cradle of civilization. The many gods of Egypt allowed for innovation and changing story telling. Islam refused it, replacing innovation with blind submission and obedience. Christianity and Judaism are not much better. Yet, trying to replace religion with government leaders failed in Russia. Stalin wanted Trotsky, Lenin, and Stalin to replace the Father, Son, and Holy Ghost. The problem was that the stories of them were too brief.

Proof of this lies in how art experience gets stored in our DNA or genetic makeup that ultimately "houses" our Soul or sense of Consciousness. Many tests prove that the melodies of the great composers (Bach, Mozart, Beethoven) and allegories and sense of comedy from immortal plays (Shakespeare, Chaucer, Aristophanes, Homer) art vision (Leonardo di Vinci) evokes emotional response and recollection in people born today. The powers of those arts to shape our souls proves true. I believe the merging of Chinese Art and Architecture, and the stories, in a superior level, with the Western advancements will show which Chinese Arts humans share.

How is this possible? Immortal Ether, storing all Information, sharing it in its beautiful web, inter-dimensionally.

Speaking of this … why do almost all people think of inter-dimensional travel as external? Because of art. My guess is the first or second dimension we travel to will be internal.

Imagine our entire being, our body, on a journey into the Fields of one of our heart cells. To make this possible, the scientific Laws our Fields on Earth obey must no longer exist. Yet, the Information shall remain constant. Looking at a strand of DNA in a microscope reminds us how unimaginably tiny it is. Yet if we walked up to it, and it towered above us like a twisted Empire State Building, each floor would be a journey.

The second way to look at inter-dimensional travel as an internal journey is …

Imagine exchanging places with Ether. Suddenly you are the Ether, and your body exists as a source of Information and Truth, that you play with.

A "different dimension" is one that does not observe the same Rules and Laws in its Fields, or it maybe be a dimension that does not recognize Fields at all. How do you visualize this? It is not as difficult as you think.

In a two-dimensional Universe, we, in 3 Dimensions, would simply appear as a black sphere growing bigger as we approach. 2-Dimensional beings would not "see" us in all our glory, or as we see ourselves. Does this limitation on viewing have a parallel in art? Yes. Two big ones. Before The Renaissance, the Church forbid painters to paint in 3 dimensions, or, as we perceive our whole reality. "Only God" can create in this completeness. Thankfully in the 1400's artists were freed from this rule, and all art after this flourished once an artificial "god-belief" was removed. Second example is film, which is 2-dimensional, yet we perceive it as 3-dimensional. In fact, the first films were so shocking to the eyes and senses that first-time viewers actually ran from their seats if a train was speeding toward them on screen. Their minds could not yet separate what they see from reality.

In an actual historic event, the Indians around Virginia had never seen a water vessel larger than a canoe. Because it was impossible to imagine, it was impossible to see. When the first explorers sailed close to shore, it took almost 3 days for the Indians to be able to "see" the ships. What they saw first was the parting of water on the bow and wakes, and working backwards, they finally saw the entire ship. And were stunned!

Touching again on the idea of our first journey to another dimension being an internal journey … many laws of our physics hold inside of an atom. However, we keep thinking that our journey into ourselves deals with matter at all. The space between atoms is like the space between planets. Huge! Like Ether. Our first true experiences with Ether seeking have been going on for millennia.

Our search for our soul or our consciousness. Imagine it as a vapor, or a ghost-like source of Information, historic and reactionary to immediate stimulus.

We would most likely lose our sensation of gravity, as Ether only holds the Laws of Nature of whatever dimension it is being experienced or sharing information in. Do the 1's and 0's in our computer care about gravity, any more than a number? Does Love care about distance? Or Time? These emanate from our consciousness or soul. Most likely, Time is not a factor in this dimension. But what is? Probably electric polarity or current, the transmission out of Ether into Matter, which would me very volatile, constantly changing shapes upon whim or need.

What matters in this reality in this dimension? Vibration. Energy transformation. Think of yourself as glass. Yes, glass, like we see every day. We all know what glass is, yet scientists still don't know what makes glass turn from an opaque solid into a clear assembly of atoms. Think about radio. Scientists still don't know how radio happens, they just know how to make it happen. Neither of them follow the limited Laws of Physics as we know them today, yet we enjoy them.

We enjoy pleasant dreams. We enjoy a sense of freedom. Yet, we also enjoy dominance over other animals and the world. The tension between the god-like traits and the soul-like traits will intensify – I mean, if you think reconciling conscience between the spiritual and the survival elements are tough now, wait until humanism gods challenge our rights for "inferiors" to discover the powers of their own souls or the freedom found when shedding the skins of 3 dimensions.

Again, it leads us back to the artistic experience, the experience of creating something out of seemingly nothing. Artists shall be the way to reveal to the simpler people their potential not as gods but as sublime translators of Information, Truth, and our purpose. They become the source of organization as leaders (wealthy and powerful) shed human limitations, thinking them unnecessary, like the heartless Mafia don thinks of murder. This should not be feared, because most likely, the Balance shall somehow tip in favor of the Artists again. For they have a gift that, like the Medicis, the rich can only stand close to, they cannot steal.

The race will then become not wealth or even health, it shall be the stealing of artistic impulse and freedom from Artists. For, as explained before, as we count on computers or merge with computers, we always must leave something behind. Our human skills of observation and innovation.

What will offer the first breakthroughs for this quantum deviation in understanding who and what we are in whatever we call reality?

Art. Almost all emerging science now was first dreamed up by writers and artists. artists will remain that breakthrough point. Vibration will play a key role, so all arts involving Harmonics play a role, and if you study deep down, all the arts include Harmonics.

In your role as artist, simply telling the Truth with the Information triggered from you is your only role. Ambitious artists may attempt to adapt a bridge or language for their creations.

The appearance of aliens would be an example of "Life Imitating Art" bypassing our nearly prehistoric method of communicating primarily with words to the Brain Brain. Perhaps they are not watching the computer coders and world leaders as much as they are watching the artists to determine we are ready for their appearance.

Actors, dancers, musicians … architects designing doors to portals to allow separation of our conscious from matter … interesting stuff, interesting responsibilities.

And do you know what is really, really exciting? You, the fans, the viewers of art – the energy to evolve as a species resides in you, even if the message of Information is created by the artist. That is what is most exciting to an artist.

Why would aliens come to Earth? Perhaps to laugh at our challenge or desire to be more like gods, when they, with supreme god-like powers, desperately want to evolve into emotional beings capable of transcending the algorithms to spontaneously create something from nothing that simply pleases our senses.

Perhaps to keep an eye on us and keep us from making the mistake of letting Artificial Intelligence from taking over and making us less human – at which point aliens might intervene. They seem to be observing all nuclear tests closely and a "Terminator" tech revolution would also wipe out humanity as we know it and the aliens might want to keep us human. To be living a life with a limited brain rather than a fully-accessible knowledge or our creation, purpose, and reality. In a sense, to live like dogs being trained step by step until suddenly dogs talk and cook their own food. The suspense must be killing it at their box office!

Religion is a "Deal" – Spirtuality is a "Journey"

What could be more diametrically opposed than science and religion. You would think that the greatest leaps forward in science happened when religion waned in power. Yet the opposite is true. The reason why lies in the goal of religion, which is organization, of desires, logic, and people. This allows the ignorant masses to focus on a target chosen by religion to take them farther from death by disease or starvation. Science served religious empires. Religion supplied the resources for scientific exploration.

Science is just one of many creative arenas. Yet, it gave hope to people following religion that they might stave off an early death and gave credit to God for giving divine guidance to physicians, especially in Western Europe, which was where almost all early cures came from. In fact, most science was simply conquering disease, starvation, and warfare. It wasn't until religion stopped persecuting discoveries as the actions of a heretic that Physicists felt free to reveal the truth about the cosmos.

Yet as powerful as science is now, does not imbue the masses with the sensations they desire, or the myriad emotions and tingling feeling, the inspirations and inner joy that gives life meaning.

People seek escape from limitations. Death, damnation – these are both severe limitations. In the 1500s, people were supremely ignorant, incapable of complex thoughts, and the stories told at church overpowered their imagination. Even if a claim was ludicrous if science came forward with a discovery that made the The Church more powerful this breakthrough was encouraged. All you have to do is look at how they persecuted Copernicus to see the penalty paid for going against religion.

As it evolved to true awe-inspiring greatness art proved humans could escape their limits. Indeed, art was the first thing mankind made repetitively that had no use in survival. It simply invoked feeling. That feeling was pride that as a human, the ancient man could tell stories or show art that lived long past his own life. Whether on cave walls or carved into marble in Florence, Art provided the sensations people relate to happiness and freedom. Yet, religion organized these feelings into obedience.

But how does the artistic process feelings compare with the feelings of those that admire it? While in art mode the body releases primitive chemicals in the Brain Brain that are addictive because it staves off anxiety over our mortality and it feeds Hope to our ego that craves idolatry and greatness.

While we think of early art as paintings or sculptures, if you think about it, the first Art was probably Music and Dance. Imagine yourself in the jungles along the Nile. Each morning, in hunter-gatherer mode, the symphony of animal sounds filled your ears and mind. The movement of Nature and your tribe in hunt mode choreographed with the sounds of Nature. In a journey to stave off mortality. This all relies on the vibration of music and the impulse of humanity to assert itself upon the world. Each discovery that led to longer lives or stronger unity among the tribes was worshipped.

Perception of Physics – what is this, what is its energy, what does it do? – altered the world. When a primary need such as fire was removed when mankind learned how to make fire rather than find fire from lightning, this freed our mind to create useless things like art.

The greater the risk the more innovative humans act, including art, and this breaking free of mortality is met with sensations of warmth, tingling, and more that we interpret as happiness. In fact this is evident in the modern world. In nations with very little need for innovation or exceptionalism because the country makes life too safe for them, the suicide and unhappiness rate increases dramatically. Singapore and Switzerland rank at the top of lifestyle and income, yet they have more suicides and less artists per capita than Laos.

What is the quotient missing? Hope. To escape limitations. Art, the personification of it. Rather than wonder if awareness of death fuels our drive to evolve ponder instead if art will disappear as we conquer age!

The fundamentals of religion give us a clue. In all religions, the follower enters into a "deal" that states, "If you give me this, I will grant you immortality and happiness in the hereafter." It is a deal. But, an artist does not begin art as a deal, not even if being paid in advance for it. It begins with a sensation. This begins a journey, from nothing to art. It evolves our independent spiritual wisdom, release, and is passed down in DNA. The idea of getting "something from nothing" or "The Void" reflects creating art. The eye for the truth in Nature delivers it (the laws of shadows, color, and light.)

The creation of art out of nothing ignites the addicting sensations that propel Artists in all fields. Watch a dancer at the end of a performance or free dance – total bliss. Those that cannot dance or sculpt or sing – they watch in awe and absorb all they can through their limited external senses. But they are a fraction of the senses that the artist felt.

Hope to keep evolving the art … and accept that the journey begins over and over.

The ultimate value of the art and its immortality quotient requires the masses of non-artists to come to an agreement - sometimes with words, other times only sensed. That agreement is "inter-subjective." There is no right or wrong, internally you decide if the art moves you. If it touches where logic cannot. This again is a journey.

Upon viewing art, in parallax view mode, this ungifted person shares the journey that broke the artist free from limitations. Proof to survive. Proof to deserve bliss.

Because of this interaction, artists and storytellers will become valuable commodities as computers merge powers with human abilities. This link to us as biological beings capable of inspiration and execution of that gives soulful meaning to the Computer Age and near future.

How does this relate to everyday people, unaware they are interacting with art? Think of Google. Its primary usefulness is that it saves us Time, which means it staves off mortality in a way. We give godlike powers to Google. It knows all, sees all, travels at the speed of visualization, saves us, and defines our world. Google is a god.

Yet the algorithm was written in a highly Free-Play creative mode by its inventors.

Now, an art for called Google is one of the most powerful "gods" on the planet. We want conclusions from Google, such as the algorithm answer. But also enjoy the side trips it takes us on, which is the journey. New ways of word association to maximize Google searches now happens a few times in a single generation all across the globe. Nothing like this ever happened before.

While the first goal might be to bring happiness, what we really are trying to do is become gods through the art of code programming.

An elegant algorithm … how shall that dance look to us? Will we soon see the interplay in code as an art form? Why is this so important?

Because advanced coding merges Science and religion in its closest ties.

THE INFINITY STONES THAT END ART

[**Note**: This chapter was written after the main chapters of this book that completed in 2017.]

"All of life's questions can be answered in the movies" is a line used often in our 6 books delivering the future merging of Edge Technology and filmmaking. In this case, The Avengers: Endgame provides the stages, questions, and answers to a pivotal era for Hollywood and all arenas of artistic free expression.

One of the great gifts given humanity is the ability to "create something out of nothing" and for this creation to not only serve a purpose as a tool, but for it to also be a thing of inspiration, beauty, and virtues. It is the codifying of cultural morality.

Barry Diller (a Hollywood movie legend/mogul) recently declared movies are dead – Big Tech is killing theatrical movies. The way to preserve them must come from within the very arena that is destroying movies. The problem is the solution cannot come from any player responsible for the death.

This is why Crypto-Integrity-Tao and Creativity First Films work together to deliver "more than the sum of their parts." So many human skills are being replaced by digital systems we need to understand how this harms the evolution of humanity.

Already they know that Kindle reading does not activate the brain the same was as reading a book. This is why most private schools where Silicon Valley Big Tech leaders send their kids to rely on textbooks: the parents know this fact and don't want their kids to lose valuable skills.

Recently, one of the biggest news stories was about an Artist that declared that NFT's are like the invention of paper. That's just insulting to our intelligence. But it remained at the top of the news because it skews reality well, and in the direction Blockchain/crypto news sources and billionaires want to go. By suggesting "the ridiculous" … "the useless" suggestions that propel investment into cryptocurrencies seem plausible. And in filmmaking and even fine art, this concept that NFT's are like the invention of paper removes the power of human interaction with others and a medium held by our hand, which develops our minds. More senses are involved which evolves us as creatures.

To more totally alter our social evolution all forces symbolizing the creative power of humans as well as what ways we interpret and live life must be working together. Don't think of them as businesses just paying each other. Think of them functioning like a hand. From this idea we identify the forces in today's world that are like the Infinity Stones, and also how the holders of the Stones are grouping to form the glove Thanos worn when he snapped his fingers to eliminate half of all the lives. While Thanos claimed he removed these lives at random clearly he chose who deserved to survive – his enablers. Like Kali, he "ate his young" when he killed Gamora in the pursuit of becoming Shiva, the great destroyer.

Across the world since ancient Egypt thousands of years ago cultures championed art and the freedom of artists in painting, sculpture, writing, dance, music, and architecture. The Renaissance, the Ming, Zhou, and Qin Dynasties, Incan Empire, Persian Empire, Egyptian era, Russian and Indian peaks in creativity, Khmer Angkor Wat, Siam's art, and of course India's art, sculpture, and architecture all captured the spirit of the times and people. It captured the Truth of the era which paved roads toward greater freedom, hope, aspirations, and spiritual development.

What we could not find the words to describe was illustrated and experienced through Art.

When nations fall into totalitarian or fanatical rule the first thing rulers do is lock up the writers because they can add emotion to their explanations of what is truly going on. Then (totalitarian) fanatics abuse human emotion to get the lower classes to begin destroying any Art that reflects the longer-term greatness of a Culture. It all must seem like a grassroots truth and movement at first. Ultimate totalitarian governments make gradual moves to replace what gives us meaning so eventually we are getting all meaning from the new leaders. Yet, this chapter is not about politics as much as it is about replacing what makes us human with technology that makes us machines serving an oligarchy of technology.

Ultimately, this is about two things: removing the inner ability to create joy … and preventing our senses from recognizing reality. Part of recognizing reality is remembering history and our family origins. The arts reflect stages of humanity and the individual's power to stand alone to create a truth that might negate the lies of centralized government and Big Tech, as well as any law agency like the KGB and FBI which today operates more like the old KGB.

The Internet has changed the means, systems, and social interaction that deliver totalitarian rule. The Internet giants alone could not force this fundamental change in humanity despite the growing temptations they offer us in exchange. These temptations include Apps that do anything for us, superiority by aligning with consensus, escape from guilt by giving up responsibility through victimhood, and trading experience for shortcuts to mastering a skill.

All of these go against the greater culture of a nation, and splinter the people into smaller groups, which results in the collapse of a nation. Unity of the people despite the temptations of technology should be the most important goal of leaders of any country now, and that means embracing long term culture and respect for what our ancestors valued. The whole national goal requires our liberties come with responsibilities, and that improved lifestyles should be based on merit and virtues.

In the Avengers' series of movies the last two were "Infinity Wars" and "Infinity: Endgame." The battle for galactic species came down to who possessed the 6 Infinity Stones. Thanos believed his greatness and vision gave him the right to remove half of every living creature in the universe… and that the survivors would be a grateful universe. To do this he needed to master and control Time, Power, Reality, Space, individual Souls, and the collective and individual Mind.

INFINITY STONES
Time – Green
Power – Purple
Reality – Red
Soul - Orange
Mind – Yellow
Space - Blue

Seen throughout the films are the fantastic art and architecture from cultures of creatures across the universe. In the first film, Thanos left the Art and enough people alive to continue civilizations, yet they remembered the past and this made them sad and un-worshipping. Thanos realizes that to create a future that is grateful for his murderous tyranny he must remove all reminders of the free will, love, ancestors, and art. Governments alone would not be able to do this because Reality, the individual human Soul, and collective universal Soul would remember over Time what gave them meaning, and how the government took meaning away from them. Time is ultimately the one Stone used to reverse the genocide Thanos thought he had the superior right to unleash upon all creatures.

Governments today would need the help of Edge Technology to alter Reality and what the human Soul values most. Which is a way of saying the centralized leaders would have to alter what gives meaning to people over Time and replace all the memories any one person or culture holds dear.

The Sanskrit wisdom said that ultimately all grand system become Shiva, The Great Destroyer. It requires the ego, selfness, and hunger for ever more power for any conglomerate of forces to put Virtues and long-term Culture above the destruction of those seeking ultimate Power. Alongside Shiva may be Kali, who eats her newborns.

Shiva meets other systems turning into Shiva; more importantly, Shivas needs reflections in other Shivas to justify genocide of people and menticide (also known as menticide of the mind.)

Menticide "occurs when you willingly
think and act out of someone else's
interpretation of reality
to their benefit
and against your survival."

The process includes indoctrination, deluding others, teaching falseness of society, propagandizing, influence peddling, removal of duties through victim association, censorship, shunning, and mob brutality that replaces order with increasing disorder.

People get numb and forget what is real inside them and outside them. Lies become truth. While this might seem to be within the domain of the Mind Stone, it also relies on the Space Stone. And this could only be possible on a global scale with today's technology.

Space is a "Great Unknown." We know as little about the far reaches of Celestial Space as we know about the depth of the oceans and the incredible connections within our bodies and minds. Because it is a Great Unknown we trust on our greatest scientists to tell us the latest discoveries. These discoveries impact changes in every culture in the world, and on every individual.

Up until the 1400s the key sciences of Astronomy and Biology were controlled by religions and royal leaders of nations or empires. Science needs debate and the freedom to question scientific discoveries or uses of medicines to evolve our understanding of the world, and how to stay healthy and alive while raising our families. Whenever we allowed free debate and challenges to the standards of our physical universe by science, the world evolved upward toward more freedom. The primary fears for survival were starvation, disease, and wars. Science helped reduce these fears and for about 150 years our fear of these mass killing events lessened.

Free of fear, societies, industrialization, innovation, and intelligence rose. But this created a problem. Royals, the governments of tyranny, powerful family dynasties, and global trade lost their key tool – fear. Science in and of itself answers questions about Space, both on the planet and in our cosmos. And it constantly challenges old understandings. That is how we harnessed raw materials to build nations and advance civilizations. But now politicians around the world use "The Science" to re-introduce us to fear to control individuals and lower expectations, while simultaneously tearing apart the very thing that mattered most to us which is raising our children to have a better life based on the reality of Nature, our emotions, and biological Epigenetic instincts and mannerisms.

"The Science" includes all Space because it dictates how we relate to everything we think of as "within our space." This includes distant galaxies, Dark Energy, our bodily health, education because this shapes the mind within a body that occupies Space at a given Time, viruses too small to see that can kill us, and motivations to create wars.

"The Science" is most obvious in our response to COVID: to question The Science was heresy. But, thousands of actual scientists did not believe the dogma of wearing masks, the vaccine, closing schools, and all for good reason as actual deaths related only to COVID was much smaller than reported. COVID used The Science to instill us with fear which makes us easier to control and reduce the individual natural ways to survive and also introduce changes in government and education.

The Science allowed us to think that boys can become girls through medicine simply by wishing it. The Science allowed the closing of anything that helped propel society forward. At the same time The Science gave us NGO's set on destroying society that is self-governed with an anointed class of elites in power. Canada turned into a nation of tyranny under Trudeau, and he is allowed to do this based on The Science. He is now going to allow children to ask their school counselor to introduce them to a doctor that will help them commit suicide without informing the parents.

The Science tells us that the very DNA of a Caucasian is racist, while certain groups or Tribes of people do not carry this damaged DNA and cannot be racist. We even have racist bridges, racist weather, and racist swimming – all backed up by The Science because this includes The Political Science.

The Science distorts all the proofs it wants to without retaliation from actual scientists or true leaders allow freedom. The Science is akin to fundamentalist governments, religions, and cults.

The Science allowed Al Gore to lie about Global Warming, make billions off of it, and force the world into a year in 2022 during which millions of people will die "right now this winter" due to their Climate Change agenda. The Climate Change agenda demands we live in fear that within 8 years the world will end, coastal cities will be underwater, and a majority of humans will perish. But none of this is coming true. The push to Green Energy will kill more people in this winter than history – all because it is using The Science, which can never be questioned, to eliminate fossil fuels.

The Science just agreed to do more gain of function on viruses within the USA and came up with a variation of COVID that has a 90% death rate in mice. Why would they allow this research? To increase one of our primal fears.

Climate Change will re-introduce starvation as we run out of fuel for farming and delivery. All that is left is war, and so we create the Ukraine War, which might end up being a nuclear war. Even if it does not, the same politicians and elites demanding Climate Change will kill us are encouraging a war the releases more of their so-called humanity-killing emissions into the air than any other manmade source.

Big Pharma prevents cures for diseases more than they introduce actual cures. They made a fortune supporting The Science told to us by politicians. The Science closed schools and resulted in more suicides of children in the USA and Europe than every in history, and at the same time set back the mental development of an entire era of children.

Yet, Obama would tell us "this is the new normal" when it can be reversed by simply refusing the "experts" forcing dogma upon us.

The ultimate misuse of the Space Stone comes when it can open a portal to the mind like it can in the dimensions of our universe. And this is the goal and ultimate use of the Space Stone – to get access to that Space in our souls and minds the government cannot touch. They want to force us to question what gives our life meaning and how we know what we know at birth, which is that we will die and must control our own morality. The Science wants to remove this from our conscious actions. They want to destroy imagination and free will through these Stones, and needs the Space Stone since it controls a fake science called The Science.

It begins with media announcing The Science we must follow (into war or into getting experimental shots) and ends with the war for our minds. The solution might be found in the book "The Janus Point" which puts limits on The Science once truth is revealed on a higher level through the merging or co-evolution of systems of higher complexity. This is one of the only ways to challenge A.I. we cannot control or equal in intelligence. A.I. will surpass the minds of the world and take over if the Glove of Thanos is completed.

The Science from the Space Stone is the only Stone that is a hydra – there is no one head to cut off to kill it. And its portals all lead to fear, fear of both the unknown, and others, by timeless methods. They must re-introduce us to the primal fears of starvation, disease, and war to end our lives. They push forward agendas that give more power to each of these three.

Watch what they do in the real world, then merge this with what you know in your bones is true and gives you meaning. Compare that to what The Science is telling us. Already nothing that Greta the screaming Climate Child predicted is close to coming true. But this was a tool used by the Space Stone to scream fear from a source that cannot be attacked (because she is a child.) This is the same as using racism or Hitler to shut down any discourse or alternative solutions, and it prevents us from seeing reality.

The reality is that most people in the United States are not racist and what we do will not cause Climate to change so much it kills off humanity. But, the wars the same group of Menticide Elites push -and the Green Agenda failing to keep us warm and feed us - WILL kill people in 2023. Fear is their goal and is not the plan to prevent a collapse of decent society and the hopes within free will.

Recent history of revolutions that resulted in far worse leaders resulting in far worse victims proves it only takes 6% of extremists with loud voices and possible violence to overthrow a nation or fundamentally change it for the worse. Once social bullies get the support of the government, and the power is spread through technology, the Soul either gives in for self-preservation or it recognizes what is happening and fights to its death. This is where we are today and the power of story and meaning are our defense and offense.

To deliver the destruction of the Infinity Stones requires a union of politics, sources of knowledge, altering memory, controlling and destroying art, limiting the mind by limiting language, forgetting the relationship of Time to truth, and ultimately controlling reality in "realities" that are so addicting and get so much superficial attention to the user that people don't want to return to the real world. Do this - and through fear or pleasure the tyrants control a human being.

Politics and technology create the mob mentality, lies, and destruction lynching the human ability to create beauty and joy from within. This requires we lose hope and accept betrayal of people and our own senses as they interpret our world. Films force us to experience life viscerally – a computer does the opposite. Life encourages communication for mutual goals and reminds us how humans have to be jacks of all trades and also specialists in a minimal number of arenas.

Films remind us of the individual journeys we all make on our path to virtues and meaning.

Quickly, Big Tech with their mobs and politicians will control what films and series we get to see. The decision to end the 12-week window for theaters is not just about money. It is because the Thanoses of today know that it is much harder to manipulate and destroy a person when he or she is in a group that reflects how much they share in common and cherish. Common sense and ways to interpret reality expand in big social scenes of storytelling especially. Alone at home, any types of control and phobias can be nurtured along with lies.

Amazon is dumbing down and numbing up movies to lower our expectations for inspiration and Art.

When one fear fades, they create new ones but always lie about the origin of the fear. The fear was created by the dynamics of the systems that are controlled by people that want to be Thanos.

Now for the first time there are shortages of products and food in the United States: this breeds fear, then helplessness, and finally capitulation. To make sure this hits everyone but the elites, the

Menticide monsters add in government mandates. Then they surround people with violence and despair – people that gave up on decent lives. The terrible people are given immunity, and the righteous people are called wicked for disapproving of the reversal of humanity and common sense.

Studios when they are bought by Amazon or telecoms want to take away the power of Free Will, meritocracy, togetherness, the power of story, group joy – and they must take away the power of individual Actors and Directors to speak the truth and show the truth about life and humanity. Minimizing it to the small screen is a first step to totalitarian control over what films we make and what we believe.

Destruction of several human traits helps this happen. Demonizing motherhood and any religion giving meaning to the gifts of life as humans experience it is first. And the idea we face enemies beyond our control like Climate Change and are born into guilt (simply be being male or Caucasian) are turned into tainted litmus tests for whether we belong in society. Fear drives actions, not human curiosity or adaptation.

Gaslighting, straw man arguments, return of caste systems, being held responsible – if a soul cannot be divided it must be destroyed – or at least what reflects the best of our soul must be.

There is such a thing as a self-fulfilling prophecy, which is when a person is told something so many times he or she becomes what he's told is inevitable. We now have a media-fulfilling prophecy to handle mass results of failure to get beyond limitations through love, ingenuity, and united culture.

Thanos would snap his fingers and half of all life existing would vanish and be wiped out, turned to dust. The argument is that the universe cannot provide enough food. This ignores accelerating intelligence, which is a human fact. Thanos was a demi-god that wanted to be all gods to all peoples in all ways.

Zuckerberg wants to be Jesus. Woke identity is the Holy Ghost breathed into someone for salvation by breathing in "The Father." The Father this time is not a compassionate God bestowing gifts of creativity, consciousness, and Free Will. "The Father" is the glove of Stones and its blessing is amnesia. Truth hurts.

From Pulp Fiction, "… Feel that sting? It's the sting of pride." And what was at stake in this film was what gave meaning to the souls of each lead character and how much it would sting if they sell this out or don't protect it.

While Thanos was a thief and demi-god declaring "I am inevitable" he was destroyed by Ironman, who was a creator, not a destroyer.

DELIVERERS OF MENTACIDE

Facebook/Meta - Soul

Politicians - Power

Mainstream Media - Mind

Big Tech - Reality

U.S.A. Education – Time

"The Science" - Space

They want to snap the fingers that dissolve half the world – the half that will not worship them.

But they failed to understand Blockchain Intersections as per "The Janus Point." The Physics of what they are attempting will catch up to those that steal the Infinity Stones -- and give the energy to the infinite forces of the infinite Mind and Universe.

The Infinity Stones exist in a state of chaos now, a state the deliverers of Menticide must embolden to function as if they are benevolent saviors. While the Deliverers follow a sort of Arrow of Time mentality, they misunderstand Time as it relates to complexity within bubbles of order. This alters the way the Time Infinity Stone works its wonders.

If you combine the essence of the Physics of "The Three Body Problem" with "The Janus Point" then our modern-day Thanos ignores a) the acceleration ability of the human mind and b) ratios that allow Truth to create Reality in an "Emergent Mind Universe."

It is the unsacrificed self that must be honored. We are not here to sacrifice our spirit. Blockchain Parallax View, which I wrote about in 2018 in "Pillars of Encryption" will allow a Space perspective that lets us better view what I expect to be a Multiverse of Media.

A Multiverse to join all the Infinity Stones will come soon and try to separate us from their Arrow of Time -- but not on the terms these new Thanos digits want: instead of being "nowhere" we will be "nowhen." And this means that the old meaning and the truth reflected in creating something out of nothing (like a newborn child or complex art) will keep returning to humanity to give people meaning.

In other words, even while "The Six" that possess the Infinity Stones seem inevitable their hold on what the Stones represent is temporary and illusionary. Within the DNA of enough people is the power to free the Stones so they go back to reflecting truth, balancing with cosmic Nature, and ultimately to preserve humanity through "creating something out of nothing."

Art is the only substitute for experience we ourselves never lived through. Art frees us to find the process of Agency of goodness and reminds each person he or she has been chosen. Beauty is to enthuse us for work, and work is to raise us up – by our own merits.

There is only one way to keep the sacred power of the Infinity Stones and remove them from the glove of power. Fortunately, that victory of transferring the Soul, Power, Mind, Time, Space, and Reality to the Free Will to create a reflection of beauty can be achieved – and quickly.

Humanity is simultaneously fragile and invincible.

THE VANITY OF BONFIRES

There are few people left alive that remember the terrific novel entitled "The Bonfire of the Vanities" by Tom Wolfe written in 1987. It later became a movie. In it you see a stellar cast: young Bruce Willis, Tom Hanks, Morgan Freeman, John Hancock, Donald Moffat, and Melanie Griffith.

Yet, the premise of the title is based upon an event that happened in 1497, when the Dominican priest Girolamo Savonarola ordered the burning of objects considered sinful, such as cosmetics, wigs, mirrors, books, and art. In what can only be defined as a perverted religious zealot power this priest destroyed literature, art, and even mirrors. Why mirrors? So we could not observe ourselves, and by removing the greatest examples of Art people could not observe the progress of mankind.

Part of this progress seen through Art was the ability of a person to put the Virtues of people into something that could be observed in a most beautiful and lasting sense: Honesty, Loyalty, Reverence, Love, Beauty, and the value of a fine social reputation.

Girolamo had an incredible appeal to what can only be named the "pliable and ignorant minds" of the people his message could reach. Many could not even read. The Church under Salvatore in Florence had spread the idea of a "fire and brimstone damnation" for sinners. It gave a power to the Catholic Church never imagined before this decade because until then the religious converts were lured by the "carrot" and not driven by the "stick" to the point they obey the edicts of Christianity and Catholicism.

While this method united nations for war and unified innovative progress, it also destroyed art and made individuals fearful of righteousness in their own image. Anyone opposing the Church was banned. Mobs attacked, bullied and even killed people deemed not zealous enough - either in bonfires, hangings, drownings, or legal inquisitions given power to torture by the mob that thought this abuse of power was about saving their own souls.

Now … in Wolf's film, a decent man tries to do what is right to protect a woman he loves. It destroys his life. And everyone makes him the scapegoat because the woman becomes very rich. She also embodies the mob mentality of the time, which crucified the rich. They labeled the rich racist and opportunists. Way back in 1987, and even in 1497, when racism was based on a caste system or indentured servants, racism was practiced openly.

To preserve what was true, the anti-hero Tom Hanks had to tell something not true. The only reason he could not tell the truth was due to the exact system by which evidence is given to a court. The tape recording of the woman admitting she was drunk and killed a man was all Hanks needed to be innocent. Revealing this tape recording of the truth proving he was innocent was admissible as evidence because the taped conversation was done by a friend. Hanks himself did not tape it. He had to choose if it was presented.

Always a moral man, Hank's character realized that not only would he go down for a crime he did not commit, but that this would lead to a series of legal actions to destroy other men of Integrity around him. And rising to the top would be the woman that used sex to lure him into the crime in the first place. Hanks had to choose which Virtue was the most important to uphold: the Virtue of protecting a sleezy woman … or honorable people. When they needed more hatred for Hanks, they made the hit-and-run as racist as possible.

I switched the title for this chapter because of the people involved in the trial that were working together to convince him to keep lying – to admit that he was driving the car that killed the man. These betraying people would gain power and money if only lies could be stronger than the truth to Hanks. These bad people used mob mentality and fake news to gain the power they wanted to force him to lie and go down as a racist murderer.

The judge for this is Morgan Freeman. When he accepted the evidence that proved Tom Hanks in the name of love was protecting the real killer Melanie Griffith all charges against Hanks for a racist crime were dropped. The entire courtroom rose up in chaos. Suddenly they pointed the finger at Freeman - the judge they loved up until that moment - and called him racist. When Freeman finally got control, he gave a speech. It's worth it to repeat it close to verbatim.

"You dare call me racist? What does it mean the color of a man's skin if the witness perjured themselves. If a prosecutor and district attorney throw a man to the mob for political gain? And men of the cloth, men of God, take the prime cuts (the money and power with the mob.) Is that justice? I'll tell you what justice really is. Justice is the law and the law is man's feeble attempt to shut down the principles of decency, and it is not a deal or an angle or a contract or a hustle. It is what your grandmother taught you. It is in your bones."

At the end of the movie, the writer of the novel capturing this fall from grace won the Pulitzer Prize, even though up to the last second the writer hated Hanks. Hanks was the sacrificial lamb who tried to please two masters. Hanks had everything the rich and powerful want, including the love of this sleezy woman, and he left it to save his soul.

Willis had nothing and sacrificed his soul for everything – defending the reputation of Hanks by standing up to the mob and writing about Virtues instead of what is trending.

In this movie they created characters symbolic of today's peddlers of hatred and victimhood as today's America but never could have seen the future we live in now. They had a religious hustler like Al Sharpton. Also included were Attorney Generals elected with money from donors that gave them an agenda to enforce. And, there was the mob. Social bullying that resulted in the divorce and destruction of Tom Hank's character – who began this fall because he wanted to protect a slut of a wife played by Griffin. In other words he tried to do the noble thing and not only take blame for driving the car that hit a man and killed him - he insisted on trying to save the guilty posing as a victim because he was guilted and intimidated into thinking it was a higher moral action.

On the course of righteous action, Hanks was used by unrighteous people to be a symbol of hate. All that mattered was destroying Hanks and what he stood for. By assuming the role of a victim, Griffin was no longer accountable for anything she does and can never be racist or unscrupulous. What

Willis realized was that this case was all about "them" -- the sinners and liars grandstanding to cover up their own lies and racism.

The vanity of the people lighting bonfires upon which we burn books, art, reputations, laws, the Declaration of Independence, history, facts, guilt, truth, and honor changes everything: it is all about THEM. Those demanding bonfires? Their virtue signaling makes tragedies all about them. The vanity of these people should be standing out as so grotesque they get removed from office and jailed for crimes proven in course through evidence. But nowadays, those starting the bonfires have too much to gain from the distortion of truth while the lower class need distortion to survive.

People burning the great books, movies, art, history, truth, education, nuclear family – this is vanity, one of the Seven Deadly Sins. It will consume them. They are naïve to think it won't.

Films will be created capturing this last ten years of an attempted destruction of what can be called The American Way. Driving the killer vehicles are the Social Justice Warriors, virtue signalers, cancel culture voices, teachers, university directors, ministers, Media leaders, Big Tech, and flash mobs run by coordinated and heavily funded militias like Antifa and Black Lives Matter.

It is all about them, their vanity, how they love seeing themselves in the mirror which today is a new sort of entertainment television making the splinter groups feel superior in followers of Tweets and Likes.

Repulsing them is true Art that symbolizes the progress of societies. They have no plan to rebuild into higher Integrity and Virtue. It is all about how to manipulate people to believe their life is a lie, or actually that their lie is their life.

Both the guilty political leaders and NGO leaders are usually guilty of what they accuse other people of doing – they just spin it and by sheer numbers alter the past and present.

In the end, Hanks lies to preserve not himself, but journalists and decent people following him. His is not an action of vanity, it is an action of humility. The mob is silenced. They lost their lamb moments before it would be sacrificed at the altar of hatred manipulated by the media.

MUTLIVERSES AND DARTH VADER

[**Note:** This chapter was written after the main chapters where written in 2017.]

Through every epoch of civilization mythology with its "Hero Journey" inspired virtues and led anyone to believe they can rise in character. Even the most poor and oppressed found value in myths and historic storytelling. It did not matter if the story began in China, or India, the Cradle of Civilization, Cambodia, Japan, Egypt, Greece, Rome, Scandinavia, Peru, Mexico, with the Aborigines, Jews, Aztecs, Incas, Mayans, or Native Americans. The most powerful original sources were in the great Classics like Homer's "The Odyssey," the religious books like the Bible, Torah, Koran, I Ching, and ancient Sanskrit literatures.

Performances on stage with elaborate sets and music and operas elevated the experiences that uplifted humanity. The earliest silent films impacted people on a level so intense people fainted in theaters. This became stronger when sound was added.

Films became by far the most powerful form of storytelling as it engaged multiple senses, adding images and cinematic techniques to make a story more emotional, visceral, and powerful. This is a great power that can uplift anyone. Films follow "The Hero's Journey" as it leads the audience to embrace their own hero within, waiting to rise to the next challenge or dream.

Now Metaverses, Augmented Reality, Artificial Reality, evolving Apps, Games, and A.I. merge in various stages of hardware and software evolution. While these speed up answers and make someone appear smarter the algorithms hijack attention and Big Tech works against any personal evolution toward being a free-minded hero.

The more important question is: what is missing from a healthy society if these merge for use primarily by someone alone that is shutting out the real world? The "group experience" that myths and plays told to a gathering of people around fires or in theaters presented confirmation by the person next to you that something was funny, meaningful, and worth sacrificing for. The ability of what you see inspiring and instilling shared the best human virtues and people could feel this as it rises in a group audience.

The clearest proof is that if the fans are watching a comedy, when the whole theater is laughing, it makes each individual's sense of humor and ways of expressing it rise. Same thing happens in a horror or suspense movie when the entire theatrical audience jumps when they get spooked, some shouting out loud, making the danger greater. This danger is also symbolic of losing the individual fan's honor, virtues, dreams, or heart and soul due to the visual experience or living vicariously through the characters. Love stories fill us with the idea we will find love if we have not found it already, and we need to make the most of ourselves to find it.

But theatrical experiences faded during the pandemic. Due to poor decisions among Hollywood studios the biggest companies produced fewer films luring big audiences to the theater. High prices forced by greedy studios didn't help when it costs over $120 for a family of four to go to the theater and enjoy snacks. (Blame the studios because they kept increasing their percentage of box office gross which forced theaters to increase the cost of refreshments.)

Then the at-home entertainment offered by Streaming, Cable, Games, TikTok, and YouTube consumed more and more hours. Bing-watching dulled minds for days in a row. So much was lost – especially personal responsibility for time, actions, and thoughts. Tech Giants learned that bad faith practices made them more money so they piled on many ways to keep users staring at smart phones or computers instead of interacting with live people experiencing life. This also gave Tech immense power in shaping the idea of Free Will, character, identity, honor, nobility, spirituality, national pride, and the value of a family.

Self-help books became the rage in the 1980s. Many of these books was beneficial but most began to cater to making the reader "belong" to a fad or deviation from social norms. These books championed victimhood or falsely warned someone is a victim if they follow traditional family values. Look deeper and it circles back to power over Free Will and money. Facebook's COO wrote in her book "Lean In" that it is more honorable to not have children and devote your meaning of life to Facebook and technology instead.

The Internet puts amazing advantages at our fingertips: search any facts, see any location, buy anything online. Yet, the Internet has a Dark Side and a Light Side.

Luke Skywalker discovered his greatest evil to battle was within himself when he learned the leader of the Dark Side was his own Father, Darth Vadar. Vadar previously was a Jedi Knight and one of the most powerful forces of good and justice, seated alongside Yoda. The Federation fought to preserve Free Will and destroy Dark Energy by bringing in the Light Force. Both the good Federation and bad Dark Side achieved incredible advancements in technology. Only one of them decried they could replace what previously existed with the more powerful decentralized universal force with a centralized leadership. Yoda and the Council he sat on knew the perils and loss of souls that the Dark Side ushers in for unsuspecting young recruits. Destroying whoever and whatever held memories of freedom and the past had to be stopped.

Thanos realized he must remove the memory of who and what we love in the film End Game and sacrificed his own daughter to be a god that creates a "more benevolent universe."

Many descriptions of Multiverses include ideals like making them more benevolent and converting Light into Dark. In "Dr. Strange" people could practice the "Dark Arts" in the Mirror Dimension so that it would not harm the reality we live in … until suddenly the Mirror Dimension became the most evil and dangerous place to control reality. This ends well in the statement, "It's not about you" meaning that the Dark Forces make it all about themselves, and this is their downfall or fatal flaw.

Now with the merging of powerful Edge Technology and our opportunity to see the future as one filled with Multiverses we can create an option that nurtures humanity and sends the most elegant messages into the universe. This evolution requires a mindset and also a remarkable proprietary innovation that can harness the Maxwell's Demon energy at Spirit and Scale to shape the future. Think of it simply by remembering the story in "Horton Hears a Who."

Recently, one person said the most important museum to create is the museum of hip-hop music. They talked about putting spotlights on underserved communities, like the New Black Panthers, and nihilistic groups like Antifa, who might simply be as misunderstood as pedophiles, who now are being considered a gender identity that should not be punished. There are so many problems with this aspect I don't know where to begin.

They talked about the importance of Thought Leadership on an "approved basis" only (with them being the sole arbitrator of what is approved.) This was followed quickly be legal cases decided with more weight put on equity and race of who owes something for a previous crime of being born into white skin. These people understand knowledge is important but said it is more important that knowledge must be inclusive, which is code for "censored."

Knowing what motivates Big Tech and the End Game of those corps with government, most likely the emotional experiences will be of two kinds: to make people suffer so they are more obedient, or to trigger a drug-like high that is enhanced either with drugs or technology that can deliver pleasure. This allows people to be distracted by emotional and sensual experiences so they do not mind or notice that living a life according to free will, borne desires and artistic skills, and the idea of joy being created from within are left behind.

You can't replace prayer with Meta-data (or Meta/Instagram) experiences because that would make an artificial god out of technology, and that god will not be a holy, giving, healing God.

Tech in the A.I. era delivers the threat of losing all innocence. Collateral damage is losing the freedom to recognize yourself as an individual, what rights that individual has, and what the individual decides is too personal to share.

"The Janus Point" is important in both the Physics sense and in the artistic visual way of presenting two faces of a person: the Light and the Dark side. Big Tech worships centralization of power and will demand soon enough we worship them. Think a bit on why they are so against religions that are not centered on submission.

Mark Zuckerberg famously said to someone in the audience that asked him about how his vision challenges the sanctity of the creation of a human being with a soul, and he laughs, and said, "You don't need Jesus, you've got me." Hiding in plain site is his agenda, his End Game. The main hubs of Social Media will tempt you that they know your desires better than you do by using A.I. to create debate points that answer all your questions. They can train you like homing pigeons to come back to roost and recharge on their data input.

On an overall analysis they are building a world that is devolving our human observation skills and making them dependent upon lazy steps to get to conclusions. Social acceptance by people you do not even know (or know if they exist) will become essential to rank high enough on their scores. They will create an oasis feeding the thirsts of the Seven Vices, and this already is commonplace.

What they don't like is when someone notices they are starving and beating to a pulp the Seven Virtues.

If someone is banished into a void that is akin to sensory deprivation, when do they have the option to be released? Will transhuman devices be inserted or broadcast to lock people in a mental prison unable to leave the online worlds? The ability for A.I. and social acceptance to hyper-program the mind will be incredible.

Now add to that both The Fed and EU with its CBDC announced it will determine equity and human choices to see if someone gets taxed for being too lucky, wealthy, or even have imagination to do a joke. This will be a humorless world. Why include jokes? Because comedy is funny because there are elements of truth in it.

With all this negative explanation can a Multiverse be wonderful and nurture life. Yes!

Right now, our universe is consumerism mixed with greed mixed with gamer curiosity and play mixed with sexiness and driven by approval from followers. Simm City had an underground deviant sex Simm City that made more money than the normal world. A hypersexual, porn and mob power Multiverse will draw in a huge mass of people if the inventor finds a way to make the experience more immersive. The only way to minimize that is to recognize attention and distraction are drugs that wants to kill you as you know yourself and offer a better experience before Meta launches.

Alex Mashinsky wisely said that Blockchain and crypto won't gain the Scale and Spirit it needs to be embraced worldwide until it solves "their" problem – meaning the problems of the players. Everything so far is only to solve the problems of the rich and powerful. Oh sure, they shouted that digital currency was to serve the under-banked but look back and not a single poor or isolated social arena was improved by what billionaires drove to highs and then dumped at lows.

Facebook/Meta would never in a million years put its users first. Part of the reason why is they are incapable intellectually to envision what this requires. The second reason is they are greedy opportunists.

And then, there is the End Game for most giants like Google, Amazon, Microsoft, Facebook, and others. Their goal will by design humiliate and torment the players, locking them into a serfdom. The first to make a true Multiverse that provides multiple forms of a Meritocracy that reflects the Truth and obeys the Laws of Nature will change everything we know in film and Social Media. Adventures into anywhere -- Games, Live Events, Fashion Shows, nightclubs, gambling oasis, farmland, places to save shelter animals, ways to clean up the environment) -- can be enjoyed through a superior avatar and Interoperable Token by visitors to the ultimate Multiverse platform. Young people are craving to find a purpose in life and some Truth about who they are without consumerism and groupthink or cancel culture distorting the truth they know in their bones.

CLUES GIVEN BY BUKOWSKI

"The Genius of the Crowd"
"Beware the preachers ...
There is genius in their hatred ...
Not being able to create art,
they will not understand art.
They will consider their failure as a creator
a failure of the world.
Their hatred will be perfect,
their finest art."

Determine a Brand that represents the fact that the players' greatest adventure in the Multiverse is a journey into themselves to discovery their own great hero within... and allow for a way for other players to notice this on a selective basis. We cannot accelerate the innovations in hardware for a Virtuous and moral Multiverse. This means that to have one now the vision and software had to begin development over ten years ago. Luckily, this vision did originate in 2016. To speed up Spirit and Scale we use a proven method: let the fans do the heavy lifting until hardware catches up. Is there an example of this winning over all competition? Yes. Google. What does fans doing heavy lifting involve? That is a proprietary secret. OnlyFans, Roblox, and Epic Games are other examples of players doing the heavy lifting. The Tech Giants' visions for ultimate power and the craving for it, the almost sexual lust to pervert society, the defining people as cattle and useless idiots, and the vision they alone deserve to survive and be gods blinds them from what will rule the A.I. Social Media worlds.

Their Siren Song ... their promise ... may sound familiar ...
Give yourself to the Dark Side.
It is the only way you can save your friends.

HEADLINE: "CREATIVE BRAINS SAVE WORLD"

[**NOTE:** This is one of the few chapters added after the original completion of this book in 2017-2018. This chapter was added in April of 2024.]

Why is it that people seem to be far more intolerant and brutal to others? Why has division come easier? Why has hatred and brutality between people that seem to only see half of history or the big picture increased? Why do both extremes of politics and religions seem incapable of consideration and empathy for other beliefs?

Is this our imagination or has something happened on top of news and politics to make violence and vitriol overwhelming? Why has polite disagreement shrunk to almost no arena? Why has merit been erased from how we evaluate people? Why have so many given up free will?

The answer engages a neurological, anthropologic, biological, socio-political, and now technological reason for this social shift.

The author and Philosopher Iain McGilchrist wrote a sensationally complete book on this, "The Matter with Things." Nihilists and materialists may use the book to enforce their arguments, while spiritual people that believe the soul, love, qualia, and mind awareness exist beyond the brain could do the same thing.

Before diving into the meat of this book it is worthwhile to point out that universal aspects of Will in "Things" without a central developed brain has already been proven. The sand dunes of the Sahara can send messages to compel distant sand formations to change shape to withstand a wind storm that will hit the Western edge of the desert weeks later. The Sahara Desert has no brain yet makes both survival decisions and ultimately carves artistic forms out of sand.

The second example is the jellyfish that is coated with a virus that glows purple when the optimum time to mate is set up: the virus has no central brain yet interprets the maturity of the jellyfish, the current, distant storms, and the density of virus on approaching jellyfish before turning purple.

The point is that a durable Multiverse universe that fits our greatest Virtues must start with Will and move toward the highest form of Elegance.

If enough minds share a vision to grow
into higher elegance – and not just human minds
but also animal, mineral, and environmental sources –
the Will of the universe will participate
revealing secrets to our existence that only

belong in a more elegant near-future reality.
And the medium must involve an art form.
Think of it as winking to the universe we reflected its message.

Materialism rejects most of what we imagine of faith, qualia, love, consciousness, compassion. "Reductive Materialists" believe there is nothing except what the brain can interpret. This means the brain creates what we imagine as a subset of the Mind. This makes it impossible to see any Virtues in any shape our eyes, mind, or sense of touch can perceive. Nature, Matter, and what we react to are "nothing but" a bundle of senseless particles. This perspective denies the concepts found in Astrophysics and the latest Physics that Will is the essence of the universe and that Dark Energy (which cannot possibly be imagined in a particle or material sense) is the most abundance form for Field in our universe. My work on finding a way for people to share qualia will completely refute Reductive Materialism.

The greater question is why so many forces (political, technological, and social media justice culture) are pushing to replace what gives us meaning with nihilism disguised as the superior treatment of marginalized people and the reduction of people to data that can be put in a caste system for generations.

See, if you strip away all the good-sounding talking points, if you ask anyone if they would choose hope and joy over Armageddon and utter futility, they would choose the former. Materialists and Accelerationists have framed their goals as pathways to acceptance and then demand acceptance is in and of itself is a joy created outside the individual by what they Virtue Signal. How have they as a massive (mostly) United States attack dog been so fully conned? As McGilchrist says, "It endangers everything that we should value."

All these influences come down to a simple point - attention – covertly funded and driven by puppet masters that hunger for power.

Most people are told in school or through wise sources that the brain is comprised of two hemispheres: the right side of the brain handles creative and emotional conclusions while the left side of the brain deals with life in black and white, automatic reactions and decisions. While this idea of the brain has useful applications, the way this is described is misleading or at least does not serve the greatest potential.

The left side of the brain responsibilities revolve primarily around survival specifically in getting food and other essentials for life. As this side of the brain starts with genetic code instructions it also learns about the world and becomes more predatorial. With only a few lessons or examples the left side that pays off in more food and control over life than the right side and lately the left side has becomes hellbent on manipulating the world. The left side is zeroed in on getting "things" which may be material but more important to the point of this article are not material in the form of food or prowling territory for more food sources.

MgGilchrist's Razor (if there were one) would argue social snobbery is a "thing."

Power in itself is not a material thing yet the people forcing a power grab make useful idiots of budding Reductionist Materialism fanatics. These fanatics in turn get small rewards (imagined or material) for towing the line. Sometimes the reward is simply a merit badge for ignorantly serving people with true power over their lives. Or, in the case of consumerism and Influencers, fans want to emulate their idols who already have a strongly-developed left hemisphere allowing it to lie, use, and abuse fans.

McGilquist would identify short-term memory as a prerequisite for Cancel Culture and Virtue Signalers because in the long term their "thing" proves to not be true and not deliver the altruistic future it promises. This makes it a rarity in our reality.

If a cause gives an icon (person or company) more attention it gives it more money and power to feed and in that feeding off food and power internal conversations about whether the pursuit of even more power mute more and more over time. This can happen very quickly, especially when managers point to the rewards of left brain thinking. You need to see that acceptance has been turned into a food source. Yet it only feeds an illusion.

Targets for predation by the social causes expand beyond food and water into taking power over an expanding hunting ground, and also in surrounding oneself with bodies to be sacrificed in case of an attack on the morals of the icon or idol. This is shown partly in cancel culture and blind Climate Change fury at protests.

In today's hunting ground we include the usual real estate, social status, and consumer flash of expensive brands with the number of minutes anyone can command attention on platforms like Instagram and TikTok. These social covets of humanity are global and not bad into and unto themselves. However, when given sole control over how we act and what we justify the left side of the brain starts being narcissistic and ultimately becomes sociopathic, psychopathic, brutal, and totally lacking in compassion.

Is the individual gaslighting himself or herself? Worth contemplating, because it ultimately must create a self-feeding dynamic to sustain the delusion.

The reason why a person led almost solely by a brain that encouraged dominance of the left side of the brain while ignoring neural impulse status of the right side of the brain becomes ruthless and acts without conscience falls upon it not having the counterbalance our Creator designed or that is reflected in our greatest source of truth: Nature. You do not see animals or plants virtue signaling.

To achieve its maximum potential the left side of the brain is narrowly focused: it ignores the distracting information about the world and thoughts about anything in it, especially in the history of that topic. The left side ignores creativity, curiosity, emotional context of what our actions result in, wonder, and most forms of art, higher capacity for joy, what gives us meaning, and spiritual growth.

Despots and tyrants destroy art forms and end education, replacing it with fear. This oddly becomes a proving point to the power whores: they'll sleep with the fat man and spread a disease if it lines their pockets.

The more the left brain gets its way, the more neural connections it activates. Like a muscle it gets stronger. If this continues unbalanced the right side of the brain has its neural impulses ignored and eventually diminished. In short-term ways the right side of the brain atrophies or is at least ignored as the left side demands control over decisions.

The rewards the left side of the brain brings us build upon themselves. If the left side of the brain blots out true compassion by replacing it with brainwashed slogans that end up not delivering compassion it can make us do terrible things without reflection.

To justify it in on our conscience level the left side of the brain points out to the few benefits and rewards it delivered while refusing to study the short and long-term impact of it taking control. Another way of saying this is if the left hemisphere demands 100 actions and 97 of them failed -- failed to deliver good things or caused bad things -- the left side ignores these and only points to the few gifts, rewards, or successes. A good example are the George Floyd riots: the made people look righteous among their peers and gave them power to hurt people that disagreed but left the cities burned down and countless people dead.

In fact, the left hemisphere of the brain simply cannot look at the big picture or alternatives that might be compassionate because it is not designed for that mental activity. This is what the right side of the brain is designed to do and we rose as a species because there was for centuries a balance between both hemispheres. Yes, there were tyrants and terrible wars or purges, but overall humanity prevailed. We grew globally as a civilization.

Art is the strongest example and action that leads to greater social elegance and truth.

History has proven that higher examples of art shape an era by feeding imagination. With stronger imagination we see more and by seeing more we apply it to more and through IQ bring it back to improve life… which is the purpose of the One Fish books.

The right hemisphere of the brain identifies and interprets what gives us on a soul level value. Much of that is not material in the slightest.

Humans were designed to be curious and communicative. The exact placement of the tongue in our mouths, our vocal cords, the placement of our eyes, the reality of emotional impact on our minds and bodies… altogether the sensory dexterity is sensational almost beyond comprehension. So very much good came and comes from this. But now it seems that a permanent imbalance has set in and the left side of the brain controls society, cultures, education, and power.

The computer and smart phone dramatically increased the left hemisphere of the brain to dominate and reduce the impulses and deep thinking of and valuing the natural focus of the right hemisphere. The reason why lies in the fact that the physical act of narrowing focus onto a small screen results in engaging the left side of the brain and its ability to laser focus on a small object and extract from it all it can while ignoring the outside world. The left side then (like any practiced sport) improves the ability to ignore the beauty of our natural world and reasons to keep it thriving.

A quick example from my personal life might help. While learning to camp in the wilderness I learned from experienced guides to look up from the trail to distant objects all around me every 3 to 5 minutes. It was proven that constantly looking down at the uneven trail every step cuts off awareness of dangers and opportunities around the hiker.

Tunnel vision in the wilds of Nature was then and is now dangerous. Awareness for distant smells and cloud warning of dangers fell. Appreciation for the beauty of the horizon was lost to the hiker. It's not complicated when seen in that activity.

In other chapters written in "One Fish Says to Another Fish" we discuss "The Final Filter" of our spiritual and mental ascension. The Technology Era is the one great leap forward that we accelerates access to what exists in the world and individual and collect Minds. We get a thrill from this acceleration. Power mongers grab the reins. To keep holding the reins of Thumos often demands we hold back or avoid holds all civilizations the ultimate elegance. The ultimate elegance would be to gain all technology gives us and apply it to human virtues, morals, and gifts given us by our Creator.

The hypothesis is that higher intelligence exists in the cosmos in biological or pure energy forms and watches our evolution as a species to see if we find some way to avoid the trap of losing the balance within the brain and Mind as the technology age accelerates what our senses perceive of our reality.

So how are we doing?

I think the capacity for compassion, incredible works of art, breakthroughs in understanding our entire human being, and observing the Fields in our universe improved us as human beings and proved we are worthy of being shown the deeper secrets of our existence. This comes from the billions of people in the world, not from the small subset of national and technological leaders in elite caste systems they usher in, because really this is an individual journey assisted by systems and inventions.

We need a megaphone held up to our right brain impulses to clarify our most valued hope and goal.

Politicians, leaders of war, drug lords, rapists, murderers, terrorists, fundamentalists of all types share the stage of giving in and letting the left side of the brain justify all actions. Isn't it unsettling to include politicians and technology leaders with drug lords and terrorists? Now, not all politicians and tech leaders are driven only by the left side of the brain to take anything it wants in its justification for "food" and power over others to remove threats to the way it works out plans. But the support staff in politics and social power uniting through technology mostly have allowed the left side of the brain to fully dominate the right side.

The left-brain-run people that gain power surrender all conflicting mental opinion and decision making. For this they gain power. Their focus is narrowed so they either simply are blind to the damage to people they think are below them, or they do not care. They move from Machiavellian to narcissists and then to psychopaths when that most sinful level is needed. People with this are the useful idiots.

To accelerate in the race for left-brain dominance over others their minds degrade to where they don't process alternatives or consequences. The removal of historic art and banning words that articulate the damage they wreck upon society are removed to reduce the temptation for self-reflections and honest consequences. Gradually, they believe the victims deserved their suffering and or deaths.

Assisted suicide for reasons as trivial as being depressed are pushed in England and Canada. Why? People in power want to cull the herd of hungry minds and mouths that do not push the Accelerationist dogma. Forced surrender of where we find joy and reflections of truth begins the process. Turning the idea into a hope is a way to appeal to the right side of the brain. Without hope the void is filled with fear. Even fear of people that don't think the same yet are peaceful and productive. The violent useful team leaders fill our minds with fear which needs the left hemisphere of the brain to survive are introduced.

Collaboration between food sources, Big Pharma, distorted news, Big Tech, and politics fills our entertainment with ways that reduce the responsibility of each of us to stay healthy. Destroy the body and its balance and it is easier for people to give up on life and due to fear rely on their own left side of the brain. The people suffering then justify violence against the people, art, religions, and sources of meaning. The useful idiots as many terrorist organizations call them lose the ability to see the wider perspective.

It is no coincidence that up to 75% of ads are for drugs, law firms, and insurance.

The United States was always the deterrent to a world full of enemies and even seemingly allied nations. Yet within allies as much as targeted foes both sides are creating an elite group at the top of the systems.

Decades ago a campaign to destroy what makes the United States so extraordinary began. As strange as it sounds the key arena turned upside down was the art world of New York City (which is a fascinating subject unto itself that is explored in other chapters.)

If all questions can be answered in the movies, the two films to refresh your memory are "Three Days of the Condor" which shows the early years of this heartless left-brain offensive, and "Die Hard with a Vengeance." In the first movie it ends with Redford discovering the truth of the assassinations of people that will fix the world.

In the Die Hard sequel the bad guy (Timothy Oliphant) creates a "Fire Sale." This is explained by stating that if any one or two systems collapse the government and American Way will recover. All systems must be overwhelmed at once.

Scholars on American Way government explain that the capitalist system only works if the lowest class does not get so big in population that the top of the pyramid of earning can pay for education and basic living for the poor. The immigrant tsunami is meant to do just that goal.

No matter how many people suffer the left-brain people in power, Democrat or Republican, cannot see or care about the damage – everything they force on us is not just acceptable it is the way to go if it brings total dominance of the left brain faster.

McGilchrist suggests the counterbalance is found in Intuition, Metaphor, and Imagination. The end result of this is the process of what is referenced in my other books as "creating something out of nothing."

The equation is propose as essential to riding the crest of the wave of our fast-moving future is "Observation skills over adaptation speed" and this equation relates to Metaphors and recognizing patterns.

People seeking power want to mute the voices of Intuition, Metaphor, and Imagination because these qualities can find the real Truth in our world beyond what we examine or get fed from the words of other people that might have the opposite agenda of sublime actions and the virtues driving them. He states that "Truth, Beauty, and Goodness" have been disparaged.

For decades deep-thinking filmmakers have stated that the beauty captured in cinematic art will save the world: McGilchrist has a more nuanced explanation for how this could happen and what we lose if we let our left hemisphere take over.

The concept of belief in God is repulsive to most left-leaning politicians and the entire governing body of the E.U. – yet they would never state it publicly. They instead attack the institutions or support opponents for the very religion that created their nations and put them in power through evolving governing rules.

My opinion is similar to McGilchrist's in that humans share in the process of God becoming God; our faiths, Intuitions, what we recognize as Beauty and Goodness merge as we contemplate God, forming him in our ideas. If there were not a grand creator, we would create one and under various names imagine him as God.

Back to the runaway demand for people to follow the left hemisphere of the brain at the expense of the Goodness in the right hemisphere…

Lying becomes the default of the politicians and tech leaders regardless of any religion. They then use words that are known to trigger emotional responses from the right side of the brain in order to make the right side think it is involved and the person cares about suffering. This short-term use of the right brain tricks it to give in until it is too late.

Climate Change surely uses the same technique. If you read in other books about using fear to forward violence you see how it can be used to destroy American energy and then create shortages, riots, sabotage of energy and delivery, and more.

All these fringe groups help the left-side brain leaders of the E.U., United Nations, Untied States, and Pentagon (as well as CIA and FBI) force destructive measures. The open border will insert people that are wholly left-brain and are terrorists and gang members. The deaths from fentanyl are ignored by both political parties. So is homelessness. Anything to destroy where we find safety so that our right side of the brain is free to create and uplift humanity is promoted. And then the leaders set up their payoffs, such as taking the money missing from the Ukraine war, electrical vehicle funding, homeless and Black Lives Matter allotments, and all other financial crimes. If the left side of the brain can point to a secret payoff it makes it harder for the right side of the brain.

A formula to consider in this order (of input from our world) not only read and understood but valued through the right side of our brain as it is felt through senses could be reduced to the following.

Look at each line as a step to starve what the right side of the brain was designed to use as a balance for our evolution to the point we focus less on power and more on the mysteries of the universe and the reality we can use to rise up levels to the point where our actions reflect elegance recognized by the universe.

1. Feed to only match a survival level
2. Removal of conscious if a cause is feverish enough

3. Removal of reflections of truth and virtues
4. Rewards for being predatorial
5. Begin controlling neural activity through drugs or vice payoffs
6. Education must revolve around indoctrination instead of intelligence
7. Tyranny from The Big is replaced by tyranny by The Small
8. Negation of biological truths and Empirical facts
9. Systems set up to reward bully groups in any merit-based arena
10. Control of family and Epigenetic examples of Virtues
11. Removal of ways to receive joy and meaning from within and without
12. Higher rewards for being so predatorial it makes news that shapes arenas
13. Surrounding the lost people with predatorial pundits
14. Education that foments division among students
15. Perversion of proven educational tools and methods as an excuse for failure
16. Greater redistribution of wealth and freedom earned through delayed gratification
17. Restriction of food and energy to left-brain followers
18. Perversion of sports
19. Spread of protecting criminals over victims
20. Mob mentality justifying moral atrocities
21. Rewards that are more deviant for being predatorial
22. Reduction of right-brain neural networks by drugs or fear
23. Use of Fast Brainwashing in cultish styles
24. Persecution of anyone questioning rhetoric decided by invisible Elitists
25. Acceptance of high crime and unemployment as the New Normal
26. Direct threats to "free thinkers" for simply not buying what's sold us
27. Starving the right side of the brain by demanding artists be persecuted
28. Government sanctioning of targeting right-brain citizens
29. Covert Elite left-brain castes and its systems being used as bait
30. Promise of utopian power I more cruelty is forced on the masses
31. Cut off food based on color or faith

32. Limit survivors to left-brain thinkers
33. Insisting that parents do not raise children preventing epigenetic traits
34. Agencies or systems raising children replacing Free Play with politics
35. A caste system of promotion based on alignment and not results
36. Removal of history to prevent the return of the Seven Virtues
37. The first wave of enabling zealots are killed or pushed aside
38. The replacements are far more radicalized
39. Politicians realize radicalized can turn against them in new leadership
40. All Art, religion, creativity is controlled, censored, banned, or dogma
41. Our only source of joy is aligning with the loudest bullies
42. We can't remember our past as well because less of our brain recorded it

Do you notice any of these steps being taken? How about a majority of these examples? Enough to form a tsunami of intimidation and mimicking?

There are ways to counterbalance the left brain's power while nurturing the right brain. To find them, think of those works of art that appear like one thing (a tiger's head) but if you look at it long enough what it really is surprises you… these surprises can hide in plain sight. Other salvations lie in what the left brain wants to stamp out: individual creativity and virtues reflected in Nature, our greatest source of truth.

**Just like water always finds a way out,
Intuition, Goodness, and Creativity find a way
to escape prisons of the mind and soul.**

Kurt Vonnegut who spent a hellish number of years in prison wrote about dystopian futures in a few books. In "Player Piano" even the activity of learning to play the piano is not allowed because it presents a source of value and joy. The lead character Tally says the revolution against the rise of technology fails because "human beings themselves are the greatest, indeed perhaps the only, impediment to human freedom and happiness." Imagine writing this 1952, when technology (automation) was vending machines and factory lines. In the book a rebellion rises up and destroys the machines… but then humans that internally love seeing machines they created work start to re-build the machines.

Instead of thinking the fate of that book is our only future remember the other quotes by Vonnegut.

"We have to continually be jumping off cliffs and developing our wings on the way down."

Which leads to his other quote.

"The arts are not a way to make a living.
They are a very human way of making life
more bearable. Practicing an art, no matter
how badly, is a way to make your soul grow,
for heaven's sake."

And finally, Kurt Vonnegut's most wonderful quote:

"There is no good reason why
good cannot triumph as often as evil does.
The triumph of anything is a matter of organization.
if there are such things as angels,
I hope they are organized along the lines of the Mafia."

The two Parts of "Accelerating Intelligence" set the moral, philosophical, and accelerating foundation for "The Age of Imagination" and this guides the new era of filmmaking presented in the second book titled "Multiverses of Entertainment."

The films, television series, and mini-series are the "Products" and the platform avatar presented in that third book bring the "Spirit and Scale" needed to aim the direction of "The Power of Story" and create the new Dynamic in the Multiverses of entertainment.

Clearly explained, the next book on films delivers the social, financial, and thrilling payoff for accelerating your philosophy so that what gives you meaning and cred remain in your human potential as it faces the onslaught of data-ism from Artificial Intelligence.

ACKNOWLEDGEMENTS

It also must be acknowledged that my
mind absorbed Information
from so many sources that I cannot always
remember what inspired thoughts and got onto
paper, yet I can mention a few key sources of
my ideas presented herein.
The books "Tools of Titans,"
"Homo Deus," "The Janus Point,"
"The Matter with Things," and many
blogs on Critical Thinking,
and books by Thomas Sowell.
The magazines including
Wired, MIT, and New Scientist Magazines.
The websites Academy of Ideas, Sixth Tone,
CoinTelegraph, blockckhain.news,
Jing Daily, WSJ, Bloomberg,
and other deep thought websites.

www.ingramcontent.com/pod-product-compliance
Lightning Source LLC
LaVergne TN
LVHW080022210926
848481LV00011B/3087